NUTRACEUTICALS AND DIETARY SUPPLEMENTS

Applications in Health Improvement and Disease Management

NUTRACEUTICALS AND DIETARY SUPPLEMENTS

Applications in Health Improvement and Disease Management

Edited by

Raj K. Keservani, MPharm

Anil K. Sharma, PhD, MPharm

Rajesh K. Kesharwani, PhD

APPLE ACADEMIC PRESS

Apple Academic Press Inc.
4164 Lakeshore Road
Burlington ON L7L 1A4
Canada

Apple Academic Press, Inc.
1265 Goldenrod Circle NE
Palm Bay, Florida 32905
USA

Exclusive co-publishing with CRC Press, a Taylor & Francis Group

International Standard Book Number-13: 978-1-77188-873-8 (Hardcover)
International Standard Book Number-13: 978-0-36782-151-7 (eBook)

Library and Archives Canada Cataloguing in Publication

Title: Nutraceuticals and dietary supplements : applications in health improvement and disease management / edited by Raj K. Keservani, MPharm, Anil K. Sharma, PhD, MPharm, Rajesh K. Kesharwani, PhD.
Names: Keservani, Raj K., 1981- editor. | Sharma, Anil K., 1980- editor. | Kesharwani, Rajesh Kumar, 1978- editor.
Description: Includes bibliographical references and index.
Identifiers: Canadiana (print) 20200255126 | Canadiana (ebook) 20200255266 | ISBN 9781771888738 (hardcover) | ISBN 9780367821517 (ebook)
Subjects: LCSH: Functional foods—Therapeutic use. | LCSH: Dietary supplements—Therapeutic use
Classification: LCC QP144.F85 N88 2021 | DDC 613.2—dc23

CIP data on file with US Library of Congress

Apple Academic Press also publishes its books in a variety of electronic formats. Some content that appears in print may not be available in electronic format. For information about Apple Academic Press products, visit our website at **www.appleacademicpress.com** and the CRC Press website at **www.crcpress.com**

About the Editors

Raj K. Keservani, MPharm
Faculty of B. Pharmacy, CSM Group of Institutions, Prayagraj, India

Raj K. Keservani, MPharm, is on the Faculty of B. Pharmacy, CSM Group of Institutions, Allahabad, India. He has more than 10 years of academic (teaching) experience at various institutes of India in pharmaceutical education. He has published over 35 peer-reviewed papers in the field of pharmaceutical sciences in national and international journals, more than 25 book chapters, two co-authored books, and 12 co-edited books. He is also active as a reviewer for several international scientific journals. Mr. Keservani graduated with a pharmacy degree from the Department of Pharmacy, Kumaun University, Nainital (UA), India. He received his Master of Pharmacy (MPharm) (specialization in pharmaceutics) from the School of Pharmaceutical Sciences, Rajiv Gandhi Proudyogiki Vishwavidyalaya, Bhopal, India. His research interests include nutraceutical and functional foods, novel drug delivery systems (NDDS), transdermal drug delivery, health science, cancer biology, and neurobiology.

Anil K. Sharma, PhD, MPharm
Department of Pharmacy, School of Medical and Allied Sciences, GD Goenka University, Gurugram, India

Anil K. Sharma, PhD, MPharm, is an assistant professor at the Department of Pharmacy, School of Medical and Allied Sciences, GD Goenka University, Gurugram, India. He has more than 10 years of academic experience in pharmaceutical sciences. He has published over 35 peer-reviewed papers in the field of pharmaceutical sciences in national and international journals as well as 28 book chapters. In addition, he has edited more than 16 books with reputable publishers. He received a bachelor's degree in pharmacy from the University of Rajasthan, Jaipur, India, and a Master of Pharmacy degree from the School of Pharmaceutical Sciences, Rajiv Gandhi Proudyogiki Vishwavidyalaya, Bhopal, India, with a specialization in pharmaceutics. He earned his PhD in pharmacy from the University of Delhi, India. His research interests encompass nutraceutical and functional foods, novel drug delivery systems (NDDS), drug delivery, nanotechnology, health science/life science, and biology/cancer biology/neurobiology.

Rajesh K. Kesharwani, PhD
Department of Advanced Science & Technology, NIET, NIMS University Rajasthan, Jaipur, India

Rajesh K. Kesharwani, PhD, is affiliated with the Department of Advanced Science & Technology, NIET, Nims University Rajasthan, Jaipur, India. He has more than seven years of research and two years of teaching experience at various institutes in India, imparting bioinformatics and biotechnology education. He has received several awards, including the NASI-Swarna Jayanti Puruskar–2013 from The National Academy of Sciences of India. He has authored over 32 peer-reviewed articles and 10 book chapters. He has been a member of many scientific communities as well as a reviewer for many international journals. His research fields of interest are medical informatics, protein structure and function prediction, computer-aided drug designing, structural biology, drug delivery, cancer biology, and next-generation sequence analysis. Dr. Kesharwani received a BSc in biology from Ewing Christian College, Allahabad, India, an autonomous college of the University of Allahabad; MSc (Biochemistry) from Awadesh Pratap Singh University, Rewa, Madhya Pradesh, India; and MTech-IT (specialization in Bioinformatics) from the Indian Institute of Information Technology, Allahabad, India. He earned his PhD from the Indian Institute of Information Technology, Allahabad, and received a Ministry of Human Resource Development (India) Fellowship and Senior Research Fellowship from the Indian Council of Medical Research, India.

Contents

Contributors

Lucrece Ahovegbe
Pharmbiotechnology and Traditional Medicine Center of Excellence, School of Medicine, Mbarara University of Science and Technology, Mbarara, Uganda
Unite de Formation et de Recherché en Pharmacie, Faculté des Sciences de la Santé, Université d'abomey-Calavi, Cotonou, Benin

Dulce Esperanza Alarcón-Yaquetto
Research and Development Laboratories (LID), Faculty of Sciences and Philosophy, Universidad Peruana Cayetano Heredia, Lima, Peru
High Altitude Research Institute, Universidad Peruana Cayetano Heredia, Lima, Peru.

Francisco J. Barba
Nutrition and Food Science Area, Preventive Medicine and Public Health Food Science, Toxicology and Forensic Medicine Department Faculty of Pharmacy Universitat de València, Avda Vicent Andrés Estellés, s/n, 46100 Burjassot València, Spain

Anamika Basu
Department of Biochemistry, Gurudas College, Kolkata 700054, India

Sabrina Bimonte
Istituto Nazionale Tumori, IRCCS—Fondazione "G. Pascale", Via Semmola, Naples, Italy

Mladen Brnčić
Faculty of Food Technology and Biotechnology University of Zagreb, Pierottijeva 6, 10000 Zagreb, Croatia

Suzana Rimac Brnčić
Faculty of Food Technology and Biotechnology University of Zagreb, Pierottijeva 6, 10000 Zagreb, Croatia

Domenico Carbone
Department of Emergency Medicine, Umberto I Hospital, Nocera Inferiore, Salerno, Italy

Marco Cascella
Istituto Nazionale Tumori, IRCCS— Fondazione "G. Pascale", Via Semmola, Naples, Italy

Arturo Cuomo
Istituto Nazionale Tumori, IRCCS- Fondazione "G. Pascale", Via Semmola, Naples, Italy

Alberto A. Escobar-Puentes
Instituto de Ciencias Biomédicas, Universidad Autónoma de Ciudad Juárez, Anillo Envolvente del Pronaf y Estocolmo s/n, CP 32310, Cd. Juárez, Chihuahua, México

Adriana García-Gurrola
Instituto de Ciencias Biomédicas. Universidad Autónoma de Ciudad Juárez, Anillo Envolvente del Pronaf y Estocolmo s/n, CP 32310, Cd. Juárez, Chihuahua, México

Gustavo F. Gonzales
Research and Development Laboratories (LID), Faculty of Sciences and Philosophy, Universidad Peruana Cayetano Heredia, Lima, Peru
High Altitude Research Institute, Universidad Peruana Cayetano Heredia, Lima, Peru

Charu Gupta
Amity Institute of Herbal Research & Studies, Amity University, Sector 125, Noida 201313, India

Amita Joshi
Department of Pharmaceutics, B.V. Patel PERD Centre, Ahmedabad, India

Rajan Kaushik
Global Institute of Pharmaceutical Education and Research, Kashipur, Uttarakhand, India

Raj K. Keservani
Faculty of B. Pharmacy, CSM Group of Institutions, Allahabad, India

Kapil Kumar
Global Institute of Pharmaceutical Education and Research, Kashipur, Uttarakhand, India

Fanuel Lampiao
Africa Centre of Excellency in Public health and Herbal Medicine, College of Medicine, University of Malawi, Chichiri, Blantyre 3, Malawi

José M. Lorenzo
Centro Tecnológico de la Carne de Galicia, Rúa Galicia No. 4, Parque Tecnológico de Galicia, San Cibrao das Viñas 32900, Ourense, Spain

Stephen Lutoti
Pharmbiotechnology and Traditional Medicine Center of Excellence, School of Medicine, Mbarara University of Science and Technology, Mbarara, Uganda

Fernando Martínez-Bustos
Departamento de Nanomateriales, Centro de Investigación y de Estudios Avanzados del Instituto Politécnico, Nacional, Unidad Querétaro, Libramiento Norponiente No. 2000, Fracc. Real de Juriquilla, 76230, Querétaro, Querétaro, México

Elena Movilla Fierro
Complejo Hospitalario, Universitario de Ourense, Ourense, Spain

Andrew G. Mtewa
Department of Chemistry, Malawi Institute of Technology Malawi University of Science and Technology, P. O. Box 5196, Limbe, Malawi, East Africa
Pharmbiotechnology and Traditional Medicine Center of Excellence School of Medicine Mbarara University of Science and Technology Mbarara, Uganda, East Africa

Paulo E. S. Munekata
Centro Tecnológico de la Carne de Galicia, Rúa Galicia No. 4, Parque Tecnológico de Galicia, San Cibrao das Viñas 32900, Ourense, Spain

Maria Rosaria Muzio
Division of Infantile Neuropsychiatry, UOMI—Maternal and Infant Health, Naples, Italy

Davies Mweta
Department of Chemistry, Malawi Institute of Technology, Malawi University of Science and Technology Limbe, Malawi

Kennedy J. Ngwira
Institute of Molecular Science, School of Chemistry, University of Witwatersrand, P/Bag 3, WITS 2050, Johannesburg, South Africa

Patrick E. Ogwang
Pharmbiotechnology and Traditional Medicine Center of Excellence, School of Medicine, Mbarara University of Science and Technology, Mbarara, Uganda

Hedmon Okella
Pharmbiotechnology and Traditional Medicine Center of Excellence, School of Medicine, Mbarara University of Science and Technology, Mbarara, Uganda

Harsh Pandey
Department of Medicinal Chemistry, Institute of Medical Sciences, Banaras Hindu University, Varanasi, India

Mirian Pateiro
Centro Tecnológico de la Carne de Galicia, Rúa Galicia No. 4, Parque Tecnológico de Galicia, San Cibrao das Viñas 32900, Ourense, Spain

Vandana Patravale
Department of Pharmaceutical Science and Technology, Institute of Chemical Technology, Mumbai, India

Emanuel L. Peter
Pharmbiotechnology and Traditional Medicine Center of Excellence, School of Medicine, Mbarara University of Science and Technology, Mbarara, Uganda
Department of Innovation, Technology Transfer & Commercialization, National Institute for Medical Research, Dar-Es-Salaam, Tanzania

Dhan Prakash
Amity Institute of Herbal Research & Studies, Amity University, Sector 125, Noida 201313, India

Swati Pund
Nanomedicine Laboratory, Department of Biosciences and Bioengineering, Indian Institute of Technology, Mumbai, India

Bhushan R. Rane
Shri DD Vispute College of Pharmacy & Research Center, Panvel 410206, Mumbai, India

Kumar Sandeep
Dr Bhim Rao Ambedkar Institute Rotary Cancer Hospital, All India Institute of Medical Sciences, New Delhi, India

Duncan C. Sesaazi
Pharmbiotechnology and Traditional Medicine Center of Excellence, School of Medicine, Mbarara University of Science and Technology, Mbarara, Uganda

Anil K. Sharma
Department of Pharmacy, School of Medical and Allied Sciences, GD Goenka University, Gurugram, India

Surya Kumar Singh
Department of Endocrinology, Institute of Medical Sciences, Banaras Hindu University, Varanasi, India

Shivani Srivastava
Department of Medicinal Chemistry, Institute of Medical Sciences, Banaras Hindu University, Varanasi, India

Cassim U. Tolo
Pharmbiotechnology and Traditional Medicine Center of Excellence, School of Medicine, Mbarara University of Science and Technology, Mbarara, Uganda

Yamini Bhusan Tripathi
Department of Medicinal Chemistry, Institute of Medical Sciences, Banaras Hindu University, Varanasi, India

Cinthya Vasquez-Velasquez
Research and Development Laboratories (LID), Faculty of Sciences and Philosophy, Universidad Peruana Cayetano Heredia, Lima, Peru
High Altitude Research Institute, Universidad Peruana Cayetano Heredia, Lima, Peru

Durgavati Yadav
Department of Medicinal Chemistry, Institute of Medical Sciences, Banaras Hindu University, Varanasi, India

Abbreviations

2VO	two-vessel occlusion
AADAT	aminoadipate aminotransferase
ABAT	4-Aminobutyrate aminotransferase
ABC	ATP-binding cassette
AD	Alzheimer's disease,
ADE	alcohol deprivation effect
AFAR	aflatoxin B1 aldehyde reductase
AGEs	advanced glycation end products
AGXT	alanine-glyoxylate aminotransferase
AGXT2	alanine-glyoxylate aminotransferase 2
ALAS1	aminolevulinate, delta-, synthase 1
ALAS2	aminolevulinate, delta-, synthase 2
ALOX5	arachidonate 5-lipoxygenase
ALS	amyotrophiclateral sclerosis
AML	acute myeloid leukemia
AP-1	activator protein
APP	amyloid precursor protein
ARs	androgen receptors
Aβ	amyloid β
BACE	β-secretase amyloid cleaving enzyme
BaP	benzo[a]pyrene
BBB	blood–brain barrier
BCAT1	branched chain amino-acid transaminase 1, cytosolic
BCAT2	branched chain amino-acid transaminase 2, mitochondrial
BCSFB	blood cerebrospinal fluid barrier
BDNF	brain-derived neurotrophic factor
BHT	butylated hydroxyl toluene
BIS	Bureau of Indian Standards
BPH	benign prostatic hyperplasia
cAMP	cyclic adenosine phosphate
CB1 and CB2	cannabinoid receptors 1 and 2
CBS	cystathionine-beta-synthase
CCBL1	cysteine conjugate-beta lyase, cytoplasmic
CCBL2	cysteine conjugate-beta lyase 2
CGNs	cerebellar granule neurons
CLL	chronic lymphocytic leukemia

CNS	central nervous system
CSAD	cysteine sulfinic acid decarboxylase
CSF	cerebrospinal fluid
CTH	cystathionase (cystathionine gamma-lyase)
Cu	copper
DATS	diallyl trisulfide
DDC	dopa decarboxylase (aromatic L-amino acid decarboxylase)
DHT	dihydrotestosterone
DOCA	deoxycholic acid
DPP-IV	dipeptidyl peptidase 4
DSC	differential scanning calorimetry
DSHEA	Dietary Supplement Health and Education Act
DSPC	1,2-distearoyl-sn-glycerol-3-phosphatidylcholine
EFSA	European Food Safety Authority
EGCG	epigallocatechin-3-gallate
ERα	estrogen receptor alpha
ERBB2	human epidermal growth factor receptor 2 or tyrosine kinase-type cell surface receptor HER2
ERs	estrogen receptors
FAAH	fatty amidic acid hydrolase
FDA	Food and Drug Administration
FFAs	free fatty acids
FSSAI	Food Safety and Standards Authority of India
FTCD	formimidoyltransferase cyclodeaminase
FTIR	Fourier transform infrared
GABA	gamma-aminobutyric acid
GAD1	glutamate decarboxylase 1 (brain, 67 kDa)
GAD2	glutamate decarboxylase 2
GCAT	glycine C-acetyltransferase
g-GCS	g-glutamylcysteine synthase
GIP	glucose-dependent insulinotropic polypeptide
GLDC	glycine dehydrogenase
GLP-1	glucagon like polypeptide-1
GLUT-2	glucose transporter
GMP	good manufacturing practice
GOT1	glutamic-oxaloacetic transaminase 1
GOT2	glutamic-oxaloacetic transaminase 2
GPT	glutamic-pyruvate transaminase (alanine aminotransferase)
GPT2	glutamic pyruvate transaminase (alanine aminotransferase) 2
GPx	GTH peroxidase
GRAS	Generally recognized as safe
GSIS	glucose-stimulated insulin secretion

GST	glutathione-*S*-transferase
GTE	green tea extract
HBV	hepatitis B virus
HCV	hepatitis C virus
HD	Huntington's disease
HDC	histidine decarboxylase
HES	hydroxyethyl starch
hi-PSC	human induced pluripotent stem cells
HIV	human immune deficiency virus
HO-1	heme oxygenase-1
HPAEC-PAD	high-performance anion-exchange chromatographic methods coupled with pulsed amperometric detection
HPLC	high-performance liquid chromatography
HPVs	human papillomaviruses
HSD17B1	hydroxysteroid (17-beta) dehydrogenase 1
hTERT	human telomerase reverse transcriptase
HTLV1	human T-cell lymphotropic virus type-1
IAPP	islet amyloid polypeptide
IL	interleukin
IMS	Infant Milk Substitutes
iNOS	inducible NOS
IRS-1	insulin receptor substrate-1
JNK	jun N-terminal kinase
KSHV	Kaposi's sarcoma herpes virus
KYNU	kynureninase
LAB	lactic acid bacteria
LUTS	lower urinary tract
MCP	modified citrus pectin
MDA	malondialdehyde
miRNAs	microRNAs
MOCOS	molybdenum cofactor sulfurase
MPM	malignant peritoneal mesothelioma
MSCs	mesenchymal stem cells
MWM	Morris water maze
n-3 PUFAs	omega-3 polyunsaturated fatty acids
nAChR	acetyl choline receptor
NADPH	nicotinamide adenine dinucleotide phosphate
NFkB	nuclear factor-kB
NFS1	nitrogen fixation 1 homolog (S. cerevisiae)
NF-κB	nuclear factor κB
Ni	nickel
NLCs	nanostructured lipid carriers

NNK	nitrosamine carcinogen 4-(methylnitrosamino)-1-(3-pyridyl)-1-butanone
NO	nitric oxide
NOS	reactive nitrogen species
NOS2	nitric oxide synthase 2
NR3A1	nuclear receptor subfamily 3, group A, member 1
Nrf2	nuclear erythroid-2 related factor 2
NS	nigella sativa
NSO	NS oil
NT/NFT	neurofibrillary tangles
OAT	ornithine aminotransferase
ODC1	ornithine decarboxylase 1
OH	hydroxyl
PAsp	poly(aspartic acid)
PBCA	poly(*N*-butylcyanoacrylate)
PCL	poly(ε-caprolactone)
PD	Parkinson's disease
pDNA	plasmid DNA
PE	phosphatidylethanolamine
PEG	polyethylene glycol
PEITC	phenethyl isothiocyanate
PEO	polyethylene oxide
PGE2	prostaglandin E2
PGlu	poly(glutamic acid)
PI3k	phosphoinositide 2 kinase
PLA	poly(lactic acid)
PLGA	poly lactic-*co*-glycolic acid
PLGA	poly(lactic-*co*-glycolic acid)
PNPO	pyridoxamine 5′-phosphate oxidase
POPC	1-palmitoyl-2-oleoylphosphatidylcholine
PPARgamma	peroxisome proliferator-activated receptor-gamma
PPARγ	peroxisome proliferator-activated receptor-gamma
Pre-MULAM	preventive medicine using locally available materials
PSAT1	phosphoserine aminotransferase 1
P-tau	phosphorylated tau
PTEN	phosphatase and tensin homolog deleted on chromosome 10
PTGS1	prostaglandin-endoperoxide synthase 1
Ptgs2	prostaglandin-endoperoxide synthase 2
PTPN6	protein tyrosine phosphatase, nonreceptor type 6
PTPRJ	protein tyrosine phosphatase, receptor type, J
PTPRN	protein tyrosine phosphatase, receptor type, N
PTPRT	protein tyrosine phosphatase, receptor type, T

PUFAs	polyunsaturated fatty acids
PUFAs	polyunsaturated fatty acids
PVA	poly(vinyl alcohol)
PYGB	phosphorylase, glycogen; brain
PYGL	phosphorylase, glycogen, liver
PYGM	phosphorylase, glycogen, muscle
QR	quinone reductase
QYD	Qing Ye Dang
RES	reticulo-endothelial system
ROS	reactive oxygen species
RTD	ready-to-drink
Sac	S-allyl cysteine
SAMC	S-allyl mercapto cysteine
SASP	senescence-associated secretory phenotype
SDS	serine dehydratase
SEM	scanning electron microscopy
SFDA	State Food and Drug Administration
SGPL1	sphingosine-1-phosphate lyase 1
SHMT1	serine hydroxymethyltransferase 1
SHMT2	serine hydroxymethyltransferase 2
SLNs	solid lipid nanoparticles
SOD	superoxide dismutase
SPP	sweet potato protein
SPTLC1	serine palmitoyltransferase, long-chain base subunit 1
SPTLC2	serine palmitoyltransferase, long-chain base subunit 2
SREBPs	sterol response element-binding proteins
SRR	serine racemase
ST	starch
STZ-T2DM	streptozotocin-induced diabetic
SWMA	Standards of Weights and Measures
T2D	type 2 diabetes
TAT	tyrosine aminotransferase
Tau	tubulin-associated unit
TEM	transmission electron microscopy
TGF-β	transforming growth factor beta1 or TGF-β1
THC	tetrahydrocurcumin
TLC	thin-layer chromatography
TNF-α	tumor necrosis factor-α
TQ	thymoquinone
TQRFNE	TQ rich-fraction nanoemulsion
TrkB	tropomyosin-related kinase B (TrkB)
Txnip	thioredoxin-interacting protein

UGT	UDP-glucuronosyl transferases
UV	ultraviolet
VEGF	vascular endothelial growth factor
WF	water fraction
WHO	World Health Organization
XRD	X-ray diffraction
αSN	α-synuclein

Preface

New dietary patterns, realistic trends, and consumption of foods have health as well as socioeconomic impacts. The modern-day world is battling maladies related to life style, for example, obesity, osteoporosis, malignant growths, diabetes, hypersensitivities, and dental issues. Though a number of bioactives are acknowledged to provide beneficial properties, the underlying mechanisms are diverse and can act exclusively or jointly with other ingredients. These products are increasingly becoming more common in the market place due to high acceptability and aggressive marketing campaigns. The role of food has much advanced from being a source of livelihood to mitigating, and treating a number of ailments as adjuvants to their synthetic counterparts. This is, of course, in concordance with Hippocrates famous quote "*Let food be thy medicine and medicine be thy food*." In certain instances, these nutraceuticals and dietary supplements have remarkably demonstrated higher efficacy than drugs. The increasing popularity among the people has prompted regulators to establish guidelines regarding their purity, safety, and quality.

The present book strives to provide requisite information to its readers relevant to herbal products. The text of this book is written by highly skilled, experienced, and renowned scientists, and researchers around the globe with up-to-date information to offer insights on herbal medicines to readers, researchers, academicians, scientists and industrialists worldwide.

This book, *Nutraceuticals and Dietary Supplements: Applications in Health Improvement and Disease Management* is comprised of 12 chapters with four sections that include an introduction, nomenclature, characteristics, and applications. The uses of nutraceuticals as agents to counter excess alcohol consumption also finds a place along with descriptions of the life-style disorder diabetes in separate section.

SECTION I: PREAMBLE OF NUTRACEUTICALS AND FORMULATIONS

Chapter 1, on functional foods, written by José M. Lorenzo and associates, gives an overview of functional foods. The authors have touched upon the main factors related to functional food development, the relation between consumers' preferences and functional food supply, and strategies to develop and establish a product in the market in this chapter.

The details of dietary supplements have been presented in Chapter 2, Nutraceutical Bioactives, written by Andrew G. Mtewa and colleagues. The authors have given elementary information of several nutraceuticals along with the pros and cons followed by the market scenario. Further, there is mention of analytical techniques used for analysis of nutraceuticals. The last segment of the chapter briefly gives an idea of manufacturing aspects.

Chapter 3, Introduction to Nutraceutical and Dietary Supplements, written by Bhushan R. Rane and Raj K. Keservani, provides an overview of nutraceuticals, embracing general features, market size, regulatory perspectivesm and some of the applications to control certain diseases. The important information has been provided as tabulated.

The multidimensional uses of herbal medicines are described in Chapter 4, Herbs in Health Management, written by Kapil Kumar and Rajan Kaushik. The chapter highlights the commonly available plants with medicinal importance. It also contains some plants that are used among the community but need to be further explored for their potential benefits to human health.

Chapter 5, Starch-Based Nanocarriers of Nutraceuticals: Synthesis and Applications, written by Alberto A. Escobar-Puentes and associates, focuses on formulations containing nutraceuticals that have starch at their core. The inherent advantages offered by nanotechnology are exploited at best in this compilation.

SECTION II: ROLE OF NUTRACEUTICALS IN CANCER TREATMENT

A general introduction to the role of dietary supplements in curbing the menace of cancer has been provided in Chapter 6, Study of Nutraceuticals in Cancer Treatment: An *In Silico* Approach, written by Anamika Basu. The author has given insights into *in silico* drug designing principles. Based on these nutraceuticals, yielding optimum benefits could be formulated and could hopefully cater to the needs of mankind in a much more fruitful manner.

Chapter 7 Nutraceuticals in Cancer Therapy, written by Andrew G. Mtewa and colleagues, gives a focused view of applications of nutraceuticals in the cure/treatment of several cancers. The probable mechanism of neoplastic cell death has been proposed, and examples are listed in tables.

SECTION III: NUTRACEUTICALS AND MENTAL ILLNESS

The applications of nutraceuticals pertaining to the treatment of mental illnesses have been discussed in Chapter 8, Exploration of Nano-Nutraceuticals in Neurodegenerative Diseases, written by Swati Pund and associates. The authors have strived to compile all nanometric nutraceutical formulations. This chapter provides an insight into applications of nutraceuticals in maintaining neurophysiology along with various scientific nanotechnological interventions to improve neuroprotective efficacy of nutraceuticals.

Chapter 9, Preclinical Perspectives of Mechanisms Underlying the Neuroprotective Effects of Selected Phyto-Constituents in Alzheimer's Disease, written by Marco Cascella and colleagues, has addressed the clinical demonstration of much-hyped dietary supplements and nutraceuticals. The studies performed in the laboratory are well correlated with *in vivo* behavior of these formulations.

SECTION IV: SPECIALIZED APPLICATIONS OF NUTRACEUTICALS

The specialized uses of nutraceuticals containing plants grown in specific regions are given by Chapter 10, The Science behind Maca, a Traditional Crop from the Central Andes, written by Gustavo F. Gonzales and associates. The chapter discusses the maca plant growing in the valleys of Andes mountains, which has been recognized as vitalizer and fertility booster since ancient times. The objective of the chapter is to cover the tradition behind the medicinal usage of this nutritional plant and the scientific research that demonstrates its properties.

Chapter 11, Nutraceutical Intervention for Treatment of Alcoholism/Drinking Problems, written by Charu Gupta and Dhan Prakash, describes alcohol dependence, which is essentially a brain disorder. The chapter has focused on the health abuse of alcoholism and the role of herbs and medicinal plants (importance, medicinal value, benefits, and the mode of action) for the treatment of alcoholism and drinking problems.

Chapter 12, Beta Cells and Diabetes, written by Shivani Srivastava and colleagues, describes the status of diabetes associated with the health of beta cells. The authors have covered the variables responsible for the development of type 2 diabetes and have suggested herbal remedies along with yoga to control the glucose levels below the threshold.

PART I

Preamble of Nutraceuticals and Formulations

CHAPTER 1

Functional Foods

JOSÉ M. LORENZO[1*], PAULO E. S. MUNEKATA[1], MIRIAN PATEIRO[1], ELENA MOVILLA FIERRO[2], SUZANA RIMAC BRNČIĆ[3], MLADEN BRNČIĆ[3], and FRANCISCO J. BARBA[4]

[1]*Centro Tecnológico de la Carne de Galicia, San Cibrao das Viñas, 32900 Ourense, Spain.*

[2]*Complejo Hospitalario Universitario de Ourense, Ourense, Spain*

[3]*Faculty of Food Technology and Biotechnology, University of Zagreb, Zagreb, Croatia*

[4]*Nutrition and Food Science Area, Preventive Medicine and Public Health, Food Science, Toxicology and Forensic Medicine Department, Faculty of Pharmacy, Universitat de València, Avda. Vicent Andrés Estellés, s/n, 46100 Burjassot, València, Spain*

[*]*Corresponding author. E-mail: jmlorenzo@ceteca.net*

ABSTRACT

Functional food is a category of food that comprises any food that is modified to provide clinically proven and documented health benefits in addition to its nutrients. This includes all food products that are either enriched, fortified, removed, or naturally enhanced of functional ingredients such as probiotics, prebiotics, minerals, fibers, fatty acids, phenolic compounds, and other bioactive compounds. The development of functional foods is a topic of important discussion as it is centered on the technological, consumer, marketing, and legal/regulatory aspects. On consumer side, a change in perception toward healthy food by means of improving control of diet composition as part of a major strategy to improve personal health status has led to the purchase intention of consumers worldwide. On the food industry side, the continuous monitoring of consumer behavior combined with the strategy of marketing tools promoted changes in the approach of food industries to meet consumer's preferences. In this complex scenario, governmental regulatory agencies and health authorities created a central point among the consumer's preferences, the continuous development of new products by food industries, and the scientific evidence that regulates (at some extent) the commercialization of functional food. In this chapter, the main factors related to functional food development, the relation between consumer's preferences and functional food supply, and the strategies to develop and establish a product in the market are overviewed and discussed.

1.1 INTRODUCTION

The definition of functional food involved complex concepts that can be misinterpreted. The European Commission proposes that important considerations must be stated regarding functional foods, in order to clarify its definition. The concept comprises the following assumptions: food is the base product; the food has either a naturally or an intentionally added compound or mixture of compounds that promotes additional health and well-being, or reduces the risk of a disease in comparison to its nutritional effects; the beneficial effects can be supported by an accredited source of information along with proper presentation and communication to consumers; and the additional health-related benefits can be achieved as part of a normal diet (Stein and Rodríguez-Cerezo, 2008). Such definition clarifies, at some extent, what can be understood as functional food. However, other important considerations can also be added such as improving biodisponibility of an active compound and targeting the general population of specific populational groups (such as patients of noncommunicable diseases, carriers of genetic disorders, and elderly population). It is important to emphasize that functional food is a concept added to certain food products rather than a well-defined group, which is in accordance with the current view about this issue: food claim and marketing regulation (Dipplock et al., 1999; Siró et al., 2008).

Functional food is a lucrative market, particularly in USA, Europe, and Japan where legislative, consumer behavior, and economic aspects, for instance, are some of the favorable factors leading to the constant growth of such market (Bigliardi and Galati, 2013). Moreover, functional foods that are reported to be mostly consumed in these and other countries are milk and dairy products (such as low fat and skim milk and yogurt), probiotics, fruit juices, and cereals. However, this market is so widely distributed and associated with several factors that no definitive profile of functional food consumer can be defined (Ozen et al., 2012).

The aim of this chapter is to describe the main steps of functional food development, explore the complex variables associated with consumer perception and their purchasing behavior to buy functional foods, and finally, exemplify additional factors and actions taken by food companies to achieve successful commercialization of functional food.

1.2 DEVELOPING A FUNCTIONAL FOOD

Developing a functional food, from a food developing point of view, is a complex and multistage task. Moreover, the complex combination of consumer desires, needs, and expectations with the current set of laws must be consider during the development of functional foods. The development of functional foods are developed either by adding either increasing a compound usually found in the food or by adding a new compound to the food matrix. This can be described by a sequence of stages: (1) the definition of a target health benefit and its progress, (2) identification and/or selection of functional ingredients and assessment of physiological effect, (3) development of a suitable food matrix, (4) bioaccessibility and/or bioavailability tests, (5) human studies, (6) approval by regulatory authorities, and last but not least (7) product announcement and commercialization (Siró et al., 2008). These steps are subject to change since each functional food development has unique details and factors (Figure 1.1).

The first stage comprises the selection of health benefit based on scientific evidence and/or consumer needs and desires. Regarding scientific evidence, studies about the potential health benefits of active compounds are the main source of information such as those related to cardiovascular diseases, for instance. Such noncommunicable disease is among the leading causes of death (accounting for around 70% of all deaths). The World Health Organization (WHO) highlights the necessity to strengthen the preventive measures for these diseases and improve the health status of general population, and care for those with noncommunicable diseases (World Health Organization, 2016). Conversely, consumers may desire for food that can improve their health status, such as those to improve mental and physical performance, in addition to functional foods that may help them reduce cardiovascular diseases (Weststrate et al., 2002).

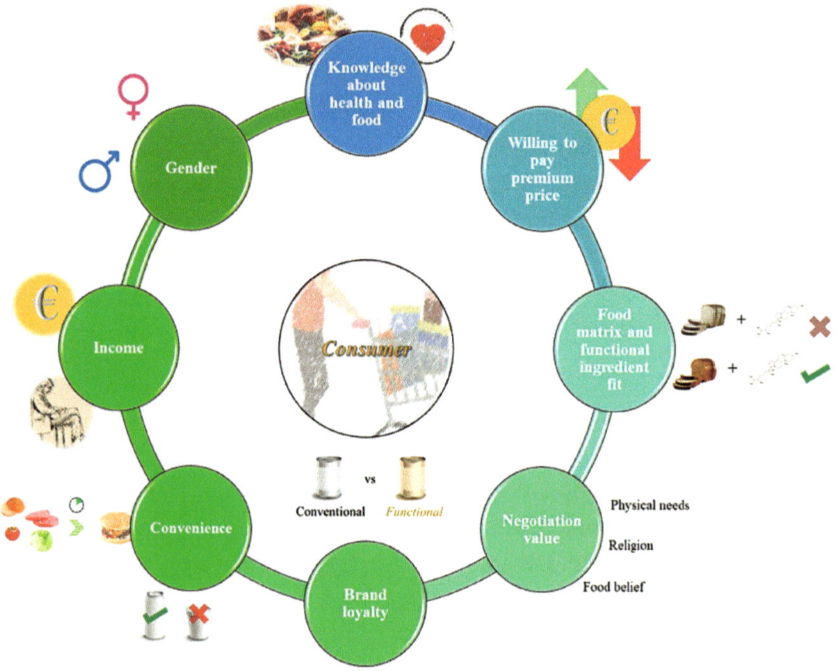

FIGURE 1.1 Functional food development stages.

The role of medical doctors and nutritionists is essential to constantly expand the level of information about new active ingredients and also to reveal potential applications of already known active ingredients. For instance, marine microorganisms are emerging as a potential source of bioactive compounds such as carbohydrates, fatty acids, vitamins, carotenoids, peptides, and other compounds that are associated with reducing cholesterol, hypotensive, or antidiabetic activity (Dewapriya and Kim, 2014). Phenolic compounds, a widely distributed and complex group, are emerging as potential active compounds to reduce the oxidative stress and also the risk of other diseases (Caleja et al., 2017).

In a similar way, the development of functional food can also consider the use of well-established functional ingredients added to conventional foods and are already available in the market. According to the European

Food Safety Authority (EFSA), the health claim of some active ingredients has already been authorized for use on food label such as calcium (European Food Safety Authority, 2016), iodine (European Food Safety Authority, 2014a), n-3 fatty acids (European Food Safety Authority, 2011), phytosterols (European Food Safety Authority, 2014b), and vitamin C (European Food Safety Authority, 2015). A similar view about the claims for oleic acid (Food and Drug Administration, 2018), n-3 fatty acids (Food and Drug Administration, 2004), and partially hydrolyzed whey protein (Food and Drug Administration, 2011) can also be observed in the Food and Drug Administration of United States laws for health claims in food labels.

It is worth mentioning that increasing food components of natural occurrence or adding active ingredients not naturally found in a specific food, at the industrial level, is considered as the main strategy used by the food industry to produce functional food. In addition, enhancing active compounds by natural ways, which can occur by special animal feed (Siró et al., 2008).

Particularly for the potential and emerging functional ingredients without health claims, the evaluation of physiological effects at the in vitro and animal level must be studied. The in vitro studies are particularly relevant to explore mechanisms and potential toxicological effects. For instance, meat peptides are promising functional ingredients with health benefits associated with cardiovascular, diabetes, mental health, and antioxidant in vitro assays (Lafarga and Hayes, 2014). In a further step, animal trails provide more information regarding the physiological effects of functional ingredients and support/disprove results obtained from in vivo studies, at some extent. For instance, the fungal bioactive polysaccharides were reported to induce protective mechanisms against cardiovascular diseases and diabetes in animal trials (Giavasis, 2014).

The ultimate target of functional ingredient physiological evaluation is the evaluation at the human level. Although important and crucial information is revealed by both in vitro and in vivo experiments, the human studies provide more realistic information about the physiological effects of selected compounds in the human body (Perel et al., 2007; van der Worp et al., 2010). An interesting strategy to conduct human studies with functional ingredients is to apply the same approach used for drug and regulatory approval by means of randomized and placebo-controlled studies. Moreover, meta-analysis studies can also be considered to strengthen the scientific evidence about a selected functional ingredient (Gregori and Gafare, 2012).

The relevance of clinical trials is also associated with data related to bioaccessibility and bioavailability of the functional ingredient. Bioaccessibility consists in the available amount of a selected compound for absorption in the gut as a consequence of digestive events on the food matrix. On the other hand, bioavailability is related to the amount of a selected compound that achieves systemic circulation and is also used in normal metabolism (Palafox-Carlos et al., 2011; Rein et al., 2013). Bioaccessibility and bioavailability are essential information due to potential impact that digestion has on the functionality of functional ingredients. The relevance of such variables is exemplified by the transformation of resveratrol during phase II metabolism. After absorption, resveratrol is ultimately converted into phase II metabolites, which can lead to the drastic reduction in parent resveratrol concentration and may induce different physiological effects (Patel et al., 2011).

The development of functional food also requires the appropriate isolation of functional ingredients and evaluation of functional food quality and stability. Considering the impact of the extraction technique on the isolation of functional ingredients, both conventional and innovative techniques have been explored to obtain selected functional ingredients. Regarding the extraction of selected compounds from vegetable sources, conventional extraction techniques such as hydrodistillation, maceration, and Soxhlet extraction are commonly applied. These techniques comprise common operations (e.g., heating, solvent reflux, and mixing) and can be easily performed. However, such techniques may cause environmental impact owing to their by-product (such as solvent disposal). Alternatively, other approaches associated with intrinsic product properties and physical phenomenon have been explored in last decades. Some innovative technologies can be cited here, such as enzyme-, microwaves-, pulsed electric field-, ultrasound-assisted extract, along with pressurized liquid and supercritical fluid extraction techniques (Azmir et al., 2013; Fogliano and Vitaglione, 2005).

The evaluation of product characteristics (e.g., physicochemical, texture, and sensory properties) and storage stability (microbial growth and sensory acceptability, for instance) during appropriate storage is crucial to estimate shelf life and carry out changes in product development and processing. Selecting the most appropriate processing technique and storage conditions can avoid the degradation of functional ingredients and prevent the reduction of health benefit effect associated with their consumption (Igual et al., 2010). It is worth mentioning that preparatory strategies can assist the preservation of health promoting activity exerted by functional ingredients. Encapsulation is an interesting technology to reduce the impact of heat, oxygen, and moisture to sensible compounds. Moreover, encapsulation can contribute to mask unpleasant taste, such as bitterness, control the release of functional ingredients, and improve the stabilization of incompatible food ingredients. This approach can be carried out in a protein, lipidic, or carbohydrate encapsulating matrix by means of spray drying, spray cooling, freeze drying, coacervation, extrusion, and liposomes, for instance (Augustin and Hemar, 2009). Moreover, studies with consumer perception and willing to buy such functional product are recommended at this stage of functional food development. Consumers perceive the combination of the food matrix and functional ingredient, which can eventually influence the product acceptance (Krutulyte et al., 2011).

The following step in functional food development is human studies. Up to this stage, all information is dispersed in well-defined groups of data such as the mechanisms and physiological impact of functional ingredients, bioaccessibility/bioavailability of functional ingredients, and characteristics and shelf life of functional products. However, in human studies, the functional food is ingested by target consumers and the evaluation of physiological effects is monitored. For instance, the fortification of blackcurrant juice with crowberry powder induced a slight improvement in the postprandial serum glycemic level in healthy subjects. The authors associated the increased polyphenolic content in functional beverage (rich in anthocyanins and polyphenols) with such effect (Törrönen et al., 2012).

It is important to mention that functional foods are part of a healthy lifestyle rather than an alternative therapeutic strategy to be used, which means that functional foods are first dedicated to healthy consumers who seek for health benefits by consuming functional foods. Although such a perspective is an important characteristic of functional foods, patients of target diseases were also subjected to human studies with functional foods wherein the health benefits and improved health status were observed. In this condition, the data obtained from such trials strengthen the role of functional food to reduce the risk of some diseases and improve the health status (Vitaglione et al., 2007).

Once the results of human studies support the consumption of a selected functional food, the approval by regulatory authorities is necessary for the proper use of health claim. Such information is indicated on food label and is regarded as crucial to prevent misinterpretation by consumers. Particularly in Europe, the Nutrition and Health Claim Regulation was created to stimulate the development of new functional food and also regulate the use of health-related information on food label. Three main categories of health claims are currently used in Europe: general functional claims from both well-accepted and new scientific evidence, reduction of disease risk claims, and health and development-related claims (European Parliament and Council of the European Union, 2006).

The evaluation of general functional claims (related to the effect of functional food on development, growth, and psychological and weight management) and other claims is carried out by EFSA that consider three consecutive steps to authorize their use. The first criterion is dedicated to evaluate if the functional ingredient has already been well characterized; the second criterion deals with the physiological effect of the functional food; and the third criterion considers the cause-effect relationship. Once all criteria are successfully fulfilled, EFSA publishes a scientific opinion regarding the use of health claim on food label. Moreover, the claim must be declared according to the specific designation pointed out by EFSA. However, failing to meet one of the criteria leads to the termination of assessment (European Food Safety Authority, 2016, 2017).

Product announcement and commercialization can take place after the last stage of product development. From this point, the relation between the food industry and consumers is expected to be more direct in order to evaluate the insertion and establishment of the new functional food in the market. Consumer acceptance and willingness to buy a selected functional food became even more relevant, but consumers are still not fully aware of functional food benefits. For instance, the concept of "functional food" remains unclear to most consumers. Moreover, important barriers remain the commercialization of functional foods. The higher rate of failure and investments than usually observed for a conventional product development are considered as some of the main specific challenges of the functional food market (Annunziata and Vecchio, 2011; Menrad, 2003). It is also relevant to consider that although the sequence proposed in this section is a reasonable order of critical steps to develop functional foods, the order of steps is subjected to change and even more than one stage can be simultaneously carried out.

Alternatively, the production of functional food from animal origin can be carried out by improving the feeding of animals with functional ingredients. One of the main functional foods of animal origin is the omega-3-enriched egg. The feeding fatty acid composition is improved by omega-3-rich sources such as fish oil, oilseeds, and microalgae (Fraeye et al., 2012). Moreover, the production of functional eggs can also include the use of vitamins and minerals such as folic acid and selenium, respectively (Surai and Sparks, 2001).

1.3 CONSUMERS AND THE DEMAND FOR FUNCTIONAL FOOD

Understanding consumers is a challenging task that is influenced by several factors such as knowledge about food and health, gender, age, willing to pay premium price, food matrix, and functional ingredient fit, negotiation value, brand loyalty, convenience, and income (Figure 1.2). It is important to mention that such variables display variations across studies and, therefore, a common functional profile cannot be defined (Aschemann-Witzel and Grunert, 2015; Brečić et al., 2014; Büyükkaragöz et al., 2014; Çakiroğlu and Uçar, 2018; Carrillo et al., 2012). The current consumer's knowledge about functional foods is considered low and demands constant attention and improvement. An example of this complex scenario is the perception of resveratrol as an active compound associated with health benefits. Danish consumers perceive functional foods as a means of general use rather than relating them with the health benefits of a particular compound present in the foods (Aschemann-Witzel and Grunert, 2015).

To improve consumer's knowledge about functional ingredients in foods, Bornkessel et al. (2014) suggested that significant predictors must be selected. In this study, the main predictors among German consumers regarding 10 selected functional ingredients (calcium, dietary fiber, iodine, omega-3-fatty acids, phytosterols, vitamin C, xylitol, and probiotics, antioxidants, and glucosamine) were health motivation and information strategies. Moreover, better communication of health benefits associated with functional food consumption with consumers is necessary. Similarly, a study with more than 1000 German consumers revealed that skepticism and reduced knowledge about functional foods are the main factors that negatively influence the intention to buy cardioprotective functional food (Jung et al., 2018).

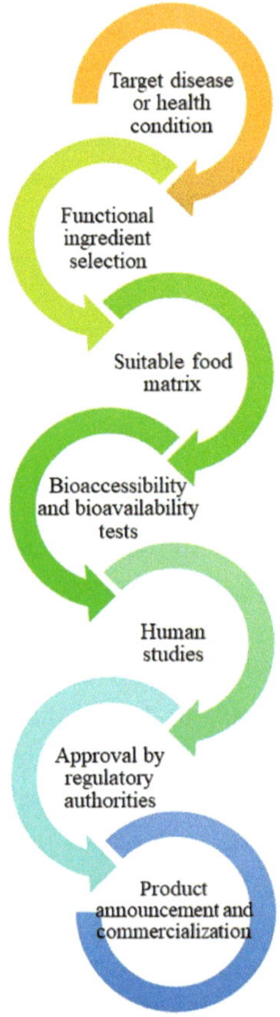

FIGURE 1.2 Factors associated with consumer choice for functional foods.

Conversely, the knowledge about the health benefits related to the consumption of lycopene was indicated as one of the main factors to increase the willingness to pay a premium price for lycopene-enriched tomatoes. Moreover, the food technology neophobia was only significant in hypothetical situations, which indicates that such phobia is more associated with personal opinion than actual consumer behavior (La Barbera et al., 2016).

The balance between the food matrix and the health-related ingredient is important for consumers. In this context, Danish consumer may consider that a better fit is perceived with less healthy food matrix rather than a healthy one (Krutulyte et al., 2011). Moreover, the influence of the food matrix-functional ingredient fit on purchase intention is particularly relevant for Canadian consumers with less knowledge about nutrition, food, and health (Lu, 2015). Another relevant result indicated by this author was the positive relation between the food matrix and health-related compound with health claim. The health claim (independent of nutritional information on the product label) improved the purchase intention of products with the low food matrix-functional ingredient fit perception.

Particularly in the group of consumers with the low nutritional knowledge level, food choice is rarely related to information and composition, as reported in a study about Spanish consumers. Such a group of consumers consider that label information is "too technical," which indicates an important barrier to attract consumer with low nutritional knowledge (Carrillo et al., 2012). Conversely, consumers with health-related attitudes tend to expect more health benefits than those who display a low level of health-related attitudes, according to a study carried out on Taiwanese consumers. Additionally, consumers with health-related attitudes are characterized by younger adults (Chen, 2011).

Consumer preference is another relevant characteristic that must be exploited. For instance, consumers skeptical about functional food prefer nonfunctional dairy products (due to the lack of belief in health-food composition relation), whereas health-conscious consumers do not prefer nonfunctional dairy products because of the lack of health benefits from such products (Bechtold and Abdulai, 2014). Another relevant aspect is the conflicting values that motivate the purchase of functional food. The association between conflicting values and purchase motivation was described by Hassan (2011), who evaluated the influence of conflicting values on the purchasing of functional foods in the three main ethnical groups of Malaysia (Chinese, Indians, and Malays). Such attitude is mainly triggered by individual physical needs, product characteristics, and cultural aspects. For example, the authors indicated that Malay consumer did not spend much time considering the consumption of functional products.

Brand is also perceived as an important factor associated with a functional food. In this context, the Italian yogurt market is a valuable example for the food industry. In this market, consumers are loyal to specific brands, which influence the success of functional dairy products. In addition, consumers of functional yogurts are less sensitive to price variation than those who consume conventional yogurts (Bonanno, 2012).

Convenience also plays a major role in the preference of consumers. For instance, Croatian consumers, particularly those who associate convenience with functional foods, buy such products as "quick fixes" (as a means to compensate for unhealthy life style). However, the authors also argued that in such a condition, consumers do not intent to improve diet quality (Brečić et al., 2014). A similar outcome was reported by De Devitiis et al. (2018) who evaluated the factors associated with the acceptance of functional fish burger. The authors observed that although the food matrix (fish) is perceived as tasty and healthy, the consumption of fish is usually limited by availability, convenience, cooking skills, and price. Conversely, fish burger is a more convenient and accessible product for the children present in the house, which is an important factor that supports its consumption. A similar outcome was observed, for instance, among elderly people and the perception of convenience (van der Zanden et al., 2014). Although such a group may be confused, skeptical, and do not associate functional food with personal health benefits, the potential consumption of protein-enriched food would occur if such improvement was carried out on food that they were already consuming.

An interesting study revealed that functional food consumers also make inferences about health claims (Grunert et al., 2011). This outcome was obtained from a survey of 720 German consumers of yogurt. The authors observed that consumers with positive attitude were the most likely group to make inferences, which may overstate the health claim. Moreover, the authors also associate such attitude with better knowledge about health and food and correct interpretation of health claims.

Regarding the effective associations, the study carried out by Bimbo et al. (2018) associated the body image dissatisfaction with the number of health-enhancing yogurt choices among Italians. The authors reported an inverse association between these factors, which was even enhanced among respondents that usually read food label and with more knowledge about health and food relationship. It is worth mentioning that such an effect was mainly observed among women.

It is worth mentioning that income can influence the consumption of functional foods. A survey with Italian consumers revealed that older consumers with low income and health condition were particularly interested in functional food (Cavaliere et al., 2015). According to the authors, this group of consumers trade expensive health care attitudes (treatments and related actions, for instance) for cheaper options such as foods with health claims. Another aspect of consumer preference regarding functional food is the taste (Verbeke, 2006). Although the health benefits motivate consumers to buy and consume functional food, consumer may not compromise taste to obtain a health benefit.

1.4 BEYOND FUNCTIONAL FOOD FOR SUCCESSFUL COMMERCIALIZATION

Achieving successful commercialization of functional foods is a complex and unique condition achieved by some functional food industries and brands. Although a well-planned food development aligned with consumer expectations and needs is the basis for functional food development, this approach is directly focused on functional food properties. A wider perception of consumer relationship with food company's image and current society trends, for instance, must be considered in this scenario to insert and establish a new functional food in the market.

This consideration was discussed by Bleiel (2010). This author explored the factors linked to functional food products that determine consumer choices such as industry adaptation to consumer perception of healthy life style. Slogans and dedicated marketing strategies to change company images from nonhealthy to natural, better lifestyle, and life enjoyment took the front seat in last decades. For instance, fast-food companies that were once associated with nonhealthy food consumption in the past are now communicating health-oriented propaganda. Another relevant aspect discussed by the authors is the inclusion of consumer perception of functional food characteristics during product development (as indicated in the previous section).

Moreover, effective communication is a key factor to attract consumer attention that eventually leads to product purchasing. An interesting example is the contrast between "added probiotics" and "better immunity," which reveals the real perception of consumers: buy a health benefit rather than a complex concept. Considering all previous information, factors, and the current characteristics of consumers, the successful commercialization of functional food can be summarized to three main pillars: *consumer need or desire to improve health status and reduce the disease risk*; *consumers must be aware of their particular health condition and status*; and *consumers have to be inclined to pay (premium prices in some cases) to consume functional foods* (Bleiel, 2010).

1.5 CONCLUSION AND FUTURE PERSPECTIVES

Functional foods are one of the main components of the food market. The development of such foods is a complex task that requires the efforts of medical doctors, nutritionists, and food technologists to be carried out at each development stage of functional foods, adapting to current trends in modern society. Understanding the consumer perception and attitude in relation to functional food consumption is a major challenge for producers of functional foods. Cognitive, behavioral, and affective factors along with the social, demographic, and economic aspects are important variables that must be constantly monitored during and after the product development. Shifting food company's image to healthy and better life style concepts and align consumer desire with health benefits by simple and direct communication must also be considered.

Although several challenges remain in the production and commercialization of functional foods, the future of functional food is promising. Continuous monitoring of consumer perception and acceptance must be carried out to expand, understand, and respond to possible variations in these aspects of consumer and functional food relationship due to the complexity of the functional food market. Focus on consumers (particularly those with less knowledge about nutrition, food, and health) to increase the amount of information and ease the selection of most appropriate functional food by balancing the several phycological, physical, socio, and demographic factors is still necessary.

Another possible aspect that can be considered in the development of functional foods is to explore more than one physiological effect. The concept of multifunctional food consists in exploring more than one functional ingredient and also inducing two or more physiological effects (Gregori and Gafare, 2012). However, this concept requires more studies to achieve appropriate methodologies to avoid bias and improve integration between studies regarding biomarkers, for instance.

AKNOWLEDGMENT

Paulo E. S. Munekata acknowledges postdoctoral fellowship support from Ministry of Economy and Competitiveness (MINECO, Spain) "Juan de la Cierva" program (FJCI-2016-29486). José M. Lorenzo and Paulo E. S.

Munekata are members of the MARCARNE network, funded by CYTED (ref.116RT0503).

KEYWORDS

- **functional ingredients**
- **product development**
- **health benefits**
- **consumer perception**
- **effective marketing**

REFERENCES

Annunziata, A.; Vecchio, R. Functional foods development in the European market: a consumer perspective. *J Funct Foods.* **2011**, 3(3): 223–228.

Aschemann-Witzel, J.; Grunert, K. G. Resveratrol and health from a consumer perspective: perception, attitude, and adoption of a new functional ingredient. *Ann NY Acad Sci.* **2015**, 1348(1): 171–179.

Augustin, M. A.; Hemar, Y. Nano-and micro-structured assemblies for encapsulation of food ingredients. *Chem Soc Ver.* **2009**, 38(4): 902–912.

Azmir, J.; Zaidul, I. S. M.; Rahman, M. M.; Sharif, K. M.; Mohamed, A.; Sahena, F.; Jahurul, M. H. A.; Ghafoor, K.; Norulaini, N. A. N.; Omar, A. K. M. Techniques for extraction of bioactive compounds from plant materials: a review. *J Food Eng.* **2013**, 117(4): 426–436.

Bechtold, K. B.; Abdulai, A. Combining attitudinal statements with choice experiments to analyze preference heterogeneity for functional dairy products. *Food Policy.* **2014**, 47: 97–106.

Bigliardi, B.; Galati, F. Innovation trends in the food industry: the case of functional foods. *Trends Food Sci Tech.* **2013**, 31(2): 118–129.

Bimbo, F.; Bonanno, A.; Van Trijp, H.; Viscecchia, R. Body image dissatisfaction and health-enhancing food choices: a pilot study from a sample of Italian yogurt consumers. *Brit Food J.* **2018**, 120(12): 2778–2792.

Bleiel, J. Functional foods from the perspective of the consumer: how to make it a success? *Int Dairy J.* **2010**, 20(4): 303–306.

Bonanno, A. Functional foods as differentiated products: the Italian yogurt market. *Eur Ver Agric Econ.* **2012**, 40(1): 45–71.

Bornkessel, S.; Bröring, S.; Omta, S. O.; van Trijp, H. What determines ingredient awareness of consumers? A study on ten functional food ingredients. *Food Qual Prefer.* **2014**, 32(C): 330–339.

Brečić, R.; Gorton, M.; Barjolle, D. Understanding variations in the consumption of functional foods—evidence from Croatia. *Brit Food J.* **2014**, 116(4): 662–675.

Büyükkaragöz, A.; Bas, M.; Sağlam, D.; Cengiz, Ş. E. Consumers' awareness, acceptance and attitudes towards functional foods in Turkey. *Int J Consum Stud.* **2014**, 38(6): 628–635.

Çakiroğlu, F. P.; Uçar, A. Consumer attitudes towards purchasing functional products. *Prog Nutr.* **2018**, 20(2): 257–262.

Caleja, C.; Ribeiro, A.; Barreiro, M. F.; Ferreira, I. C. F. R. Phenolic compounds as nutraceuticals or functional food ingredients. *Curr Pharm Design.* **2017**, 23(19): 2787–2806.

Carrillo, E.; Varela, P.; Fiszman, S. Influence of nutritional knowledge on the use and interpretation of Spanish nutritional food labels. *J Food Sci.* **2012**, 77(1): H1–H8.

Cavaliere, A.; Ricci, E. C.; Banterle, A. Nutrition and health claims: who is interested? An empirical analysis of consumer preferences in Italy. *Food Qual Prefer.* **2015**, 41: 44–51.

Chen, M. F. The joint moderating effect of health consciousness and healthy lifestyle on consumers' willingness to use functional foods in Taiwan. *Appetite*. **2011**, 57(1): 253–262.

De Devitiis, B.; Carlucci, D.; Nocella, G.; Viscecchia, R.; Bimbo, F.; Nardone, G. Insights for the development of a functional fish product: drivers and barriers, acceptance, and communication of health benefits. *J Aquat Food Prod T.* **2018**, 27(4): 430–445.

Dewapriya, P.; Kim, S. K. Marine microorganisms: an emerging avenue in modern nutraceuticals and functional foods. *Food Res Int.* **2014**, 56: 115–125.

Dipplock, A.; Aggett, P.; Ashwell, M.; Bornet, F.; Fern, E.; Robertfroid, M. Scientific concepts of functional foods in Europe: consensus document. *Br J Nutr.* **1999**, 81(1): 1–27.

European Parliament and Council of the European Union. Regulation (EC) No 1924/2006 of the European Parliament and of the Council of 20 December 2006 on nutrition and health claims made on foods, **2006**. https://eur-lex.europa.eu/LexUriServ/LexUriServ.do?uri=OJ:L:2006:404:0009:0025:EN: PDF (accessed October 24, 2018).

European Food Safety Authority. Scientific opinion on the substantiation of health claims related to docosahexaenoic acid (DHA), eicosapentaenoic acid (EPA) and brain, eye and nerve development (ID 501, 513, 540), maintenance of normal brain function (ID 497, 501, 510, 513, 519, 521, 534, 540, 688, 1323, 1360, 4294), maintenance of normal vision (ID 508, 510, 513, 519, 529, 540, 688, 2905, 4294), maintenance of normal cardiac function (ID 510, 688, 1360),"maternal health; pregnancy and nursing" (ID 514), "to fulfil increased omega-3." *EFSA J.* **2011**, 9(4): 1–30.

European Food Safety Authority. Scientific opinion on the substantiation of a health claim related to iodine and contribution to normal cognitive development pursuant to Article 14 of Regulation (EC) No 1924/2006: scientific opinion. *EFSA J.* **2014a**, 12(1): 1–10.

European Food Safety Authority. Scientific opinion on the modification of the authorisation of a health claim related to plant sterol esters and lowering blood LDL-cholesterol; high blood LDL-cholesterol is a risk factor in the development of (coronary) heart disease pursuant to Article 14 of Regulation (EC) No 1924/2006, following a request in accordance with Article 19 of Regulation (EC) No 1924/2006: scientific opinion. *EFSA J.* **2014b**, 12(2): 1–14.

European Food Safety Authority. Vitamin C and contribution to the normal function of the immune system: evaluation of a health claim pursuant to Article 14 of Regulation (EC) No 1924/2006: scientific opinion. *EFSA J.* **2015**, 13(11): 1–9.

European Food Safety Authority. Calcium and contribution to the normal development of bones: evaluation of a health claim pursuant to Article 14 of Regulation (EC) No 1924/2006: (Scientific Opinion). *EFSA J.* **2016**, 14(10): 1–9.

European Food Safety Authority. Scientific and technical guidance for the preparation and presentation of a health claim application (revision 2): scientific opinion. *EFSA J.* **2017**, 15(1): 1–31.

Fogliano, V.; Vitaglione, P. Functional foods: planning and development. *Mol Nutr Food Res.* **2005**, 49(3): 256–262.

Food and Drug Administration. Letter responding to health claim petition, dated November 3, 2003 (Martek Petition): omega-3 fatty acids and reduced risk of coronary heart disease (Docket No. 2003Q-0401), **2004**. http://wayback.archive-it.org/7993/20171114183727/https://www.fda.gov/Food/IngredientsPackagingLabeling/LabelingNutrition/ucm072932.htm (accessed October 25, 2018).

Food and Drug Administration. 100% Whey-protein partially hydrolyzed infant formula and reduced risk of atopic dermatitis, **2011**. http://wayback.archive-it.org/7993/20171114183703/https://www.fda.gov/Food/IngredientsPackagingLabeling/LabelingNutrition/ucm256731.htm (accessed October 25, 2018).

Food and Drug Administration. Oleic acid and coronary heart disease (Corbion Biotech Petition), **2018**. https://www.fda.gov/downloads/Food/LabelingNutrition/UCM624755.pdf (accessed October 26, 2018).

Fraeye, I.; Bruneel, C.; Lemahieu, C.; Buyse, J.; Muylaert, K.; Foubert, I. Dietary enrichment of eggs with omega-3 fatty acids: a review. *Food Res Int.* **2012**, 48(2): 961–969.

Giavasis, I. Bioactive fungal polysaccharides as potential functional ingredients in food and nutraceuticals. *Curr Opin Biotech.* **2014**, 26: 162–173.

Gregori, D.; Gafare, C. E. Multifunctional food: medical evidence and methodological notes on substantiating health claims. *Int J Food Sci Nutr.* **2012**, 63(Suppl.1): 29–36.

Grunert, K. G.; Scholderer, J.; Rogeaux, M. Determinants of consumer understanding of health claims. *Appetite.* **2011**, 56(2): 269–277.

Hassan, S. H. Managing conflicting values in functional food consumption: the Malaysian experience. *Brit Food J.* **2011**, 113(8): 1045–1059.

Igual, M.; García-Martínez, E.; Camacho, M. M.; Martínez-Navarrete, N. Effect of thermal treatment and storage on the stability of organic acids and the functional value of grapefruit juice. *Food Chem.* **2010**, 118(2): 291–299.

Jung, F. U. C. E.; Luck-Sikorski, C.; Krüger, M.; Wiacek, C.; Braun, P. G.; Engeli, S.; Riedel-Heller, S. G. Eat healthy? Attitudes of the German population towards industrially produced cardioprotective food. *Nutr Metab Card Dis.* **2018**, 28(5): 486–493.

Krutulyte, R.; Grunert, K. G.; Scholderer, J.; Lähteenmäki, L.; Hagemann, K. S.; Elgaard, P.; Nielsen, B.; Graverholt, J. P. Perceived fit of different combinations of carriers and functional ingredients and its effect on purchase intention. *Food Qual Prefer.* **2011**, 22(1): 11–16.

La Barbera, F.; Amato, M.; Sannino, G. Understanding consumers' intention and behaviour towards functionalised food: the role of knowledge and food technology neophobia. *Brit Food J.* **2016**, 118(4): 885–895.

Lafarga, T.; Hayes, M. Bioactive peptides from meat muscle and by-products: generation, functionality and application as functional ingredients. *Meat Sci.* **2014**, 98(2): 227–239.

Lu, J. The effect of perceived carrier-ingredient fit on purchase intention of functional food moderated by nutrition knowledge and health claim. *Brit Food J.* **2015**, 117(7): 1872–1885.

Menrad, K. Market and marketing of functional food in Europe. *J Food Eng.* **2003**, 56(2–3): 181–188.

Ozen, A. E.; Pons, A.; Tur, J. A. Worldwide consumption of functional foods: a systematic review. *Nutr Rev.* **2012**, 70(8): 472–481.

Palafox-Carlos, H.; Ayala-Zavala, J. F.; González-Aguilar, G. A. The role of dietary fiber in the bioaccessibility and bioavailability of fruit and vegetable antioxidants. *J Food Sci.* **2011**, 76(1): R6-R15.

Perel, P.; Roberts, I.; Sena, E.; Wheble, P.; Briscoe, C.; Sandercock, P.; Macleod, M.; Mignini, L. E.; Jayaram, P.; Khan, K. S. Comparison of treatment effects between animal experiments and clinical trials: systematic review. *Brit Med J.* **2007**, 334: 197.

Rein, M. J.; Renouf, M.; Cruz-Hernandez, C.; Actis-Goretta, L.; Thakkar, S. K.; da Silva Pinto, M. Bioavailability of bioactive food compounds: a challenging journey to bioefficacy. *Brit J Clin Pharmaco.* **2013**, 75(3): 588–602.

Siro, I.; Kápolna, E.; Kápolna, B.; Lugasi, A. Functional food. Product development, marketing and consumer acceptance: a review. *Appetite.* **2008**, 51(3): 456–467.

Stein, A. J.; Rodríguez-Cerezo, E. Functional food in the Europe Union, **2008**. https://ec.europa.eu/jrc/en/publication/eur-scientific-and-technical-research-reports/functional-food-european-union (accessed Nov. 1, 2018).

Surai, P. F.; Sparks, N. H. C. Designer eggs: from improvement of egg composition to functional food. *Trends Food Sci Tech.* **2001**, 12(1): 7–16.

Törrönen, R.; McDougall, G. J.; Dobson, G.; Stewart, D.; Hellström, J.; Mattila, P.; Pihlava, J. M.; Koskela, A.; Karjalainen, R. Fortification of blackcurrant juice with crowberry: impact on polyphenol composition, urinary phenolic metabolites, and postprandial glycemic response in healthy subjects. *J Funct Food.* **2012**, 4(4): 746–756.

van der Worp, H. B.; Howells, D. W.; Sena, E. S.; Porritt, M. J.; Rewell, S.; O'Collins, V.; Macleod, M. R. Can animal models of disease reliably inform human studies?. *PLoS Med.* **2010**, 7(3): e1000245.

van der Zanden, L. D.; van Kleef, E.; de Wijk, R. A.; van Trijp, H. C. Knowledge, perceptions and preferences of elderly regarding protein-enriched functional food. *Appetite.* **2014**, 80: 16–22.

Verbeke, W. Functional foods: Consumer willingness to compromise on taste for health?. *Food Qual Prefer.* **2006**, 17(1–2): 126–131.

Vitaglione, P.; Fogliano, V.; Stingo, S.; Scalfi, L.; Caporaso, N.; Morisco, F. Development of a tomato-based food for special medical purposes as therapy adjuvant for patients with HCV infection. *Eur J Clin Nutr.* **2007**, 61(7): 906–915.

Weststrate, J. A.; Van Poppel, G.; Verschuren, P. M. Functional foods, trends and future. *Brit J Nutr.* **2002**, 88(S2): S233–S235.

World Health Organization. Assessing national capacity for the prevention and control of noncommunicable diseases: Global survey. *World Health Organization*, **2016**, pp. 10–87.

CHAPTER 2

Nutraceutical Bioactives

ANDREW G. MTEWA[1,2*], KENNEDY J. NGWIRA[3], STEPHEN LUTOTI[2], DAVIES MWETA[1], PATRICK E. OGWANG[2], and DUNCAN C. SESAAZI[2]

[1]*Department of Chemistry, Malawi Institute of Technology, Malawi University of Science and Technology, Limbe, Malawi, East Africa*

[2]*Pharmbiotechnology and Traditional Medicine Center of Excellence, School of Medicine, Mbarara University of Science and Technology, Mbarara, Uganda, East Africa*

[3]*Institute of Molecular Science, School of Chemistry, University of Witwatersrand, Johannesburg, South Africa*

[*]*Corresponding author. E-mail: amtewa@must.ac.mw; andrewmtewa@yahoo.com*

ABSTRACT

Nutraceuticals are a group of products, natural or synthetic, that contain both nutritious and pharmaceutical agents. The products are more of a gray area between foods and drugs, but their bioactives essentially possess properties that fall well within both categories. Nutraceuticals present a field that is yet to be exhausted and developed. Every year, seasons almost oversupply a bulk of biomass of fruits and their residues which are mostly underutilized and find themselves in dumpsites. If deliberate investments were focused on these, a robust nutraceutical industrial revolution at a professional scale would contribute to a massive recovery of nutraceutical bioactives and the production of their derivative products to improve health and well-being of the general populace. An improved health status of a population is a key recipe for socioeconomic development. The future with nutraceuticals is rich with potential specialized nutraceutical products in various fields of life and for the exploration of a relatively new field of "nutraceuptides" where potential combination effects and therapies of nutraceuticals and peptides from animal venom would be the center of focus.

2.1 INTRODUCTION

In 1989, the term "nutraceuticals" was derived by Dr Stephen De Felice from a combination of "nutrition" and "pharmaceuticals." Nutraceuticals are plant sources containing chemicals that are primarily sought for their nutritive value and contain significant health beneficial properties (Sarin et al., 2012). These properties could be immune modulation, potential antioxidants, and other important pharmacological activities against diseases (Park et al., 2016). A related term, nutraceuptides, adds the use of animal venom and/or peptides to nutraceutical to treat or manage diseases and health (Mtewa, 2018; Mtewa and Kumphanda, 2018). Besides household benefits, natural product nutraceuticals have revolutionalized the food and pharmaceutical industries by showing potential of replacing or improving synthetic health-related chemical groups by natural nutraceuticals. Some of the synthetic chemical groups that have been affected are antioxidants and vitamin supplements. This has impacted the commercial sector as the newer natural product derivatives are more preferred by consumers than synthetic ones with a perception and belief that natural products are generally safer for consumption than synthetics. One of the most economically significant benefits of natural nutraceuticals over synthetics for manufacturers is the low-cost sourcing and production of the same right from raw materials to the market. In general, nutraceuticals are classified into functional foods, medicinal foods, dietary foods, and pharmaceutical foods.

2.2 SOME COMMON NUTRACEUTICAL BIOACTIVES

Different nutraceuticals have been proven to be important for health management as shown in Table 2.1, which is discussed in what follows.

Many botanical foods and spices are known to possess health benefits. These include vegetables, fruits, tubers, and roots. Some research reports that phenolic compounds, which are also present in botanical species and most of their products, are responsible for most biological properties (Chauhan et al., 2013). Antioxidant potential of phenolic compounds is associated with such bioactivities as anticarcinogenicity (Mtewa et al., 2017; Prakash et al., 2011), anti-allergenicity, anti-aging, and antimutagenicity, which have also been reported for synthetic and natural antioxidants. Some of the most common sources of nutraceuticals are: garlic, ginger, and turmeric used as spices to meals and beverages. Turmeric roots are rich in *curcumin* that is an effective anti-inflation and antioxidant agent.

TABLE 2.1 Some Commonly Accessible Nutraceuptides

Serial Number	Nutraceutical	Health Management Benefit
1	Tannins	Prevents and reduces cardiovascular diseases
2	Probiotics	Controlling bacteria levels in the alimentary system
3	Fatty acids: O-3	Has anti-inflammatory properties preventing chronic diseases
4	Fatty acids: α-lipoic acid	Reduces nerve pain, prevents diabetes, facilitates weight loss and wound healing
5	Mono saturated fatty acids	Prevents coronary diseases
6	Sterols	Reduce cholesterol
7	Lutein	Prevents some cancers
8	Curcumin	Prevents inflammation
9	Glucosinolates	Prevents most cancer development
10	β-Glucan	Reduces fats in the colon and prevents colon cancers
11	Vitamins (D3)	Facilitates teeth and bone development and has immune modulatory properties
12	Sulfides and thiols	Maintains a healthy immune system
13	Fiber	Maintains a healthy digestive systems and prevents colon cancers
14	Phenolics	Prevents most cancers and nephritic disorders

Plants that are rich in phenolics are: *flavonoid*, *lignans*, *catechins*, or *anthocyanin*; these are potentially very good nutraceuticals. These groups of compounds are responsible for the neutralization of free radicals, thereby reducing risks of the development of cancers and the prevention of nephrosis. Some of the plants that are rich in these compounds are: tea, red wine, cocoa, and bilberries

Plant *sterols* like *stanol esters* have a potential of reducing cholesterol in blood through cholesterol-absorption inhibition. *Tannins* present in most plants have an ability to reduce risks of cardiovascular diseases and also to maintain the health status of the urinary tract. *Ketones* from raspberry are said to have thermogenic properties, and they effectively work as a good fat-blocking agent. *Probiotics* help in maintaining a balance in the required levels of bacteria in the body, which sustains appropriate conditions in the gut and intestines. *Glucosinolates* abundant in cauliflowers are reported to be anti-cancer agents (Zu et al., 2010).

Carotenoid-rich plants like tomatoes and carrots among others are significant natural plants that protect us against diseases. *Lycopene* and *lutein* are the most important and effective antioxidant carotenoids active against most diseases (Radhika et al., 2011). *Lutein* has been reported active against breast, lung, and cervical cancers and also has protective effects on muscular degeneration that is age dependent. *Lutein* is mostly found in green vegetables.

Thiols and *sulfides* found largely in vegetables are important in maintaining an immune system (Khan et al., 2014). β-*Glucan* was reported to be an effective reducer of cholesterol in the body and also active against colon cancers when available in a soluble form (Radhika et al., 2011). *Tocotrienol* was reported to be functioning as an antioxidant and also has anticancer properties.

Some *vitamins* also play a vital role as nutraceuticals. Vitamin D_3 helps in bone and teeth development. It also has immune-modulatory properties particularly when used together with calcium. Vitamin B_{12} controls metabolic processes and is used to enhance levels of energy (Radhika et al., 2011).

Dietary fiber, both soluble and insoluble, from corn and wheat bran, oats, beans, and some nuts was reported to have reducing effects on colon cancers and maintaining the functioning of a digestive system (Khan et al., 2014). Green tea compounds are usually used as nutraceutical supplements due to their common effective antioxidant properties. These are largely extracted from phenolic compounds of the plant that are present in abundant.

Fatty acids, mainly O-3, present in flaxseed oil are vital for body functioning, but they cannot be manufactured by the body; they can only be supplemented to the body. Their role is to inhibit inflammation, which makes them able to reduce risks of chronic diseases such as cancers, arthritis, and cardiac ailments (Radhika et al., 2011). They also improve vision and mental functions, and reduce cardiovascular risks. Alpha-lipoic acid, another fatty acid and an antidiabetic, has been used to help in weight loss, wound healing, nerve pain, and in the treatment of rheumatoid arthritis. Monosaturated fatty acids from most nut trees are reported to reduce the risk of coronary diseases (Khan et al., 2014).

An isolated *alkylated piperazine* from *Arum palaestinum* showed significant cytotoxicity against tumor cells *in vitro* (El-desouky et al., 2007). An isolated compound from *Boswellia carterii*, acetyl-11-keto-b-boswellic acid, was reported to inhibit prostate cancer promoting pathways (Ahmed et al., 2015; Yuan et al., 2008).

2.3 ADVANTAGES OF NUTRACEUTICAL BIOACTIVES

Advantages of nutraceutical bioactives are as follows:

1. They are important agents for nutritional deficiencies' treatment. In this role, they can be used as supplements to food and good groups. They provide essential nutrients and minerals that are required by the body for the improvement and maintenance of health.
2. They possess protective effects against diseases, for example, work as antioxidants against potential progenesis of cancers.
3. They are advantageous as one can use them at will without necessarily seeking prescription.
4. Their sources are usually readily available. One may not be required to be on medical schemes or be on a highly paying job for them to access, at least basic natural nutraceutical bioactives. Getting a fruit or some vegetables would avail the bioactives to consumers.

2.4 DISADVANTAGES OF NUTRACEUTICAL BIOACTIVES

Some of the disadvantages of nutraceutical bioactives are as follows:

1. Processed nutraceuticals are sold at higher prices than normal food in supermarkets. This could be so due to the underdevelopment

of nutraceutical products, leading to an economical imbalance where supply is low and demand is high. The development of the products would decrease the costs of processed nutraceutical products.

2. Most nutraceutical products are being sold out with little or no professional research and tests. In this way, people may be recommending a health hazard supplemented to other foods just because there is no enough data on the side effects of the same. For example, *Gingko biloba*, a blood thinner, is sold as an ingredient to supplements. The prolonged use of this would lead to adverse health effects.
3. The availability of nutraceuticals has the potential of disrupting prescribed medical dosages for a particular disease. It has already been noted in various areas where patients first combine prescription drugs with supplements, and then get too comfortable to use them and completely stop taking the prescribed drugs thinking that the supplements are covering it all up. Civic education in communities is required on how to use nutraceuticals responsibly.

2.5 LOST OPPORTUNITIES IN NUTRACEUTICAL RECOVERY

Year in year out, there are huge quantities of biomass from around the world from wasted fruits and their other parts that are not consumed. This is largely seen in developing countries where economic challenges largely hamper sound and sustainable industrialization. If one thinks of tones of citrus fruits, pawpaws, bananas, mangoes, and other such fruits in season, it is clearly an opportunity for producing countries to design production strategies that can see the extraction and production of natural nutraceuticals and their derivatives right on site or at least close by. Available fruit peels, fruit remains from seasonal over supply, fruit pulp, and seeds, among other potential sources are the best raw materials available all over the world in a particular season of the year. However, this is not the case on the ground. In most developing countries, this self-presenting opportunity is disappointingly lost and one would see mounds of wasted fruits from seasonal oversupply, fruit wastes, and other residues lying around in market and community dumpsites—all gone because people have failed to eat. These, according to the authors of this chapter, are not merely wastes and residues; they are an unfortunate magnitude of a great opportunity for a community and/or a nation at large, slipping away from the hands, leaving the fore-bearer still wanting and begging in the next season.

2.5.1 A CHANCE TO RECLAIM THE LOST OPPORTUNITY

Having reported above an opportunity that has culturally slipped away from the populace means that no one is yet working on these plant wastes. This still remains a good news as it provides a chance for some to fill the gap and take this section of a book as a viable business idea that can save millions of state budgetary allocations on nutraceuticals and their derivative products for internal use or for export. However, there could be others who have ever tried to process nutraceuticals at an industrial level but failed to sustain the industry. There are challenges that first need to be identified and then addressed before venturing into such an industrial setup and operation. The following are only some of the main challenges that, if well addressed, can push for a good startup in the industrialization of nutraceutical from botanical species, particularly in the developing world:

1. *Unconcerned mindset*: Most of the societies that have some of the bulk biomass of remains from fruits, wasted fruits and vegetables, and seasonal oversupply of the same see it as a normal thing to keep dumping these precious raw materials onto dumpsites. The flying houseflies and lack of dumping space due to the huge size of the "wastes" do not seem to bother many, and no one dares to ask themselves what can be done to put all this to a stop. Many are comfortable with such eyesores and the next production season is usual and familiar. Such things can only come from a mindset that has been conditioned (likely subconsciously) to think that there is nothing that can be done about it. However, one needs to rise up and realize that there is a great potential to work on the "wastes" into a viable nutraceuticals industrialization business. This will make communities cleaner and more of a profit than a loss from expenditure. Mindset change needs to come first and inculcated in the masses.
2. *Lack of infrastructure*: Most community setups do not have proper infrastructures to house and manage both the raw materials and processing plants (Babbar et al., 2015). There is need for investors

to collaborate with processors as well as research technocrats in order to develop appropriate facilities required for specific intended products depending on their chemical properties.

3. *High production costs*: Any production setup is relatively costlier than taking in raw stuff. For instance, a professionally high-standard packaging plant alone could be at a required startup of between $10,000 and $20,000. This could only be a department in a professional industry that is working to produce nutraceuticals and their derivative products. The cost for the running of the plant alone is also a huge investment (Babbar et al., 2015). This could be one of the reasons why most financially challenged parts of the world keep dumping away nutraceutical raw materials.
4. *Lack of political will*: Governance drives development and does so only if deliberate political policies and strategies are pointing toward the same. Where there is a political will, it is easy for governments to lobby for, finance, and incentivize such projects as pharmaceutics in drug development and designing, as well as in the development of nutraceuticals and their derivative products to high standards. This will mark a development milestone but will also boost the economic and research potential for a nation.
5. *Pharmaceutical industrial politics*: Without overemphasizing the blow that the development of nutraceuticals from natural products and all over the world, existing multi-billion dollar pharmaceutical investments are likely to see their profits significantly reduced due to a paradigm shift where consumers may buy more natural products than synthetics. As a normal response, some may be forced to do the unthinkable; striking deal conditions with potential natural nutraceutical funders to slow down on partnering with the natural product world, state interventions to boost their local businesses, propaganda against natural medicine exploration, antagonistic influences in global health and medicines policies and guidelines, corruption with certifying agencies, among other potential responses. Currently, there are only speculations that natural medicine development cannot be advanced, which is evident from its centuries of use with no meaningful progress, but no formal evidence for such speculations and perceptions has been brought forward yet.

2.6 NUTRACEUTICAL ANALYSIS

2.6.1 CHEMICAL ANALYSIS

Chemical analysis of nutraceuticals uses both traditional and more advanced analytical tools, methods, and techniques. Of the most critical importance is the understanding of their underlying properties as individual molecules or as groups of compounds to determine best methods and tools of analysis to avoid degrading the active nutraceuticals and also to reasonably maximize yield.

Just as for medicinal and pharmaceutical compounds, nutraceuticals are extracted using appropriate solvents and analyzed through suitable chromatographic and spectroscopic techniques for purification and identification (Park et al., 2016). The sections below briefly describe fundamental steps in the chemical analysis for nutraceuticals.

2.6.1.1 EXTRACTION OF NUTRACEUTICALS FROM LOW-COST SOURCES

Extraction processes are the same for nutraceuticals as are for general phytochemicals. Usually, solvent-dependent extraction is the best where one needs to understand the polarity of the active or suspected active compounds and then use a solvent of a similar polarity in order to collect the right group of compounds (Mtewa et al., 2018). Where the polarity of the actives cannot be predetermined, serial extraction using different solvents of different polarities should be undertaken, starting from a nonpolar end to the most polar end of the solvents available. One would have hexane through dichloromethane, then ethyl acetate, and end up with methanol and then water. After extraction, proceed with each and every extract obtained from subsequent steps.

2.6.1.2 FRACTIONATION OF NUTRACEUTICALS

Fractionation basically aims at further separating the compounds that are in the extracts in order to obtain pure compounds. Fractionation can be solvent–solvent, where the extract is put in a separating funnel, missed up, and shaken together with two immiscible solvents. After a short stay, the solvents will separate, each carrying with it some compounds that are as polar as the solvents. The other method is to use a fractionating column, where a polar stationary phase, for example, silica gel (chromatographic grade), is either first added to a column, followed by a nonpolar mobile phase, and then a sample, followed again by a lot of the mobile phase, or the stationary phase, first mixed up with a little mobile phase to form a slurry (Better way than the former) and then proceed as described above. The collected solutions at different time intervals contain various compounds. All principles of column packing need to be followed (not detailed in this chapter). Fractionation can also help in the isolation of pure compounds, but this can be confirmed using thin-layer chromatography or high-performance liquid chromatography.

2.6.1.3 IDENTIFICATION OF POTENTIAL NUTRACEUTICAL ISOLATES

A set of tests on spectroscopic machines will help in determining the compounds isolated above (Mtewa et al., 2018). Nuclear magnetic resonance spectroscopy will give relative positions of carbons and protons in the isolates; infrared spectroscopy will provide details of vibrations and functional groups present. Mass spectroscopy provides masses of various chunks of compounds in the isolate. All these sets of data need to be interpreted and combined in order to come up with a structure and name of the compound therein. A qualified scientist can assist in this or, if new with these machines, proper guidance from a qualified analyst helps protect the equipment, but practice is the one that develops skills with time.

2.6.2 BIOLOGICAL ANALYSIS

Bioactivities of compounds need to be tested on various types of cells, tissues, as well as any other preparation depending on the intended bioactivity investigation. *In vitro* and *in vivo* test results need to be specified separately as they will have different interpretations when the research advances. *In silico* analyses can also be undertaken now that the compound structure is fully known.

2.6.3 PHARMACEUTICAL CONSIDERATIONS

Pure isolates that are promising for the development of a further product need to be pharmacologically complaint and fit for the purpose (Sardana, 2012). Of a critical importance is the safety profile of every isolate or any combination of the same. Data confirming the safe standing of every stage of development has to be collected and controlled using approved protocols and kept well for subsequent submissions to regulatory and approving bodies. General and other specific toxicological tests need to be undertaken. Appropriate animal models, once the stage is reached, should be followed and documented.

2.7 NUTRACEUTICAL BIOACTIVES' PRODUCTION

After the isolation and purification of compounds, tested for their bioactivities starting from cell line to animal and patient studies (Park et al., 2016), derivatives may be modified if necessary in order to improve the bioavailability of the compounds to biological targets. This is basically the chemistry of the compounds that will determine their absorptivity across membranes, distribution in the body, their metabolism, and excretion which are fundamental characteristics of a compound that must function in the body. This process takes a long time and requires huge monetary and time investments. Where products are to be labeled as drug supplements or functional foods, regulatory requirements need to be followed, (Sardana, 2012) to avoid fines and other charges for ethical and other misconducts later.

2.8 PREVENTIVE MEDICINE USING LOCALLY AVAILABLE MATERIALS (PRE-MULAM)

The future that natural bioactive nutraceuticals hold is industrialization, both at the small and large scales. With some challenges highlighted in this chapter, measures can be undertaken to address them one by one or in compound policies and approaches. One of the approaches that would work well is strategic collaboration. Strategic collaboration is one that does not only include investors due to their possession of money but also because they may easily invest in a white elephant. There is a need for a collaboration that has a combination of potentials, skills, and knowledge, such as research scientists, funders, laboratory directors, analysts, government agencies, and local communities. The inclusion of local communities is key to the sustainability of infrastructure as they feel part of the investment in their area. The presence of "development partners" in most developing countries could be handy in such types of developmental projects. As of today, there are a lot of untapped reservoirs of raw materials and opportunities to make the most and the best out of botanies—both for the health sector and the economic sector. If well implemented and managed, this could be a new future of preventive medicine using locally available materials. This concept will be easy to manage malnutrition and diseases at the same time as people will be fed on natural prophylaxes as diet, with supplementation from time to time as part of their daily life. Supplementation of foods using natural products could have fewer or no adverse effects than synthetic supplements that are flooding the markets today.

Apart from being food for the future, the future of nutraceuticals is bright. A few hurdles need to be cleared and regular deliberate engagement amongst stakeholders should be sought if success has to be attained. All stakeholders need to have a win in any development deals thereof. Apart from pre-MULAM, the future of nutraceutical is likely to cover specialized areas of military combat diets, supplements for specific disease prescriptions, physical activity special legal supplements, presurgery supplements, surgical support drops, postsurgery supplementations, and also combination of therapies and supplementations using nutraceutical derivatives and animal peptides from venoms, which is yet another area that requires robust research. Newer group of medicinal supplementary compounds called "nutraceuptides" can be attained, perhaps with better bioactivities as well as suitable chemistry for improved bioavailability, chemistry, and general pharmacology.

2.9 CONCLUSIONS

Nutraceuticals have always been there but research was limited. This remains an area that requires more attention, particularly when it comes to product development from nutraceuticals. Progress in research is undoubtedly increasing and progressing well, but there seems to be an asymptote to the growth of development and designing of better products that can help in reducing disease and health burdens in human societies. Raw materials are there; every year, tones are being dumped away that feels normal for some. These are lost opportunities that can be reclaimed by strategic collaborations for the good of societies. In future, peptides from animal venoms in combination with botanical nutraceuticals can be explored to come up with what we term "nutraceuptides."

KEYWORDS

- **nutraceuticals**
- **nutraceuptides**
- **nutraceutical bioactives**
- **preventive medicine**
- **functional foods**
- **probiotics**

REFERENCES

Ahmed, H. H.; Abd-Rabou, A. A.; Hassan, A. Z; Kotob, S. E. Phytochemical analysis and anti-cancer investigation of *Boswellia serrata* bioactive constituents in vitro. *Asian Pac J Cancer Prev*. **2015**, 16: 7179–7188.

Babbar, N.; Oberoi, H. S.; Sandhu, S. K. Therapeutic and nutraceutical potential of bioactive compounds extracted from fruit residues. *Crit Rev Food Sci Nutr.* **2015,** 55(3): 319–337.

Chauhan, B.; Kumar, G.; Kalam, N.; Ansari, S. H., Current concepts and prospects of herbal nutraceutical: a review. *J Adv Pharm Technol.* **2013,** 4(1): 4–8.

El-Desouky, S. K.; Kim, K. H.; Ryu, S. Y.; Eweas, A. F.; Gamal-Eldeen, A. M.; Kim, Y.-K. A new pyrrole alkaloid isolated from *Arum palaestinum* Boiss. and its biological activities. *Arch Pharma Res*. **2007**, 30: 927–931.

Hui-Qing, Y,; Feng, K,; Xiao-Ling, W,; Young, Y. F. C, Xiao-Yan, H.; Hong-Xiang, L. Inhibitory effect of acetyl-11-keto-b-boswellic acid on androgen receptor by interference of Sp1 binding activity in prostate cancer cells. *Biochem Pharmacol*. **2008**, 75: 2112–2121.

Khan, R. A.; Elhassan, G. O.; Qureshi, K. A. Nutraceuticals: in the treatment and prevention of diseases—an overview. *Pharma Innov J.* **2014,** 3(10): 47–50.

Mtewa, A.; Mwatseteza, J. F.; Biswick, T. Antioxidant potential and total phenolic changes in Malawian herbal formulations stored in clay pots and plastic bottles. *J Free Radic Antioxidants.* **2017,** 144: 479–494.

Mtewa, A. Nutraceuptides: perspectives into the future of supplementary food sciences and therapeutic technology. *Online J Complement Alt Med.* **2018**, 1(1): 1–2.

Mtewa, A. G.; Deyno, S.; M., K. F.; Amanjot, A.; Sesaazi, D. C. General extraction, isolation and characterization techniques in drug discovery. *Int J Sci Basic Appl Res.* **2018,** 38(1): 10–24.

Mtewa, A. G.; Kumphanda, J. Nutraceuptides: a potential future in the supplementary food sciences and therapeutic technology. *Curr Trend Complement Alt Med.* **2018**, 110: 1–2.

Park, P.; Wang, T. T.; Kim, E.; Zhong-Ji, Q. Nutraceuticals: recent advances of bioactive food components. *J Chem.* **2016,** 2016(5874726): 1–2.

Prakash, D.; Upadhyay, G.; Pushpangadan, P.; Gupta, C. Antioxidant and free radical scavenging activities of some fruits. *J. Complement Integr Med.* **2011,** 8: 1–9.

Radhika, P. R.; Singh, R. B. M.; Sivakumar, T., Nutraceuticals: an area of tremendous scope. *Int J Res Ayurveda Pharm.* **2011,** 2(2): 410–415.

Sardana, S. Herbal drug development from natural sources. *J Adv Pharm Technol Res.* **2012,** 3(2): 82.

Sarin, R.; Manisha, S.; Singh, R.; Kumar, S. Nutraceuticals: a review. *Int Res J Pharm.* **2012,** 3(4): 97–99.

Zu, Y,; Yu, H,; Liang, L. Activities of ten essential oils towards *Propioni* bacterium acnes and PC-3, A-549 and MCF-7 cancer cells. *Molecules.* **2010**, 15: 3200–3210.

CHAPTER 3

Introduction to Nutraceuticals and Dietary Supplements

BHUSHAN R. RANE[1*] and RAJ KESARWANI[2]

[1]*Department of Pharmaceutics, Shri DD Vispute College of Pharmacy and Research Center, University of Mumbai, New Panvel, Raigad 410206, Mumbai, Maharashtra, India*

[2]*Faculty of B. Pharmacy, CSM Group of Institutions, Allahabad, India*

[*]*Corresponding author. E-mail: rane7dec@gmail.com*

ABSTRACT

Nutraceuticals are the products that come under the food category that is beneficial in health and medical aspects. The drug shows some of the adverse/side effects; thus, nutraceuticals/food products/supplements are useful in the improvement of health. Nutraceutical products used as diet food, dietary supplements or nutrients, genetically modified foods, herb products, and co-processed foods or food products, for example, soups, beverages, and cereals. In the treatment of cancer or heart-related disorders, nutraceuticals are used. The use of nutraceuticals depends on the interest of consumers if they are cautious about health. In recent scenarios, worldwide growth, some of the development-related research studies, setup of standards, awareness of society or to marketing, quality assurance, and recent regulations will definitely write a new success story of nutraceuticals.

3.1 INTRODUCTION

In the 18th century, a great philosopher Hippocrates states that "let food be your medicine and medicine your food." The term "nutraceutical" was coined by Dr Stephen L. DeFelice, the Chairman and Founder of the Foundation for innovation. A nutraceutical is defined as a food product with physiological benefits that is isolated or purified from food items and ready for sale in the market in the form of a medicinal product. The nutraceutical is a combination of two terms: nutrition and pharamaceutical.

Nutraceuticals are nutritional food products that are beneficial for health and useful in the treatment of some diseases (Sapkale et al., 2012) (Figure 3.1). These are also termed as foodstuff (i.e., a prepared food or supplemental diet food) that is beneficial for health. Nutraceuticals are used to support wellness; these prevent the propagation of cancerous cells and thus control the symptoms and are also useful in therapies that are nonspecific and biological. Nutraceuticals come under the food product category only, so cannot be considered as a pharmaceutical product (Deore et al., 2014). Phytochemicals are nutraceuticals that are helpful in the treatment of cancer, diabetes, hypertension, and some heart diseases. Antioxidants are also a specific type of nutraceutical. Research envisaged and proved that consuming foods with some phytochemicals may be helpful in the treatment and cure of diabetes, cancer, some cardiovascular diseases, for example, carotenoids that are isolated from carrots. As per the Food and Drug Administration (FDA regulations), the manufacturers have to submit the clinical trial (Phases I, II, and III) data in relation to therapeutic efficacy for NDA approval, but in the case of nutraceuticals, such regulations are not too stringent in old days. Recently, manufacturers must have to prove that the composition of food products does not cause any disease. Different nutraceutical products have been accepted for the treatment of various diseases and disorders such as cancer and some diseases related to the cardiovascular system. Nutraceutical products have some health benefits like decreasing the threat and curing some issues related to hypertension, balancing cholesterol levels, maintaining the weight and, sometimes, osteoporosis, controlling sugar, as well as treating some health problems like arthritis, insomnia, cataract, and female menopause symptoms. As per the Dietary Supplement Health and Education Act of 1994 (DSHEA), the definition of nutraceuticals has been amended and expanded to add some things such as minerals, amino acids, vitamins, herbs/herbal compounds, botanicals, and some other dietary supplements; after the expansion of the term nutraceutical, the use of nutraceuticals has increased worldwide. In recent years, more interest has been shown by consumers toward the nutraceutical products in relation to health issues and diseases (Sapkale et al., 2012).

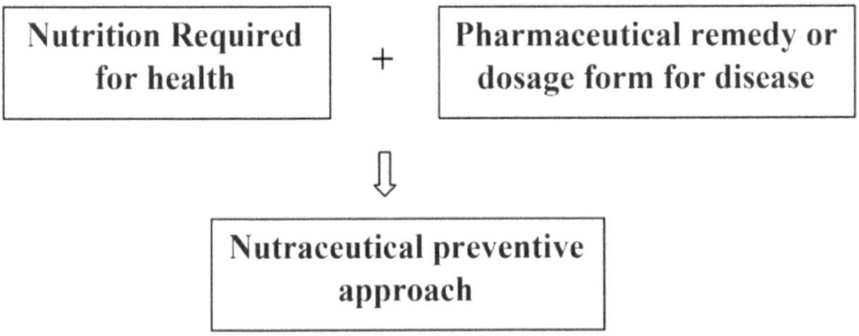

FIGURE 3.1 Concept of nutraceuticals (Lakshamna Prabhu et al., 2012).

3.1.1 *WHY TO USE NUTRACEUTICALS?*

- It is impossible to get adequate nutrition from our daily food (Lakshamna Prabhu et al., 2012).
- We are surrounded by a highly polluted and toxic environment as well as the high use of pesticides in farming that results in improper functioning of the body (Lakshamna Prabhu et al., 2012).
- Nowadays, some new ailments such as fatigue (chronic), lupus, and so forth disturb our body's regular functions (Lakshamna Prabhu et al., 2012).
- Awareness that prevention is better than cure (Deore et al., 2014).
- Chronic diseases without effective medicines (Deore et al., 2014).
- A gradual shift toward natural ingredients (Deore et al., 2014).
- The high cost of health care with an expensive disease-treatment approach (Deore et al., 2014).
- Increased work stress, busy schedule, and agriculture pollution that makes it unable to provide proper nutrition (Deore et al., 2014).
- People, especially of the younger generation, are becoming progressively more health-conscious (Deore et al., 2014).
- Drugs are foreign bodies for our system that show a lot of side effects, while nutritional supplements give a proper supply of nutrients to our body to strengthen our immunity and vitality.

- The use of nutraceuticals is the most strategic approach to uplift our body's normal functioning and also make an environment to increase the effectiveness of antibiotics against which resistance has developed (Lakshamna Prabhu et al., 2012).
- Malnourished population (Deore et al., 2014).

Thus, there is a great demand for alternative beneficial products like dietary supplements, functional food, and nutraceuticals by people as well as researchers to explore therapeutic values (Deore et al., 2014).

3.1.2 WHY NUTRACEUTICALS SEEM ATTRACTIVE? (LAKSHAMNA PRABHU ET AL., 2012).

- Nutraceuticals contain high percentage of phenolic components that are useful for human beings; thus, their proportion of daily consumption is increased.
- The chance of side effects in the case of nutraceuticals is rare.
- Their half-life is long in comparison to those of drugs.
- After administration, they are easily absorbed through the intestine.
- These come under nonprescription products and do not require any recommendation from a health specialist.
- Most people think that the use of nutraceuticals is the best approach to treat diseases/ disorders instead of using drugs.
- Supplemental foods are beneficial for the complete wellbeing of the body, and they also avoid the entry of pathogens in the body.
- Many people use nutraceuticals in certain cases when they came to know that specific treatments are failed to treat diseases.

3.1.3 RATIONALE FOR USE OF NUTRACEUTICALS

Nowadays, the use of nutraceuticals is increased as it shown a key role in the treatment of certain diseases or disorders. The use of some supplemental foods and dietary products as medicine is an old concept to treat diseases, for example, iodine is added in salt to treat iodine deficiency and iron or folic acid are mixed with wheat flour anciently to treat anemia. Likewise, for the treatment of vitamin A deficiency, the uptake of vitamin-A-rich food is the most preferable and cost-effective way (Gupta et al.,

2013). Imbalanced diet resulted in some of the chronic diseases and the progression of diseases to increase mortality and morbidity. Out of all osteoporosis and cancer cases diagnosed in many patients, 50% are due to improper diet.

3.2 CLASSIFICATION OF NUTRACEUTICALS

Nutraceuticals are food products that contain a labeled amount of nutrients and are prepared from various food materials having additional health benefits in comparison to normal food products (Tables 3.1–3.5). This is based on a claim that nutraceuticals improve health, treat or prevent chronic diseases, reduce the process of aging, increase the expected life span, help in the proper functioning of human body systems.

TABLE 3.1 Classification of Nutraceuticals Based on Their Nomenclature (Deore et al., 2014)

Class	Examples
Dietary supplements	Vitamins, minerals, herbs, amino acids, and organ tissues
Functional food	It can be any ordinary food that has components or ingredients added to give it a specific medical or physiological benefit, other than purely nutritional effects
Medical foods	These include probiotics and antioxidants manufactured in liquid, tablet, capsule forms
Farmaceuticals	The term is a combination of words farm and pharmaceuticals. These refer to medically valuable compounds produced from modified agricultural crops or animals

TABLE 3.2 Classification of Nutraceuticals Based Upon Food Source (Agrawal and Paridhavi, 2007)

Food Source	Example
Plants	β-Glucan, ascorbic acid, quercetin, cellulose, lutein, gallic acid, daidzein, allicin, lycopene, lignin, and minerals
Animals	Conjugated linoleic acid, lecithin, calcium, coenzyme Q10, selenium, and zinc
Microbes	*Saccharomyces boulardii*, *Bifidobacterium bifidum*, *B. longum*, *B. infantis*, *Lactobacillus acidophilus*, and *Streptococcus salvarius*

TABLE 3.3 Classification of Nutraceuticals Based on Ingredients (Deore et al., 2014)

Class	Example
Minerals	Calcium: maintaining bone strength
	Iron: metabolism and energy production
	Magnesium: for nerve and muscle function
	Phosphorous: part of the genetic material
	Chromium: with insulin help to convert carbohydrates and fats into energy
	Cobalt: essential components of vitamin B12
	Copper: part of hemoglobin and collagen production
	Iodine: essential for the proper functioning of thyroid
Vitamins	Vitamin A: antioxidant
	Vitamin E: antioxidant
	Vitamin K: essential for blood clotting
	Vitamin D: bone development and calcium absorption
	Vitamin C: antioxidant
	Vitamins B1, B2, and B3: energy production and healthy eyes and skin
	Folic acid: RBC formation, protection against heart disease, and prevention of birth defects
Antioxidants	Antioxidants are widely used in dietary supplements and have been investigated for the prevention of diseases such as cancer, coronary heart diseases, and also altitude sickness. An antioxidant is a molecule that inhibits the oxidation of other molecules. Oxidation is a chemical reaction that transfers electrons or hydrogen from a substance to an oxidizing agent. Oxidation reactions can produce free radicals. In turn, these radicals can start chain reactions. When the chain reaction occurs in a cell, it can cause damage to or death of the cell. Antioxidants terminate these chain reactions by removing free radical intermediates and inhibit other oxidation reactions. They do this by being oxidized themselves, so these are often reducing agents such as thiols, ascorbic acid, or polyphenols

TABLE 3.3 *(Continued)*

Class	Example
Polyunsaturated fatty acids (PUFAs)	Polyunsaturated fatty acids are fatty acids that contain more than one double bond in their backbone. This class includes many important compounds, such as essential fatty acids. Essential fatty acids, or EFAs, are fatty acids that humans and other animals must ingest because the body requires them for good health but cannot synthesize them. The term "essential fatty acids' refers to fatty acids required for biological processes and not those that only act as fuel. Only three EFAs are known for humans: alpha-linolenic acid (an omega-6 fatty acid), lauric acid (a saturated fatty acid), and palmitoleic acid (a monosaturated fatty acid). Essential fatty acids play a part in many metabolic processes, and there is evidence to suggest that low levels of essential fatty acids may be a factor in a number of illnesses, including osteoporosis
Prebiotics	Probiotics are nondigestible food ingredients that stimulate the growth and/or activity of bacteria in the digestive system in ways claimed to be beneficial to health. Traditional dietary sources of probiotics include soybeans, inulin sources, raw oats, unrefined wheat, unrefined barley, and yacon. Some of the oligosaccharides that naturally occur in breast milk are believed to play an important role in the development of a healthy immune system in infants
Probiotics	Probiotics are live microorganisms that when administered in adequate amounts confer a health benefit to the host. Lactic acid bacteria (LAB) and *Bifidobacteria* are the most common types of microbes used as probiotics, but certain yeasts and bacilli may also be used. Probiotics are commonly consumed as part of fermented foods with specially added active live cultures, such as in yogurt, or as dietary supplements
Fibers	Dietary fiber is the indigestible portion of plant foods having two main components: soluble (may be prebiotic and/or viscous) fiber, which is readily fermented in the colon into gases and physiologically active byproducts, and insoluble fiber. Bulking fibers absorb water as they move through the digestive system, easing defecation. Fermentable insoluble fibers mildly promote regularity, although not to the extent that bulking fibers do, but they can be readily fermented in the colon into gases and physiologically active byproducts
Phytochemicals	Polyphenols are useful to prevent and control arterial diseases. Flavonoids are useful to block the acetyl choline esterase (ACE) and strengthen the tiny capillaries that carry oxygen and essentials nutrients to all cells. Glucosamine and chondroitin sulfate are used against osteoarthritis and regulate gene expression and synthesis of nitric oxide (NO) and prostaglandin E2 (PGE2)
Health drinks	Health drinks are either just water fortified with some kinds of flavors or tea, milk, and juices enriched with healthy ingredients. Dairy drinks (19%), general health drinks (13.4%), ready-to-drink tea and coffee (11.2%), and sports and energy drinks (10%–12%) are famous types of health drinks

TABLE 3.4 Classification of Nutraceuticals Based Upon the Mechanism of Action (Agrawal and Paridhavi, 2007)

Anticancer	Antioxidant Activity	Anti-inflammatory	Osteogenetic or Bone Protective
Capsaicin	Ascorbic acid	Linolenic acid	CLA
Genestein	β-Carotene	EPA	Soy protein
Daidzein	Polyphenolics	DHA	Genestein
	Tocopherol	GLA	Daidzein
	α-Tocopherol	Capsaicin	Calcium
	Ellagic acid	Quercetin	Quercetin
	Lycopene	Curcumin	Inulin

TABLE 3.5 Classification of Nutraceuticals Based Upon the Chemical Nature (Agrawal and Paridhavi, 2007)

Class/Component	Source	Potential Benefits
Carotenoids	Carrots	Neutralizes free radicals that may cause damage to cells
Collagen hydrolysate	Gelatin	May help improve some symptoms associated with osteoarthritis
Dietary fiber	Wheat bran	May reduce the risk of breast and/or colon cancer
Insoluble fiber	Psyllium	
Soluble fiber		Reduces the risk of CVD
Fatty acids	Tuna fish and marine oils	May reduce the risk of CVD and improve mental and visual functions
Omega-3-fatty acids—DHA/EPA		
Flavonoids	Tea	Neutralizes free radicals and may reduce the risk of cancer
Catechins	Citrus	
Flavanones		Neutralizes free radicals and may reduce the risk of cancer
Glucosinolates, indoles, and isothiocyanates	Cruciferous vegetables	Induces detoxification enzymes and may reduce the risk of cancer
Plant sterols	Corn, soy, wheat, and wood oils	Lowers blood cholesterol levels by inhibiting cholesterol absorption
Stanol ester		
Probiotics/probiotics	Onion powder	May improve gastrointestinal health
Fructo-oligosaccharides	Yogurt and other dairy products	May improve gastrointestinal health
Lactobacillus		

TABLE 3.5 *(Continued)*

Class/Component	Source	Potential Benefits
Saponins Saponins	Soybeans, soy foods, and soy protein-containing foods	May lower LDL cholesterol and possess anticancer activity
Soy protein Soy protein	Soybeans and soy-based foods	25 g per day may reduce the risk of heart diseases
Phytoestrogens Isoflavones—daidzein and genistein	Soybeans and soy-based foods	May reduce menopause symptoms, such as hot flushes
Tannins Proanthocyanidins	Cranberries, cranberry products, cocoa, and chocolate	May improve urinary tract health and may reduce the risk of CVD

3.3 FUNCTIONAL FOOD

Functional food is defined as follows:

- Food or dietary supplement that may give basic nutritional benefits as well as additional health benefits to human beings.
- Foods that contain active components and provide basic nutrition as well as some health benefits.
- Daily uptake of nutritional food gives basic nutrition and keeps the body away from chronic diseases.
- Foods enriched with functional nutrients improve the health of human beings, for example, cereals, drinks used during sports activities, and so forth.
- Daily used food always have some proved and identified value that provides health benefits as well as reduces the risk of disease toward person who consumes it.
- It contains all fruits, leafy and nonleafy vegetables, suckers and tubers, antioxidants, fortified foods, and drink beverages (ready to drink).
- For improvement of health if anything that is consumed comes under nutraceutical food.
- Some of the functional and diet foods also come under nutraceuticals.

The scientific approach is used for the preparation or in the process of cooking food, when you may know why and how the food is used is called functional food. Every human being has to take carbohydrate-, vitamin-, fat-, protein-, and amino-acid-enriched diet daily, which are available in functional foods. All of the functional foods are considered as nutrient-enriched foods that provide essential bodybuilding or growth components to human beings; consumers remain healthy after the daily intake of these foods in their diet. It is scientifically proved from the data available through various trials (i.e., randomized study, intersubject study, and placebo-controlled or intervention trials) conducted on human volunteers that recent health issues in human beings totally depend on the consumption of daily food only. Researchers suggest taking functional foods or dietary supplements daily to avoid health issues (Hasler, 2002). Foods have a specific aroma, taste, and certain nutritive value and they are also functional. Nowadays, all of the foods are evaluated for their nutritive values and various physiological benefits to check whether they reduce the chance of disease or to improve health (Hasler, 2002). Sometimes various functional foods are used for the treatment of diseases or for the prevention of disorders except anemia, and these foods are termed as nutraceuticals, for example, fortified or enriched nutraceuticals such as various dairy products as well as citrus fruits (Sapkale et al., 2012).

3.4 DIETARY SUPPLEMENTS

DSHEA formally defined "dietary supplement."

A dietary supplement

- is a fortified product (other than tobacco) that is intended to be used as a supplementary food and contains dietary components, for example, vitamins, minerals, herbs or other herbal components, proteins, amino acids for human beings;
- is used as supplemental diet food by adding it into daily food;
- is available in different forms, for example, tablets, capsules, ready-to-drink beverages, pills, or in the liquid form;
- is not ingested or taken as a conventional food in more quantity as a part of a complete food; and
- sometimes includes recently approved novel drugs, approved antibiotics, or licensed biological products that were marketed and sold

under label dietary supplement or maybe a food before its approval, certification, or license (unless the Secretary of Health and Human Services waives this provision) (Kalra, 2003).

Such above-mentioned foods can be termed as a "dietary supplement."

Nutraceuticals are differentiated from dietary supplements on the following grounds:

- Nutraceuticals are the fortified foods or nutritional products that are used as supplemental food as well as used to prevent or treat diseases and/or disorders.
- Nutraceuticals are used as a conventional food item as a sole for meals (Kalra, 2003).

3.4.1 FOOD OR DRUG

Many people are still confused and not able to distinguish nutraceuticals from other products due to the difficulty in their determination. Feed is defined as an edible product or substance for animals that provides energy or some nutrients if used in the diet. Animal feed claims about all the nutrients available in it for its physiological functions. Nutraceuticals comes under category foods that are required for normal functioning and complete wellbeing of humans. Mainly nutraceuticals have established nutritive value, while a feed may be required to have a nutritive value; this is a basic difference between both products. According to the safety point of view, feed is generally recognized as safe (Sapkale et al., 2012).

3.5 CHARACTERIZATION OF NUTRACEUTICALS

All the nutraceuticals, supplements, and fortified food products must be characterized for their labeled components as a part of regulation. Many people consume various functional foods, but they may not know about their nutritional value and available components; however, they still consume them because such products are recognized good for health. In various countries, such products are available without prescription due to less stringent regulations on them. Actually, manufacturers or suppliers must have to monitor the raw materials and finished products for their safety and quality characteristics. Regulatory agencies like FDA must have to establish good manufacturing practice (GMP) guidelines for the manufacturing of nutraceuticals to monitor and regulate the engaged industries to ensure the quality (Lakshamna Prabhu et al., 2012).

Many parameters to be characterized in the case of nutraceuticals are as listed as follows:

a. calory testing;
b. total carbohydrate testing;
c. trans fat content;
d. total fat available (cholesterol);
e. total protein available (amino acid analysis);
f. multivitamin testing (A, C, and E);
g. percentage of moisture;
h. heavy metals and other metals such as Na, K, Ca, and Mg testing;
i. nutraceuticals available;
j. vitamin E;
k. aerobic microbial count;
l. aflatoxins;
m. yeast and mold count;
n. coliform count;
o. *Escherichia coli* test; and
p. *Salmonella* test.

To characterize functional food as well as nutraceuticals, specific, highly sensitive, and reproducible analytical methodologies are needed. Innovative high-performance liquid chromatography methods employing reversed-phase narrow bore column and high-performance anion-exchange chromatographic methods coupled with pulsed amperometric detection (HPAEC-PAD) are specific for carbohydrate analysis. Also, high-performance thin layer chromatography methods are available (Corradini et al., 2002).

3.6 BENEFITS OF REGULATIONS

Nutraceutical foods require other testing and regulations that are not similar to the testing and regulations used in pharmaceuticals. If a food product that comes under the nutraceutical class and is able to show some health benefits to human beings, then this product must be required to comply with all the regulatory requirements that are available nowadays for products with medicinal benefits. Nutraceutical products must satisfy the requirement of medicinal products as safe, efficacious, and have quality as well as follow all procedures for the marketing authorization (Deore et al., 2014). As per the Food Safety and Standards Authority of India (Food Safety Standard Act, 2006), food and food products are classified as follows:

- newly modified foods;
- genetically modified foods
- irradiated foods;
- organic foods;
- dietary foods;
- nutraceuticals;
- functional foods; and
- health supplements (Deore et al., 2014).

Few decades ago, FDA regulated supplemental dietary food with an aim to ensure that it is safe and wholesome as well as the labeled information is available on the container or packaging material of such products, which is true and not misleading the consumer (Chauhan et al., 2013). Worldwide, the regulatory agencies are able to know the needs of consumers that change day by day, so they work with an aim to provide the products that satisfy the customers. The different region-specific regulatory agencies are listed in Table 3.6.

The regulatory guidelines available in some countries need to be revised, and worldwide regulatory authorities should have to make common guidelines for all.

3.7 INDIAN REGULATORY ASPECTS OF NUTRACEUTICALS

In India, olds laws are available such as the Prevention and Food Adulteration Act, 1954, which needs to be revised. As well as in year 2006, the Government of India passed the FSSAI law to regulate foods, nutraceuticals, and supplemental foods (dietary) (Chauhan et al., 2013). Two main objectives of the FSSAI law are

- to regulate foods and food products under a single guideline, and
- to develop food-processing industries in India (Chauhan et al., 2013).

Apart from these above-mentioned laws, some other laws to be followed by food industries in India are as follows:

TABLE 3.6 Different Countries and Their Regulations (Deore et al., 2014).

Country	Body Known As	Body Governed By	Definition Covers
Canada	Natural Health Products, in 2004	FDA	Vitamins, minerals, homeopathic, Chinese medicine, amino acids, and essential fatty acids
EU	Food Supplements, in 2002	Food Safety Authority	Concentrated sources of nutrients and substances with nutritional or physiological effects
Russia	Biologically Active Food, in 1997	Ministry of Health and Social Development	Vitamins, amino acids, dietary fibres, and para-pharmaceuticals (bioflavonoids, polysaccharides)
USA	Dietary Supplement, in 1994	FDA	Vitamins, minerals, botanicals, amino acids, and herbal extracts
Australia	Complementary Medicines, in 1994	Department of Health and Ageing	Herbal medicines, vitamins, minerals, and nutritional supplements
Japan	Foods for Specific Health Issue, in 1991	Japan Health and Nutrition Food Association	Functional foods that can have three functions: nutrition, sensory satisfaction, and physiological improvement
India	Foods for Special Dietary Use, in 2005	Food Safety and Standards Act, passed in 2006 and yet to be implemented	Vitamins, minerals, botanicals, and herbal extracts
China	Health Food	State Food and Drug Administration (SFDA), in 2003	Any finished product or raw material intended for people to eat or drink, as well as any product that has traditionally served as both food and medication except for products used solely for medical purposes

- Standards of Weights and Measures Act, 1976;
- Standards of Weights and Measures (Packaged Commodities) Rules, 1977 (SWMA)
- Infant Milk Substitutes, feeding Bottles and Infant foods (Regulation of Production, Supply and Distribution) Act, 1992 and Rules, 1993 (IMS);
- Edible Oils Packaging (Regulations) Order, 1998;
- Fruit Products Order 1955 (FPO);
- Meat Food Products Order 1973;
- Milk and Milk Products (Regulation) Order 1998 (VOP);
- Atomic Energy Act, 1962 and Atomic Energy (Control or Irradiation of Food) Rules 1996;
- Consumer Protection Act 1986 and the Consumer Protection (Amendment) Act, 2002 and Rules 1987;
- Environment Protection Act, 1986 and Rules 1986;
- Agricultural Produce (Grading and Marketing) Act, 1937 (as amended up to 1986) General Grading and Marking Rules 1986 and 1988 (AG mark); and
- Bureau of Indian Standards (BIS) Act 1986 (Lakshamna Prabhu et al., 2012).

3.7.1 IMPORTANCE OF REGULATIONS

- Regulations ensure legal security and expected environment.
- They motivate and support development in the field of food and drink products with a new technology or invention.
- They avoid unfair competition between the suppliers or manufactures that is based on misleading/wrong claims.
- A regulatory agency does not permit any manufacturer to create negative claims of others' products if they do not have any positive claims in relation to their products (Chauhan et al., 2013).

3.7.2 REGULATORY CHALLENGES

To achieve success in the development of nutraceuticals, researchers must have basic knowledge about physicochemical properties of all the excipients used in the formulation, criteria and rationale behind selection of excipients, and actual technique of manufacturing as well as data of stability for study of actives to add required overages within products at the time of manufacturing. Globally, the environmental climate is different, so researchers have to prepare and optimize best suitable formula of products for different regions or countries. Different regulations are available in each part of the globe, and time for approval also varies with a lot of scrutinizing factors (Raut, 2016).

The main issue appears to be at the level of quality and claim parameters in the case of regulatory aspects. It is essential to govern the disorganized sector with a rational attitude Ayurvedic and Unani medicines are herbomineral based. Modern medicine has certain parameters that are unreasonable to validate nutraceuticals. Moreover, the marketability of medicinal plants and their products in the country is today unorganized due to several problems. Current practices of harvesting are unsuitable, and many studies have highlighted the depletion of the resource base. The guidelines should, therefore, be similar to regulatory requirements for Ayurveda, Siddha, Unani, or homeopathy (Shah and Seth, 2010).

3.8 NUTRACEUTICALS IN MANAGEMENT OF DISEASES

1. **Various Cadiac Diseases**

   Nutraceuticals are useful in various cardiac diseases such as hypertension, coronary heart issue, cerebrovascular disease, heart problems, etc. The survey gives information that up to year 1999 the proportion of deaths due to cardiovascular diseases is more throughout the world, while in 2015 the proportion increases very rapidly, mainly in developing countries. Research envisaged that in the treatment of cardiovascular diseases the diet enriched with fruits and vegetables plays a protective role. Nutraceutical products that contain antioxidants, some dietary fibers, omega-3 polyunsaturated fatty acids (n-3 PUFAs), multivitamins, and required minerals are recommended to avoid cardiovascular diseases (Agrawal and Paridhavi, 2007).

   Nutraceutical products enriched with polyphenolic compounds of molecular weight more than 30,000 Da; some anthocyanins, stilbenes, and condensed tannins available in grapes and old wine; tetrahydro-β-carbolines, serotonin, indoleamines, and melatonin available in different plant foods provide health benefits (Agrawal and Paridhavi, 2007).

Apart from the above-mentioned compounds, flavonoids are also beneficial for health; these are available in cherry and berry fruits, onion, different varieties of grapes, cruciferous vegetables, wine and apple (Agrawal and Paridhavi, 2007).

Intake of phytosterols in daily diet is useful for reducing problems associated with cardiovascular diseases. The seeds of *Fagopyrum esculentum* Moench, that is, sweat buckwheat, contain flavones, flavonoids, phytosterols, and some proteins having the ability to bind with thiamin and so forth; buckwheat proteins are useful in constipation, obesity, and, more importantly, lowering the blood cholesterol levels, ultimately reducing blood pressure (Agrawal and Paridhavi, 2007).

Rice bran is also useful in lowering cholesterol as it has a laxative property.

Animal-derived foods, that is, eggs and milk contain important nutraceuticals such as proteins, fatty acids, and fats (polyunsaturated) (Agrawal and Paridhavi, 2007).

2. **Obesity**

Obesity is a serious issue in developing countries due to today's lifestyle; it is reported that many herbal stimulants such as caffeine, ephedrine, and green tea are useful to treat obesity as they help in reducing body weight. Some other nutraceuticals also responsible for body weight loss, for example, capsaicin, conjugated linoleic acid, *Momordica charantia*, and psyllium fiber (Agrawal and Paridhavi, 2007).

3. **Diabetes**

Many people are suffering from diabetes and it is a serious issue globally. In the treatment of diabetes, psyllium-containing dietary fibers play an important role. Psyllium fibers have the ability to control glucose and lipid levels. Nutraceuticals enriched with calcium, magnesium, and vitamin D lower the risk of diabetes and increase insulin sensitivity (Agrawal and Paridhavi, 2007).

4. **Cancer**

It is reported that antioxidants have anticancer property; thus, flavonoids present in citrus fruits are beneficial to prevent cancer (Agrawal and Paridhavi, 2007).

Some other active anticancer components are β-carotene (available in leafy, orange, and green fruits); lycopene (chemically a caratenoid) obtained from various fruits, that is, tomato, papaya, watermelon, and so forth; and ellagic acid (Agrawal and Paridhavi, 2007).

In the treatment of prostate cancer, modified citrus pectin (MCP), that is, a new form of citrus pectin, is very useful. It inhibits the process of adhesion of cancerous cells to normal cells in the human body.

Phenolics also have anticancer activity, for example, ferulic, caffeic, and gallic acids obtained from tannins and curcumin obtained from *Curcuma longa.* Sulforaphane present in broccoli is responsible for anticancer activity as it produces D-glucarolactone and useful in the treatment of breast cancer as it acts as a potent phase 2 enzyme inducer (Agrawal and Paridhavi, 2007).

3.9 RECENT RESEARCH IN NUTRACEUTICALS

In the recent era, researchers deal with herbal nutraceuticals that are responsible for various therapeutic activities as well as have health benefits. Some herbal extracts have the ability to prevent chronic diseases in different ways, which is to be investigated by new-generation researchers. It gives useful information about its safety and efficacy to patients and aware physicians about the better treatment of such chronic diseases (Lakshamna Prabhu et al., 2012).

The most critical issue in treating diseases is a lack of significant clinical data about the use of nutraceuticals. Various nutraceutical-associated departments are needed to be promoted such as research and development, actual manufacturing, packaging, sales and marketing with a motto to raise the sales of nutraceutical's globally as well as aware consumers for the regular use of nutraceuticals (Lakshamna Prabhu et al., 2012). Hope, in coming years, the more relevant data and information are available for the successive growth of nutraceuticals.

3.10 GLOBAL DEMAND FOR NUTRACEUTICAL PRODUCTS

Globally, the United States and Japan are the most developed markets for nutraceuticals due to the consumer acceptability achieved in these regions. India, China, and Brazil are developing nations that show huge potential for the nutraceutical market. India and China are rich sources of flora and fauna, and in coming years, both countries will be the biggest suppliers of raw materials in the whole world for the production of herbal nutraceuticals. The world market for nutraceuticals is projected to reach US$250 billion by 2018, driven primarily by the growing affinity among the general populace toward adopting a healthy lifestyle.

Nutraceutical foods occupied the largest area in the market in 2007, worth $39.9 billion. It is expected to be increased up to $56.7 billion in 2013 (Deore et al., 2014). Nutraceutical supplemental foods occupied the second largest market share in the last decade, generating approximately $40.0 billion in 2007, which will reach more than $100 billion at the end of 2020.

Worldwide demand for ingredients required in nutraceuticals has grown 5.8% annually through 2010. Best prospects include probiotics, soy additives, lycopene, lutein, stero-based additives, green tea, glucosamine, chondroitin, and coenzyme Q10. At least, a dozen large companies currently produce and market nutraceuticals today. South India-based Parry Nutraceuticals, a leading player, produce a well-known product, spirulina. Some of the other major companies are Wockhardt, Lupin, Morepen, Laboratories, Dabur, and Himalaya. The largest Indian pharmaceutical company Ranbaxy is also planning to enter into this segment in a big way. According to recent industrial growth calculations, the current market of nutraceuticals in India is estimated at up to Rs. 2000 crore with a fruitful growth rate of around 20%. In the next few years, it will be increased but totally depend on the actual growth rate, availability of standards and regulations, and current market scene.

Nutraceutical market is not too old; it developed in early 1990s, and it is assumed that it will grow very quickly. Between 1999 and 2002, the industry developed at an average rate of 7.3% each year, but in the beginning of 2003, the rate has more than doubled to 14.7% (Chauhan et al., 2013). Currently, the global nutraceutical market has an estimated international value of US$117 billion. Some nutraceutical products are listed in Table 3.7.

The reigning current trends are in product customization and specialization. This is especially so in more developed markets. Investment in R&D focuses on to discover and develop new innovative products and approaches to treat conditions and promote health, verifying that the claims made by nutraceutical brands and market research are key strategies for players in this exploding industry. The primary force behind this enormous growth is huge and growing consumer demands. Consumers want to achieve lifestyles and optimum nutrition to prevent diabetes, hypertension, and obesity. While the US, European, and Indian markets are the world's largest currently, China is expected to surpass them by 2030.

TABLE 3.7 List of Marketed Nutraceutical's (Sarin et al., 2012)

Marketed Nutraceutical	Category
Weight smart	Nutritional supplement
Omega woman	Immune supplement
Rox	Energy drink
Proteinex	Protein supplement
PNer Plus	Neuropathic pain supplement
Mushroom Optimizer	Immune supplement
Chaser	Hangover supplement
Calcirol D-3	Calcium supplement
Appetite Intercept	Appetite suppressant
Red Bull	Energy drink
Pediasure	Nutritional supplement
Glowelle	Beauty drink
BrainSpeed Memory	Brain health

3.11 FUTURE PROSPECTIVES

Recent globalization of all the nutraceutical and functional food industries face a large number of challenges, such as differences in regulations throughout the world as well as no evidence of medicinal benefits of the products. Insights into GMP manufacturing techniques and GMP, standardized, and validated analytical techniques in relation to safety, quality, and efficacy of products must be revised (Deore et al., 2014).

Recent lifestyle and daily intake of food are responsible for metabolic syndromes that are overcome by daily intake of nutraceuticals. More issues for nutraceuticals are as follows:

- Unavailability of standards for the scientific assessment of the prevention of diseases.
- Development of an assessment system for the prevention of diseases by conducting primary human trials.
- Startups to be developed to launch new products in the market to prevent diseases.

The use of nutraceuticals and their role in the prevention of diseases do not depend on only one component but a combined effect of many components. It is required to develop methods for the study of the role of food in the prevention of various diseases; for that, some biomarkers are required for which research is required. As well as, measurement methods and standardization procedures for the biomarkers need to be developed (Lakshamna Prabhu et al., 2012).

3.12 CONCLUSIONS

Nutraceuticals are the most required fortified food products that are available in the market. All the nutrients required for human health are present in an isolated form in such products such as dietary supplements, specific diets, some herbal products, genetically modified foods, and processed foods. Statistical information reveals that nutraceuticals are a rapidly developing segment of the food industry. Nutraceutical products and dietary supplements give sufficient energy and all required nutrients to human beings that involved in daily requirements of the body for complete wellbeing and good health. The intake of nutraceuticals in the daily diet prevents diseases like cancer as well as changes the mood and works against insomnia. Regulatory agencies are required to establish some stringent guidelines for maintaining the quality, safety, and efficacy of nutraceutical products. Delegates from different countries should have to come on one platform and need to finalize the strategy through consensus for the healthy growth of the nutraceutical segment. Hope that after the successful growth of nutraceutical, they will definitely bring big changes in the lifestyle of human beings and will give a healthy and quality life to society.

KEYWORDS

- **nutraceuticals**
- **nutrients**
- **dietary food supplements**
- **functional food**
- **anticancer**

REFERENCES

Agrawal S. S.; Paridhavi M. *Herbal Drug Technology*, 2nd edn; Orient Blackswan: New Delhi, India, 2007; pp 711–717.

Chauhan B.; Gopal K.; Kalam N. Ansari S. H. Current concepts and prospects of herbal nutraceutical: A review. *J. Adv. Pharm. Technol. Res*. **2013**, *4*(1), 4–8.

Corradini C.; Galanti R.; Nicoletti I. Characterization of nutraceuticals and functional foods by innovative HPLC methods. *Ann. Chem.* **2002**, *92*(4), 387–396.

Deore S. L.; Khadabadi S. S.; Baviskar B. A. Pharamcognosy and Phytochemistry: A Comprehensive Approach; PharmaMed Press: India, **2014**; pp. 737–742.

Gupta S. K.; Mali Y.S.; Patil S. M. Nutraceutical—A bright scope and opportunity of Indian healthcare market, *Int. J. Res. Dev. Pharmacy Life Sci.* **2013**, *2*(4), 479–478.

Hasler C. M. Functional foods: Benefits, concerns and challenges—A position paper from the American Council on Science and Health. *J. Nutr.* **2002**, *132*(12), 3772–3781.

Kalra E. K. Nutraceutical—Definition and introduction, *AAPS PharmSci*, **2003**, *5*(3), 1–2.

Lakshamna Prabhu S.; Supriyaprakash T. N. K.; Dinesh Kumar C.; Suresh Kumar S.; Ragavendram T. Nutraceuticals: a review. *Elixir Int. J.* **2012**, *46*, 8372–8377.

Raut S. *Formulation of Nutraceuticals and Dietary Supplements: Formulation and Regulatory Challenges*, Proceedings of the 8th International Conference and Exhibition, Madrid, Spain, Mar 07–09, **2016**, *7*(1), 68.

Sapkale A.; Thorat M.; Vir Prasad R.; Singh M. C. Nutraceuticals—Global status and applications: a review. *Int. J. Pharm. Chem. Sci.* **2012**, *1*(3), 1166–1181.

Sarin R.; Sharma M.; Singh R.; Kumar S. Nutraceuticals: a review, *Int. Res. J. Pharmacy* **2012**, *3*(4), 95–99.

Shah B.; Seth A. K. *Textbook of Pharmacognosy and Phytochemistry,* 1st edn; Elsevier: India, **2010**, p 479.

CHAPTER 4

Herbs in Health Management

KAPIL KUMAR* and RAJAN KAUSHIK

Global Institute of Pharmaceutical Education and Research, Kashipur 244713, Uttarakhand, India

**Corresponding author. E-mail: kapil5november@gmail.com*

ABSTRACT

Health is an important aspect of life, and it must be guarded at any cost. The present chapter focuses on the origin, distribution, chemical composition, and general, traditional, and pharmacological uses of various household and easily available herbs that can be employed in the management of health. The herbs are not only nutritional foods but also remedies; when utilized in a proper way, they can help in achieving a fulfilling healthy lifestyle. The herbs are becoming popular in public than the synthetic medicines because these are devoid of many of the adverse effects linked with synthetic drugs and thus provide a safer view about them. The information supplied in this chapter is written in a lucid and easy language so that even laymen can understand it. Herbs are the gift of nature for all and we are one of them; they do not only act as remedies but also as the functional foods. The content of this chapter will help us in attaining good health through the justified use of the herbs.

4.1 INTRODUCTION

In botany, herbs are defined as the plants that lack woody stems. Herb is a common term used for many flowering garden plants or potherbs, some of which have medicinal values and some are pests, for example, holy basil, mint, turmeric, coriander, cumin, and so forth.

Herbs have been a part of many cultures for different objectives around the world since antiquity. These are used for various purposes, such as food, to flavor, in cosmetics, in medicines, and for fragrance. Herbalism is the study or science of herbs.

Health is an important dimension of life; we can remind its importance through a famous quote, that is, "If wealth is lost, nothing is lost; if health is lost, something is lost." Herbs being rich in phytochemicals, their micro- and macronutrients can be employed to manage the different health-related issues. The World Health Organization (WHO) estimates that 80% of the population of some Asian and African countries relies upon herbal medicine for some aspect of primary health care (WHO, 2004).

Some of the widely used herbs for the management of health are as follows.

4.2 CUMIN

4.2.1 BOTANICAL NAME

Cuminum cyminum

4.2.2 INDIAN NAME

Jeera

4.2.3 ORIGIN, DISTRIBUTION, AND CHEMICAL COMPOSITION

It is a kind of annual herb having a smooth surface and slender stem. When cultivated in favorable conditions, it can attain a height of up to 45 cm. The stem bears long deep greenish leaves and white or rose-colored flowers. The seeds, known as cumin seeds, are tiny, 0.5 cm in size, yellowish-brown in color, and possess a strong aromatic odor and bitter taste. An Indian kitchen is incomplete without this herb, as it is one of the prime ingredients of curry powder. The cumin plant is native to Syria, Egypt, and Turkey. Now, it is grown in Africa, India, and China also. Major cumin-producing states of India are Rajasthan and Gujarat

(Shivakumar et al., 2010; Zohary and Hopf, 2000). Cumin seeds contain aldehydic compounds, amino acids, fats, and a yellow-colored volatile oil. Cuminaldehyde is the active chemical constituent of the volatile oil. The fragrant odor of the cumin seeds is due to its volatile oil. The major chemicals present in cumin seeds are cuminaldehyde, thymol, α- and β-pinene, linalool, and so forth.

4.2.4 HEALTH BENEFITS

4.2.4.1 INSOMNIA

Powder of fried seeds mixed with the pulp of a ripe banana is used to treat sleeplessness.

4.2.4.2 COLD

Decoction of cumin seeds is an antiseptic beverage and used to alleviate the common cold.

4.2.4.3 LACTATION

Due to the high content of iron in seeds, it is a rich source of iron for lactating mothers and women undergoing menses.

4.2.4.4 DIGESTIVE DISORDER

Decoction of seeds with one teaspoon juice of fresh coriander leaves and a little quantity of common salt can be used to cure diarrhea.

4.2.4.5 SCORPION STING

To reduce the oftenness of upbeats, a seed paste blended with onion paste is applied to the place of bite.

4.2.4.6 SKIN DISORDERS

Cumin has an excellent amount of vitamin E. So, its powder can aid in getting a glowing and youthful skin.

4.3 BLACK PEPPER

4.3.1 BOTANICAL NAME

Piper nigrum

4.3.2 INDIAN NAME

Kaali Mirch

4.3.3 ORIGIN, DISTRIBUTION, AND CHEMICAL COMPOSITION

Pepper is known as the "king of spices" (Menghani et al., 2010). It is one of the ancient condiments. Pepper is a flowering vine cultivated for its fruits. It was first cited by Theophrastus in 372–287 BC. It was used by the ancient people of Greece and Roman. Black pepper is the whole dried fruit, generally black in color with a strong aromatic odor and pungent taste. The fruits are generally oblong and 0.5–1.0 cm in size. It is native to the Western Ghats in India but is now grown in many of the tropical and subtropical countries.

Pepper contains chemicals like α-terpineol, acetophenone, nerol, nerolidol, citral, 1,8-cineol, α-pinene, and alkaloid piperine (Bajad et al., 2002). Chemical analysis of black pepper shows that it consists of moisture, proteins, fats, minerals, fibers, and carbohydrates. The calorific value of pepper is 304 kcal.

4.3.4 HEALTH BENEFITS

4.3.4.1 DIGESTIVE DISORDERS

Pepper is a stimulant in nature. It improves the secretions of the salivary glands and stomach. It is used to ameliorate the appetite (Chopra and

Chopra, 1959). One tablespoon of pepper powder mixed with cow's milk is used to reduce the feeling of heaviness in the stomach.

4.3.4.2 AMNESIA

Ground pepper when used with honey works as a potent remedy to treat amnesia.

4.3.4.3 COUGH AND COLD

Turmeric and pepper powder decoction is used to treat cough and cold. Ayurvedic, Unani, Siddha, and folklore medicines in India used pepper and pepper-containing preparations for the treatment of neuritis, cold, pain, intermittent fever, and throat diseases.

4.3.4.4 ANTICANCER

Black pepper is believed to act against cancer development directly. The principal phytochemical present in the pepper, that is, piperine, has the property to inhibit some of the pro-inflammatory cytokines that are produced by tumor cells. A study by a cancer institute in Michigan revealed that the pepper along with turmeric can help in preventing the growth of cancerous stem cells in breast tumors.

4.3.4.5 PAIN

Special oil manufactured by boiling pepper powder with sesame oil can be used to treat rheumatic pain and myalgia.

4.4 LICORICE

4.4.1 BOTANICAL NAME

Glycyrrhiza glabra

4.4.2 INDIAN NAME

Mulethi

4.4.3 ORIGIN, DISTRIBUTION, AND CHEMICAL COMPOSITION

Licorice is a well-known popular drug used in Ayurvedic and Unani systems of medicine for the cure of respiratory ailments and is used as a flavoring agent. It is a tall herb that grows up to a height of 1.5 m. The name finds its origin in Greek words *Glycyrrhiza* meaning "sweet wood" and *Glabra* meaning "smooth." However, it is thought to be a native of southern Europe and Pakistan. Its various species are now cultivated in different parts of the world. It grows wild in sub-Himalayan tracts and Baluchistan. It is cultivated on a commercial scale in Spain and Yorkshire of England. In India, it is cultivated in Baramulla, Srinagar, Jammu, Indore, and Bengaluru. The chief chemical constituent of licorice is a tri-terpenoid saponin recognized as glycyrrhizin (6%–10%). Its sweet taste is attributed to the presence of this compound. Glycyrrhizin is 50 times as sweet as sugar. The other important categories of chemicals present in the drug are coumarin derivatives, flavanoid glycosides, glucose, mannitol, and 20% starch (Dastagir and Rizvi, 2016).

4.4.4 HEALTH BENEFITS

- Glycyrrhiza has demulcent and expectorant properties (Kamei et al., 2003).
- Licorice is utilized to mask the taste of bitter drugs in pharmaceutical formulations such as quinine (A cinchona alkaloid), aloe, and ammonium chloride.
- It contains glycyrrhetinic acid, which possesses mineralocorticoid activity, and hence, it is employed in the treatment of inflammation, rheumatoid arthritis, and Addison's disease.

4.4.4.1 SORE THROAT

The herb is a well-recognized home remedy for the problem of sore throat. Small pieces are chewed to relieve the soreness.

4.4.4.2 COUGH

Powder of licorice along with honey is used in the treatment of cough and cold-like conditions (Usmanghani, 1997).

4.4.4.3 CONSTIPATION

Licorice acts as a good laxative. In the case of constipation, a little powder along with jaggery and water can be taken to get relief from constipation.

4.4.4.4 ULCERS

Liquid extract of licorice is used to heal peptic ulcers.

4.4.4.5 SCALDS AND WOUNDS

A paste prepared from clarified butter and powder can be applied to heal the wounds. The poultice prepared from leaves is used to treat scalds of the head.

4.4.4.6 FLAVORING AGENT

It is employed as a flavoring agent for chewing tobacco, beverages, and confectionary, in the candy industry (Anilkumar et al., 2012), and a foam stabilizer in beers.

4.5 INDIAN SENNA

4.5.1 BOTANICAL NAME

Cassia angustifolia

4.5.2 INDIAN NAME

Bhumiari, Senai-Ki-Patti

4.5.3 ORIGIN, DISTRIBUTION, AND CHEMICAL COMPOSITION

Senna is one of the traditional remedies used by Arab physicians as a cathartic. Indian Senna is a small bushy plant of 1.5 m height. It consists of the dried leaflets of *Cassia angustifolia* belonging to family Leguminosae. Senna is native to Southern Arabia and Western Asia. In India, the herb is cultivated on a large scale in Tinnevelly, Ramnathpuram, and Madurai districts of Tamil Nadu. The powdered leaves are mainly used as a laxative; this action is due to the presence of anthraquinone glycosides that include dianthrone glycosides and sennoside A, B, C, and D. The other chemicals present in the drug include flavanols such as kaempferol, isorhamnetin, rhein, aloe emodin, and so forth (Agarwal and Bajpai, 2010).

4.5.4 HEALTH BENEFITS

4.5.4.1 CONSTIPATION

Senna acts as a potent purgative that can be taken by children, weak, or elderly people. It is popular as it is devoid of astringent after action as observed in the case of rhubarb. The sennosides found in senna improves the colonic transit rate and colonic peristalsis. It is one of the components of famous ayurvedic formulation "Kayam Churna."

4.5.4.2 INTESTINAL WORMS

Senna is used to treat helminthiasis.

4.5.4.3 SKIN PROBLEMS

Paste of leaves prepared using acetic acid can be applied to treat skin problems like acne, eczema, and pimples.

4.6 HOLY BASIL

4.6.1 INDIAN NAME

Tulsi

4.6.2 *BOTANICAL NAME*

Ocimum sanctum

4.6.3 *ORIGIN, DISTRIBUTION, AND CHEMICAL COMPOSITION*

Tulsi plant is herbaceous found throughout India. In Hindu religion, it is considered as a sacred plant and worshipped, the reason why it is called holy basil or sacred basil (Ghosh, 1995). The plant bears light green oval-shaped leaves with reddish flowers. The plant possesses aromatic odor, and leaves are generally bitter. The plant is found naturally in India but now has spread in many other countries. Tulsi plant consists of fresh or dried aerial parts of the plant *Ocimum sanctum* belonging to family Labiatae. The leaves have about 0.1%–0.9% volatile oil, which consists of around 70% eugenol, 3% cavacrol, and 20% eugenol methyl ether. The extract of leaves and stem yields some phenolic compounds that possess antioxidant activity, such as cirsiliniol, cirsimaritin, apigenin, and so forth. The seeds contain an appreciable quantity of fixed oil. They also contain alkaloids, glycosides, saponins, tannins, vitamin C, and traces of maleic, citric, and tartaric acids.

4.6.4 *HEALTH BENEFITS*

Tulsi is called "Elixir of Life" because it promotes longevity. It is a popular remedy of Ayurvedic and Siddha systems of medicine for the cure of general ailments like sore throat, bronchitis, asthma, colic pain, fever, insomnia, stress, scorpio sting, snake bite, and so forth. Nowadays, different brands of distilled tulsi water are available in the market that claim to provide antioxidant, antiaging, antibacterial (Mishra and Mishra, 2011), antiviral, antiseptic, stimulant, and diuretic properties.

4.6.4.1 STRESS

Tulsi is an adaptogen or antistress herb. Many studies on animals revealed that it has antianxiety and antistress activity similar to modern medicines like diazepam. It is worth mentioning that 12 fresh leaves are beneficial to chew in the case of stress.

4.6.4.2 SORE THROAT

Sore throat is a very common problem due to viral and bacterial infections. In such conditions, fresh leaves of Tulsi are boiled with water and decoction is used either as a drink or for gargle. This water can also be drink during bronchitis and asthmatic conditions.

4.6.4.3 FEVER

Tulsi leaves have antibacterial and antipyretic properties. During rainy days, when malaria, dengue, typhoid, and viral fever are prevalent, their decoction in combination with cardamom and clove can help to reduce the fever. A tea made from clove, ginger, and tulsi is also a good home remedy for fever.

4.6.4.4 EYE DISORDERS

Tulsi is considered to improve eyesight. Honey mixed with oil of tulsi can be used to instill in eyes to improve the eyesight (Patil et al., 2011). In Ayurvedic system of medicine, eye drops are prepared using tulsi leaf juice and triphala (harad, bahera, and amwla) to treat glaucoma, conjunctivitis, and other painful eye problems.

The juice of leaves has stimulant, aromatic, spasmolytic, and diaphoretic properties.

4.6.4.5 DENTAL PROBLEMS

Powder of dried Tulsi leaves is used to make a paste with mustard oil. This paste can be used to get rid of bad mouth odor and dental diseases like pyorrhea.

4.7 SANDAL

4.7.1 INDIAN NAME

Chandan

4.7.2 BOTANICAL NAME

Santalum album

4.7.3 ORIGIN, DISTRIBUTION, AND CHEMICAL COMPOSITION

Sandal is a small- or medium-sized evergreen semiparasitic tree indigenous to India. It grows naturally in the Karnataka state of India. In Karnataka, Hassan, Coorg, Mysore, Kolar, and Bengaluru are sandal-producing districts. It has now been naturalized in many parts of Uttar Pradesh, Madhya Pradesh, and Rajasthan. Sandal consists of the heartwood of stem and roots of the tree *Santalum album* belonging to family Santalaceae. A plant that has attained the age of 30 years is generally selected for the distillation of sandalwood oil. The tree is uprooted, the bark is removed from the stem, and logs are cut into billets. The wood has a pale yellow to red color and pleasant odor. The wood is distilled to get fragrant volatile oil. It contains 2.5% of essential oils. Sandalwood oil contains a mixture of two isomeric sesquiterpene alcohols (95%), α and β santalol. Other chemical constituents present are santene, α and β santalene, santalone, santanone, and α and β santalic acids.

4.7.4 HEALTH BENEFITS

4.7.4.1 AROMA AND PERFUMERY

Sandalwood oil is employed as a base to create costly aromas and perfumes. It finds a significant place in aromatherapy. It is also used as a scent in the incense sticks. The essential oil is generally utilized in fragrance diffusers to make the surroundings pleasant during meditation sessions.

4.7.4.2 GENITOURINARY DISEASE

The paste prepared from sandalwood powder and milk is used to cure the symptoms of gonorrhoea. The oil is also used to treat conditions like dysuria and cystitis. Scientific studies reveal that the oil possesses appreciable antiviral activity against the herpes virus.

4.7.4.3 SKIN DISEASES

The paste wood is cooling and antibacterial in nature. The oil is used to treat scabies. Sandalwood oil and mustard oil (1:2) are used to cure pimples. During summer, the paste is applied on the whole body to provide a cooling and refreshing effect.

It is believed to stimulate the pineal gland, the reason why its paste is applied on the forehead.

4.7.4.4 CONTRACEPTIVE

With neem oil, it acts as a contraceptive.

It is an ingredient of the famous Mysore Sandal Soap and many other cosmetic items.

4.8 INDIAN GOOSEBERRY

4.8.1 INDIAN NAME

Amla

4.8.2 BOTANICAL NAME

Emblica officinalis

4.8.3 ORIGIN, DISTRIBUTION, AND CHEMICAL COMPOSITION

It is considered as the king of all medicinal plants. It is a famous drug mentioned in Ayurvedic texts. Its plant is a small- or medium-sized deciduous tree that has smooth, greenish-gray, and exfoliating bark. The leaves are feathery, small, narrowly oblong and arranged pinnately. The fruits are depressed, round to globose, fleshy, and obscurely six-lobed, and have six trigonous seeds. The fruits are green when unripe and change to yellowish-red on ripening. The taste of fresh green fruit is sour and astringent but the after taste is sweet. Its fruits are being used in traditional practices from the time of Sushruta, a great authority in Ayurveda. Sushruta considered the fruit to be best of all the acidic fruits. The plant is naturally found in India and one of the ingredients of the famous Ayurvedic preparation Chyawanprash. In India, it is found in wild in deciduous forests. It is also found in Sri Lanka and Myanmar. Anand, Kanchan, Chakiya, NA6, and NA7 are its famous agricultural varieties. Amla is a very good dietary source of vitamin C, amino acids, and minerals. Its edible fruit tissues contain three times more protein and 160 times more vitamin C when compared to apple. The amino acids found in the fruit are glutamic acid, proline, alanine, aspartic acid, and lysine (29.6%, 14.6%, 5.4%, 8.1%, and 5.3%, respectively) of the total amino acids. Fruit also contains phyllemblin and curcuminoids. The fruit has 482.14 units of superoxide dismutase/g fresh weight and exhibited antisenescence activity. The seed oil contains 64.8% linolenic acid that closely resembles with linseed oil.

4.8.4 HEALTH BENEFITS

4.8.4.1 ANTIAGING

Amla is a superfruit. It contains a very high amount of vitamin C, a natural antioxidant. This helps in reducing cell damage. It scavenges free radicals and thus helps in slowing the aging process.

4.8.4.2 SORE THROAT

A combination of fresh amla juice, ginger juice, and honey is a proven remedy to treat pesky sore throat.

4.8.4.3 RHEUMATISM

One teaspoon of the powder of dry fruits along with jaggery can be helpful in relieving inflammation of joints due to rheumatism.

4.8.4.4 EYE

It is a proven fact that those who eat an amla fruit daily have the best eyesight. The fruits are beneficial in the treatment of conjunctivitis and glaucoma. A cup of amla juice when taken with honey is a good remedy to reduce intraocular tension.

4.8.4.5 SCURVEY

Being very rich in vitamin C, it is a perfect remedy for the treatment of scurvy.

4.8.4.6 CANCER

Superoxide dimustase, a potent antioxidant, is present in the fruit, which scavenges free radicals and thus helps against cancer.

4.9 CARDAMOM

4.9.1 INDIAN NAME

Elaychi

4.9.2 BOTANICAL NAME

Elettaria cardamomum

4.9.3 ORIGIN, DISTRIBUTION, AND CHEMICAL COMPOSITION

It is a perennial plant. When grown in favorable conditions, it can attain a height up to 4 m. It bears dark green color leaves. The fruits are capsules having three sides, and each chamber has two rows of seeds. The plant consists of the dried ripe fruits of *Elettaria cardamomum* belonging to family Zingiberaceae. Theophrastus described about cardamom in the 4th century BC. India is trading this herb since 1000 AD to Western countries. The plant is indigenous to South India. In India, it is cultivated on a commercial scale in Karnataka, Kerala, and Tamil Nadu. In 1965, the Cardamom Act was passed by the Government of Karnataka to encourage the cultivation of cardamom in the state. It is also naturalized in Sri Lanka, Myanmar, and Malaysia. Guatemala is the world's largest producer of this herb. Naturally, the fruits are green to yellow and possess strong aromatic odor and taste. The seeds are hard, round, and brown. Chemical analysis of the seeds revealed that seeds contain 2%–8% of volatile oil. Cineole is the chief active chemical of the volatile oil. The other chemical constituents present in the oil are α-pinene, β-pinene, myrcene, α-phellandrene, limonene, 1,8-cineole, p-cymene, terpinolene, linalool, linalyl acetate, terpinen-4-oil, α-terpineol, α-terpineol acetate, nerol, geraniol, methyl eugenol, and trans-nerolidol (Korikontimath et al., 1999). Seeds also have a significant amount of fixed oil, starch, and proteins.

4.9.4 HEALTH BENEFITS

4.9.4.1 DIGESTIVE PROBLEMS

Cardamom is carminative, that is, it expels gases from the stomach and relieves from flatulence. From a scientific study, it has been found that it stimulates the secretion of bile and thus helps in digestion. A tea made from cardamom seeds is the best remedy to treat the headache caused by indigestion.

4.9.4.2 ORAL HEALTH

Cardamom seeds are the best mouth freshener. In India, it is a custom to eat flavored seeds after a meal. This helps to get rid of bad mouth odor as well as halitosis. Cineole present in the seed works as an antiseptic and antibacterial and thus promotes good oral health.

4.9.4.3 RESPIRATORY DISORDER

According to Ayurveda, it is excellent for balancing *Kapha*, particularly in the stomach and lungs. Cardamom seeds are expectorant in nature and thus help in clearing mucus from the respiratory tract. They also help in preventing the symptoms of cold and flu. Cardamom tea is a good remedy to get relief from the symptoms of bronchitis.

4.9.4.4 ANTI-INFLAMMATORY PROPERTY

Cardamom like ginger and turmeric has good anti-inflammatory properties. The powder of cardamom seeds can be used to treat inflammation of the mouth, mucus membranes, and throat (Gadekar and Jadhav, 2017).

4.9.4.5 FLAVORING AGENT

It is not only used to flavor food but also in many pharmaceutical formulations for children especially. Nowadays, many antacid formulations are available in cardamom flavor (Ravindran *et al.*, 2002).

4.10 CHIRAYTA

4.10.1 INDIAN NAME

Chirata, Bhoonimba

4.10.2 BOTANICAL NAME

Swertia chirata

4.10.3 ORIGIN, DISTRIBUTION, AND CHEMICAL COMPOSITION

Chirata as a herb falls in the category of bitters. It is an annual herb indigenous to temperate Himalaya. It consists of entire plant of *Swertia chirata* (Scartezzini and Speroni, 2000) belonging to family Gentianaceae. The drug grows in the Himalayan region, from Kashmir to Bhutan at elevations between 1200 and 3000 m and in Naga and Khasi hills. It is also found growing in Uttarakhand, Sikkim, and Arunachal Pradesh. The plant bears alternate leaves and small flowers in a raceme arrangement. In medicinal practice, whole plants or broken pieces of the drug are employed. The plant has indistinct odor and a bitter taste. Xanthones are the main chemicals found in the Swertia species. Xanthones are present as parent polyhydroxylated compounds, but most of them present in the drug belong to mono- or polymethyl ethers or are found as glycosides (Hostettmann and Miura, 1977). The drug contains bitter principle ophelic acid, chiratin, and amarogentin. It also contains alkaloids, lignans, flavonoids, terpenoids, iridoids, and so forth. The pharmacological efficacy of *S. chirayita* can be attributed to the presence of major phytoconstituents that include amarogentin (Ray et al, 1996), swertiamarin (Lie et al., 1982), mangiferin, swerchirin (Ariño et al., 1997), sweroside, amaroswerin, gentiopicrin, and so forth.

4.10.4 HEALTH BENEFITS

4.10.4.1 INTESTINAL WORMS

Chirayta possesses anthelmintic activity and is thus useful for the treatment of helminthiasis.

4.10.4.2 DIGESTIVE PROBLEMS

Being bitter, the drug has a direct action on the salivary and gastric cells. Secretion of saliva and gastric juices helps in the prevention of nausea, bloating, hiccups, and indigestion.

4.10.4.3 DIABETES

Animal studies on the drug showed that the drug normalizes the elevated sugar level and thus helps in the management of diabetes (Saxena et al., 1991, 1993; Sekar et al., 1987).

4.10.4.4 MALARIA

Infusion of the herb is a traditional remedy used as a preventive measure during malaria epidemics.

4.10.4.5 GENERAL TONIC

The drug acts as a general tonic for the heart, liver, and eyes.

4.11 FENUGREEK

4.11.1 INDIAN NAME

Methi Dana

4.11.2 BOTANICAL NAME

Trigonella foenum graceum

4.11.3 ORIGIN, DISTRIBUTION, AND CHEMICAL COMPOSITION

Fenugreek is an erect, scented, and robust annual herb attaining a height of up to 30–80 cm; it consists of dried seeds of *Trigonella foenum graceum* belonging to family Fabaceae. This herb is native to the Mediterranean region, and in India, it is cultivated as a vegetable and for seeds. In culinary practice, seeds are used as a spice to give an aromatic taste to curry. It bears light green color compound leaves that are 1.5–2.5 cm in length. During summer, white color flowers appear that finally develop into long, slender, yellow pods containing seeds. Seeds are dark yellow to buff, hard, and angular. Upon soaking in water, seeds gets swell due to the presence of mucilage. Seeds have characteristic aromatic odor and bitter mucilaginous taste. Its leaves are used in the fresh form as a vegetable and are a good source of moisture, proteins, calcium, fibers, and carbohydrates. *T. foenum graecum* seeds are a rich source of protein (20%–30%), high in tryptophan and lysine; free amino acids (arginine, lysine, 4-hydroxyisoleucine, and histidine) (25.8%); fat (6.53%); ash content (3.26%), crude fibers (6.28%); energy (394.46 kcal/100 g seed); and moisture (11.76%). The seeds contain a steroid diosgenin, a raw material for the synthesis of many steroidal compounds. The other important sapogenins present in the seeds include yamogenin, gitogenin, tigogenin, and neotigogen. Fenugreek seeds also contain alkaloids, including trigonelline, gentianine, and carpaine compounds. The fiber fenugreekine is also reported in the seeds, a component that may have hypoglycemic activity.

4.11.4 HEALTH BENEFITS

4.11.4.1 POLYURIA

Polyuria, also called frequent urination, is a condition generally associated with diabetes. In such cases, 2–3 g powder of fenugreek seeds can be consumed with milk to regulate urination.

4.11.4.2 DIABETES

The seeds contain trigonelline that is reported to exert hypoglycaemic activity; thus, the seeds can be used in the treatment of type-I and type-II diabetes. For this purpose, around 500 mg of seeds is soaked overnight and consumed by chewing slowly on an empty stomach in the morning (Sharma et al., 1990).

4.11.4.3 INFLAMMATION AND PAIN

Seeds are made into poultice, warm, and applied directly on the affected part to treat local pain, muscle pain, inflammation, lymphadenitis, gouty pain, and so forth (Chauhan et al., 2010).

4.11.4.4 ARTERIOSCLEROSIS

This condition is called as "hardening of arteries." Fenugreek seeds contain steroidal saponins (diosgenin, yamogenin, tigogenin, and neotigogenin) that are reported to inhibit cholesterol absorption and synthesis, hence indicating its potential role in arteriosclerosis.

4.11.4.5 RHINITIS, COUGH, AND COLD

Around 2–3 g of seeds is soaked in water for 5–6 h, boiled, and filtered. The prepared decoction is used to get rid of conditions like rhinitis, cough, and cold. This decoction should be consumed at least 3 times a day.

4.11.4.6 MOUTH ULCERS

An infusion of leaves helps to cure recurrent mouth ulcers.

4.11.4.7 GASTROPROTECTIVE

The biological effect of fenugreek seeds (*T. foenum graecum*) compared to that of omeprazole was studied on ethanol-induced gastric ulcers in animal models, and it was revealed that the aqueous extract and a gel fraction isolated from the seeds have significant ulcer-protective properties.

The leaves boiled and fried in butter help to remove biliousness.

4.12 BELERIC MYROBALAN

4.12.1 INDIAN NAME

Bahera, Bibhitaki

4.12.2 BOTANICAL NAME

Terminalia belerica

4.12.3 ORIGIN, DISTRIBUTION, AND CHEMICAL COMPOSITION

Terminalia belerica is a large deciduous tree, 50 m tall and up to 3 m in diameter with a round crown. The plant is native to India; its more than 100 species are distributed throughout the world. This tree is grown as an avenue tree and can be found growing in lower hills and plains of Southeast Asia. In India, it is found in Madhya Pradesh, Uttar Pradesh, Punjab, and Rajasthan. The plant bears dark green, large, and glabrous alternate leaves. The plant has small, solitary, and greenish-white flowers with honey-like odor. The generic name "*Terminalia*" comes from a Latin word "terminus" or "terminalis" (ending) and refers to the habit of the leaves being crowded or borne on the tips of the shoots. The fruits are drupes, gray-brown in color, ovoid, and have a sweet kernel inside. Its fruits are similar to *Terminalia chebula* but without ridges. The fruits are an ingredient of famous Ayurvedic formulation Triphala. In Ayurveda, the fruits are believed to nourish the dhatus and pacify the tridosha. Fruits are rich in tannins like gallic acid, ellagic acid, and chebulagic acid. They also contain mannitol, glucose, galactose, fructose, and rhamnose (Saroya, 2011).

4.12.4 HEALTH BENEFITS

4.12.4.1 EYE DISORDERS

Traditionally, the powder of fruits is mixed with water and left overnight; in the morning, the liquid is strained and is used to treat epiphora. A paste of the bark is made with honey and is used as an eyeliner to treat eye pain.

4.12.4.2 LOSS OF APPETITE

The powder of bahera fruits alone or in combination with amla and harad helps in strengthening the stomach. After the meals, up to 5 g of fruit powder can be consumed to improve digestion and help in the loss of appetite.

4.12.4.3 COUGH

It is a household remedy for cough. Traditionally, the fruit is fried in cow ghee and covered with wheat flour and roasted again. This recipe is used to cure cough and catarrh.

4.12.4.4 CONSTIPATION

The pulp of half-ripe fruits can be used to treat constipation; however, dried fruits should be avoided as they have a completely opposite effect (Ghani, 2003).

4.12.4.5 DIARRHEA

Traditionally, roasted fruits of bahera are used to treat diarrhea.

4.12.4.6 RHEUMATISM

Seed oil is applied to swollen joints to treat rheumatic inflammation.

4.12.4.7 PHARMACOLOGICAL PROPERTIES

It possesses analgesic, antidiarrhoeal, antihypertensive, anti-Salmonella, antispasmodic, bronchodilator, and antimicrobial activities as reported in various scientific studies.

4.13 ISAPGOL

4.13.1 INDIAN NAME

Isabgol

4.13.2 BOTANICAL NAME

Plantago ovata

4.13.3 ORIGIN, DISTRIBUTION, AND CHEMICAL COMPOSITION

Isabgol plant is a small, hairy annual herb. Isapgol word is a combination of two different words Isap meaning horse and gol/ghula meaning ear because the seeds have the shape of an ear of a horse. The plant is indigenous to the Mediterranean region and West Asia. It is also cultivated in Spain, Cuba, and France. At present, India is the largest exporter of psyllium seeds. In India, the plant is cultivated for commercial purposes in Gujarat, Rajasthan, Punjab, and Maharashtra. The factory for psyllium husk is situated in Sidhpur, Gujarat. It consists of cleaned, dried, and ripe seeds of *Plantago ovata* belonging to family Plantaginaceae. RI-87, 89, GI-1, 2, HI-1, 2, and MIB-121 are some of the famous varieties of *Plantago*. The seeds are small, 1–2 mm in length, reddish-brown, and odorless but with a bland and mucilaginous taste. The outer seed coat contains about 30% mucilage (hydrocolloidal polysaccharide), fixed oil, steroids, tannins, proteins, and iridoid glycoside acubin. It is mainly composed of xylose, arabinose, and galacturonic acid with rhamnose and galactose. Two polysaccharide fractions are isolated from its mucilage. Of which, one is soluble in cold water and on hydrolysis produces D-xylose (46%), aldobiouronic acid (40%), L-arabinose (7%), and insoluble residue (2%), while the other fraction is soluble in hot water forming a highly viscous solution that sets to a gel on cooling and yields on hydrolysis D-xylose (80%), L-arabinose (14%), aldobiouronic acid (0.3%), and traces of D-galactose.

4.13.4 HEALTH BENEFITS

4.13.4.1 CONSTIPATION

Psyllium husk or commonly called Isabgol Ki Bhusi is one of the common remedies for constipation. It is generally soaked in water or curd before its use. It works as the osmotic laxative by increasing the mass of stool and thus facilitating its passage. The action is mechanical more than physiological (Voderholzer et al., 1997).

4.13.4.2 RHEUMATISM

An emollient poultice of crushed seeds is applied to rheumatic and glandular swelling.

4.13.4.3 NOCTURNAL EMISSION

According to Ayurveda, the nature of the drug is cold, which has a cooling effect on the body and thus reduces the excitement.

4.13.4.4 DRY COUGH AND SORE THROAT

Pudding made up of psyllium husk produces instant relief in the scratchy throat due to its emollient action and reduces the frequency of dry cough.

4.13.4.5 OBESITY AND CHOLESTEROL LOWERING

Psyllium husk is one of the richest sources of dietary fibers. Intake of dietary fibers can help in lowering the concentration of LDL in plasma. Psyllium intake has consistently shown significant reductions in plasma low-density lipoprotein cholesterol levels ranging from 10% to 24%, according to a scientific study (Anderson et al., 1987).

Isabgol preparations are given after colostomy to assist in the production of smooth solid fecal mass. The mucilage is not digested by enzymes and intestinal bacteria in the gut and comes out unchanged. Jelly-like mucilage absorbs irritating products and toxins of the gut and they are expelled from the body.

4.14 HERBS AS NUTRACEUTICALS

The word nutraceuticals is a combination of two words, namely, nutritive and pharmaceutical, means a food stuff (as a fortified food or dietary supplement) that provides health benefits. According to Defelice, "nutraceuticals are food or part of a food that provides medical or health benefits including the prevention and/or treatment of a disease" (Trottier et al., 2010). Nutraceuticals or functional foods are classified according to their natural sources, pharmacological categories, and/or chemical constituents present in the products.

Nutraceuticals are nonspecific biological therapies used to promote wellness, prevent malignant processes, and control symptoms. These can be grouped into the following three broad categories (Hathcock, 2001):

1. Substances with established nutritional functions, such as vitamins, minerals, amino acids, and fatty acids—nutrients.
2. Herbs or botanical products as concentrates and extracts—herbals.
3. Reagents produced from other sources (e.g., pyruvate, chondroitin sulfate, and steroid hormone precursors) serving specific functions, such as sports nutrition, weight-loss supplements, and meal replacements—dietary supplements.

Herbs or their extracts can be used as functional foods, that is, food products that can be consumed as a part of the common diet to have beneficial effects that go beyond basic nutritional function.

The most rapidly growing segment of nutraceuticals is herbal/natural products followed by dietary supplements. The generation of scientific research linked foods of plant origin and health has resulted in understanding that plant bioactive compounds have antioxidant and other health-promoting properties (Dahiya, 2013).

Table 4.1 shows some of the herbs used as nutraceuticals.

TABLE 4.1 Some of the Herbs Used as Nutraceuticals

Common Name	Biological Source	Nutraceuticals	Applications
Turmeric	Rhizomes of *Curcuma longa*; Zingiberaceae	Curcumin	Antioxidant, anticancer, anti-inflammatory, antimicrobial, antidiabetic, antiviral, cardiovascular diseases, and so forth
Licorice	Dried stem and roots of *Glycyrrhiza glabra*; Leguminosae	Glycyrrhizin and Liquiritin	Anti-inflammatory, expectorant
Bhring Raj	Fresh plants of *Eclipta*; Compositae	Ecliptine, Wedelolactone	Hepatic tonic, deobstruent, and anticancer
Ginger	Rhizomes of *Zingiber officinale;* Zingiberaceae	Zingiberene, Shogaols, and Gingerols (Semwal et al., 2015)	Anticancer, anti-oxidant, antimicrobial, anti-inflammatory, and antiallergic
Bael	Unripe fruits or slices of *Aegle mermelos* Corr.; Rutaceae	Marmelosin (Bhattacherjee et al., 2016; Bramhachari and Reddy, 2010)	Antibacterial, antihistaminic, antiarthritic, digestive, demulcent, stomachic, nutritive, and antidiabetic
Garlic	Dried bulbs of *Allium sativum*; Liliaceae	Allin and Allicin (Touloupakis and Ghanotakis, 2010)	Antimicrobial, antiparasitic, antifungal, antiviral, anti-inflammatory, cholesterol-lowering, and antioxidant
Onion	Dried bulbs *of Allium cepa*; Liliaceae	Allin and Allicin	Improves blood circulation, antibacterial, and anticancer
Ashwa-gandha	Dried roots of *Withania somnifera Dubal*; Solanaceae	Withanoloide A (Akhoon et al.; 2018)	Cancer
Punarnava	Herbs of *Boerhaavia Diffusa*; Nyctaginaceae	Punarnavine (Manu and Kuttan, 2009)	Immunomodulator
Shatavar	Dried tuberous roots of *Asparagus racemosus* Willd; Liliaceae	Asparagine (Veena et al., 2014)	Anti-oxidant, immunomodulator

KEYWORDS

- **health**
- **health management**
- **herb**
- **herbalism**
- **nutraceuticals**
- **basil**
- **cumin**

REFERENCES

Agarwal, V.; Bajpai, M. Pharmacognostical and Biological studies on Senna and Its Products: An Overview. *Int. J. Pharm. Bio. Sci.* **2010**, *1*(2), 1–10.

Akhoon, B. A.; Rathor, L.; Pandey; R. Withanolide A Extends the Lifespan in Human EGFR-Driven Cancerous *Caenorhabditis elegans*. *Exp. Gerontol.* **2018**, *104*, 113–117.

Anilkumar, D.; Hemang, J.; Nishteswar, K. Review of *Glycyrrhiza glabra* (yastimadhu)—A Broad Spectrum Herbal Drug. *Pharm. Sci. Monit.* **2012**, *3*(4), 3171–3195.

Ariño, A.; Arberas, I.; Leiton, M. J.; Renobales, M. D.; Dominguez, J.B. The Extraction of Yellow Gentian Root (*Gentianalutea* L.). *Z. Lebensmitteluntersuchung Forschung* **1997**, *205*, 295–299.

Bhattacherjee, A. K.; Dikshit, A.; Kumar, P. C.; Pandey, D.; Tandon, D.K. Profiling Nutraceuticals in Bael [*Aegle marmelos* (L.) Correa] at Various Stages of Fruit Development. *J. Hortic. Sci. Biotechnol.* **2016**, *91*(2), 169–174.

Chopra, R. N.; Chopra, I. C. Indian Council of Medical Research; New Delhi, India. Review of Work on Indian Medicinal Plants. *Spec. Rep. Ser.* **1995**, *1*, 99–107.

Dahiya, K. Nutraceuticals and Their Impact on Human Health. *J. Plant. Biochem. Physiol.* **2013**, *1*(4), 1–4.

Dastagir, G.; Rizavi, M. A. *Glycyrrhiza glabra* L. (Liquorice). *Pak. J. Pharm. Sci.* **2016**, *29*(5), 1727–1733.

Gadekar, V. S. J.; Jadhav, G. V. A. U. Review of Spices as a Medicine. *World J. Pharmacy Pharm. Sci.* **2017**, *6*(8), 413–425.

Ghosh, G. R. Tulasi (N.O. Labiatae, Genus—Ocimum). *New Approaches Med. Health (NAMAH)* **1995**, *3*, 23–29.

Hathcock, J. Dietary Supplements: How They Are Used and Regulated. *J. Nutr.* **2001**, *131*, 1114–1117.

Kamei, J.; Nakamura, R.; Ichiki, H.; Kubo, M. Antitussive Principles of *Glycyrrhiza radix*, A Main Component of Kampo Preparations Bakumondo-to (Mai-men-dongtang). *Eur. J. Pharmacol.* **2003**, *69*, 159–163.

Manu, K. A.; Kuttan, G. Immunomodulatory Activities of Punarnavine, An Alkaloid from *Boerhaavia diffusa*. *Immunopharmacol. Immunotoxicol.* **2009**, *31*(3), 377–387.
Menghani, E.; Jain, S. C.; Jain, R. Scientific Validation of *Piper nigrum* by HPLC and Antioxidative Assay Markers. *Asian J. Biotechnol.* **2010**, *2*, 133–138.
Mishra, P.; Mishra; S. Study of Antibacterial Activity of *Ocimum sanctum* Extract against Gram Positive and Gram Negative Bacteria. *Am. J. Food Technol.* **2011,** *6*, 336–341.
Patil, R.; Patil, R.; Ahirwar, B.; Ahirwar, D. Isolation and Characterization of Antidiabetic Component (Bioactivity Guided Fractionation) from *Ocimum sanctum* L. (Lamiaceae) Aerial Part. *Asian Pac. J. Trop. Med.* **2011**, *4*, 278–282.
Ravindran, P. N.; Johny, A. K.; Nirmal, B. K. Spices in Our Daily Life. *Satabdi Smaranika* **2002**, *2*, 102–105.
Saxena, A.M.; Bajpai, M.B.; Mukherjee, S.K. Swerchirin induced blood sugar lowering of streptozotocin treated hyperglycemic rats. *Ind. J. Exp. Biol.* **1991**, 29, 674–675.
Saxena, A. M.; Bajpai, M. B.; Murthy, P. S.; Mukherjee, S. K. Mechanism of Blood Sugar Lowering by a Swerchirin-Containing Hexane Fraction (SWI) of *Swertia chirayita*. *Indian J. Exp. Biol.* **1993**, *31*, 178–181.
Scartezzini, P.; Speroni, E. Review on Some Plants of Indian Traditional Medicine with Antioxidant Activity. *J. Ethnopharmacol.* **2000**, *71*, 23–42.
Sekar, B. C.; Mukherjee, B.; Chakravarti, R. B.; Mukherjee, S. K. Effect of Different Fractions of *Swertia chirayita* on the Blood Sugar Level of Albino Rats. *J. Ethnopharmacol.* **1987**, *21*, 175–181.
Semwal, R. B.; Semwal, D. K.; Combrinck, S; Viljoen, A. M. Gingerols and Shogaols: Important Nutraceutical Principles from Ginger. *Phytochemistry* **2015**, *117*, 554–568.
Sharma, R. D.; Raghuram, T. C.; Rao, N. S. Effect of Fenugreek Seeds on Blood Glucose and Serum Lipids in Type I Diabetes. *Eur. J. Clin. Nutr.* **1990**, *44*, 301–306.
Touloupakis, E.; Ghanotakis, D. F. Nutraceutical Use of Garlic Sulfur-Containing Compounds. In *Bio-Farms for Nutraceuticals: Functional Food and Safety Control by Biosensors*, Springer: Boston, MA, 2010, Chap. 9; pp 110–121.
Trottier, G.; Boström, P. J.; Lawrentschuk, N.; Fleshner, N. E. Nutraceuticals and Prostate Cancer Prevention: A Current Review. *Nat. Rev. Urol.* **2010**, *7*, 21–30.
Usmanghani, K. *Researches on Materia Medica*; Department of Pharmacognosy, Faculty of Pharmacy, University of Karachi: Karachi, Pakistan, 1995; pp 29–35.
Veena, N.; Arora, S.; Kapila S.; Singh, R. R. B. *et al.* Immunomodulatory and Antioxidative Potential of Milk Fortified with *Asparagus racemosus* (Shatavari). *J. Med. Plan. Stud.* **2014**, *2*(6), 13–19.
Voderholzer, W. A.; Schatke, W.; Muhldorfer, B. E. *et al.* Clinical Response to Dietary Fiber Treatment of Chronic Constipation. *Am. J. Gastroenterol.* **1997**, *92*, 95–98.
WHO (World Health Organization), New WHO Guidelines to Promote Proper Use of Alternative Medicines. Online News Release [Online] 2004.
http://www.who.int/mediacentre/news/releases/2004/pr44/en/ (accessed March 24, 2020).
Zohary, D.; Hopf, M. *Domestication of Plants in the Old World*. Oxford University Press: Oxford, UK, 2000; 206 p.

CHAPTER 5

Starch-Based Nanocarriers of Nutraceuticals: Synthesis and Applications

ALBERTO A. ESCOBAR-PUENTES[1], ADRIANA GARCÍA-GURROLA[1], and FERNANDO MARTÍNEZ-BUSTOS[2*]

[1]*Instituto de Ciencias Biomédicas, Universidad Autónoma de Ciudad Juárez, Anillo Envolvente del Pronaf y Estocolmo s/n, CP 32310, Cd. Juárez, Chihuahua, México*

[2]*Departamento de Nanomateriales, Centro de Investigación y de Estudios Avanzados del Instituto Politécnico Nacional, Unidad Querétaro, Libramiento Norponiente No. 2000, Fracc. Real de Juriquilla, 76230, Querétaro, Querétaro, México*

[*]*Corresponding author. E-mail: fmartinez@cinvestav.mx*

ABSTRACT

Currently, the application of nutraceutical compounds has been on the rise due to their beneficial properties for human health. However, being highly sensitive molecules have low stability during food production and during human metabolism, possess low bioavailability and cellular absorption, limiting their bioactivity. In response, nanoencapsulation is a promising nanotechnology tool for the efficient protection and transport of bioactive compounds. In this respect, starch nanoparticles, as a biodegradable polymer, have a promising use for the protection and controlled release of various nutraceuticals due to their good controlled-release profile, high bioavailability, economic cost, and nontoxicity. The present chapter shows the most recent works published on the use of starch nanoparticles as encapsulating materials of various nutraceuticals, such as vitamins, essential oils, phenolic acids, and polyphenols. For the first time, the following classification is proposed: nonesterified starch nanoparticles, esterified starch nanoparticles, adsorbent nanoparticles, hollow starch nanoparticles or nanoreactors, hybrid starch–magnetic nanoparticles, and ternary nanoparticles. With the employ of starch-based nanocarriers, it has been possible to increase the functionality of the nutraceuticals under adverse conditions. The enrichment of a dairy product with vitamin D has been successfully achieved, without compromising the stability and sensory properties of the food system. Vitamin E nanocapsules embedded in nanofibers were proposed with promising results in the regeneration of skin tissues. Moreover, the nanoencapsulation of the several nutraceuticals has increased their stability against simulated digestive conditions, cellular absorption, and the delivery of active substances in cancer cell lines. Finally, scientific evidence was compiled and reported concerning to nutraceutical–nanocarrier molecular interaction. In conclusion, scientific results showed the effective protection of nutraceuticals through their encapsulation in starch-based nanovehicles, increasing their bioactive properties, opening a wide range of applications in biomedicine and food industries.

5.1 INTRODUCTION

Nowadays, there is a rising interest in the investigation of molecules with biological activity, such as polyphenols, vitamins, and essential oils, due to their potential properties for maintaining and promoting human health. However, nutraceutical molecules have low stability, which limited their bioactive properties and nutraceutical applications. In response to these challenges, nanoencapsulation is a nanotechnological solution, wherein it is possible to protect sensitive compounds from degradation in food systems and human metabolism, increasing their molecular stability, solubility, and cellular absorption.

Recently, nanotechnology has acquired great recognition in the fields of medicine and food to such a degree that it has been cataloged as "nano-medicine" (Wang et al., 2014) and as "nanochemo-prevention" and "nano-chemotherapy" in the oncology field (Santos et al., 2013). In this regard, as a branch of nanotechnology, nanoencapsulation has been established as an innovative approach with potential applications in nutraceutical research. The nanomaterials employed as encapsulating carriers exhibit high cellular penetration by endocytosis due to their submicron size (10–1000 nm) and high surface-to-volume ratio, resulting in efficient release systems for active compounds and drugs. At present, have been reported successful cases of nanoencapsulation of nutraceuticals, such as green tea polyphenols and curcumin, which may confirm their usefulness in "nanochemo-prevention" and "nanochemotherapy" (Li et al., 2015; Santos et al., 2013; Wang et al., 2014, 2018).

The obtention and application of nanomaterials from biodegradable sources have gained much attention in the food and pharmaceutical industries due to their characteristics of null toxicity, controlled release, stability, bioavailability, and the property to administer several active ingredients in several food systems and human beings. Starch is a biodegradable polymer cataloged as a polyglucoside and an alternative widely used during the last years to obtain nanoparticles and perform nanoencapsulation of medicines and nutraceuticals due to the advantages of large-scale obtention and low-toxic properties (de Oliveira et al., 2018; Simón-Vázquez et al., 2012). Starch is the main carbohydrate biopolymer due to its low cost, availability, renewability, and biodegradability (Sadeghi et al., 2017). It is found in nature as microgranules composed of amylose (linear) and amylopectin (branched) molecules (Sadeghi et al., 2017), and it is architecturally structured by semicrystalline nanoblocks, which can be obtained by several procedures, such as nonsolvent precipitation (Qin et al., 2016), high-pressure homogenization (Young et al., 2016), high-energy milling (Patel et al., 2016), sonication (Boufi et al., 2018), reactive extrusion (Song et al., 2011), and the combination of acid hydrolysis and sonication (Kim et al., 2013) methods that have been carefully scrutinized by Kim et al. (2015). In the present chapter, information regarding the nanoencapsulation of different nutraceuticals such as phenolic acids, polyphenols, vitamins, and essential oils in starch nanoparticles was reviewed in detail. Likewise, recent scientific studies have shown that nanoencapsulation with starch nanoparticles improves the bioactive and controlled-release properties of nutraceuticals.

5.2 NUTRACEUTICALS AND THEIR LIMITATIONS

During the last decade, dietary habits have changed progressively, and functional and nutraceutical foods have received an increasing demand from consumers due to their positive effects on health (Oliveira et al., 2018).

Nutraceuticals are the result of combining human nutrition with medicine, a term deriving from hybridization between a nutrient and a pharmaceutical, and they are defined as "food ingredients that have health benefits and inhibit the progress of diseases" (Grumezescu et al. 2016; Khorasani et al., 2018; Oliveira et al., 2018). Several fruits and vegetables are postulated as the foremost source of these bioactive compounds (Quintín et al., 2018).

Vitamins are proficient molecules for the maintenance of human health, contribute to the development of human beings, and involved in human metabolism (Katouzian and Jafari, 2016). Unfortunately, the most of these bioactive molecules are synthesized in human beings in small quantities or are not produced (Katouzian and Jafari, 2016), evidencing the importance of its consumption. Otherwise, aromatic hydrophobic liquids such as essential fatty acids are widely employed in food systems and medicines, as they contribute to health. These compounds inhibit the expansion of neoplasm and possess antiasthmatic, antioxidant, and antiobesity properties (Quintín et al., 2018), all of these due to a considerable amount of fatty acids and, moreover, to their being a good source of phenolic compounds that modulate cell signaling (Kozowska et al., 2016). However, they are easily degraded by oxidation, volatilization, heating, and light; in addition, their low solubility in water and their strong odor limit their applications. To overcome these drawbacks, several encapsulation techniques have been studied (Asbahani et al., 2015).

Undoubtedly, the most substantial nutraceutical field is that represented by the polyphenolic compounds (Quintín et al., 2018), which are secondary metabolites and very abundant in many fruits, vegetables, and teas. Their structure consists of an aromatic ring of benzene with one or more hydroxyl substituents, representing a variety of more than 8000 compounds (Zhu et al., 2018). Among phenolic compounds, we find phenolic acids, stilbenes, hydroxycinnamates, and flavonoids; the latter represents the most representative group and is basically composed of 15 carbons, with two aromatic rings conjugated by three carbon bridges (C6–C3–C6). Within the classification of flavonoids, we find flavonols, anthocyanidins, flavan-3-ols, flavanones, and isoflavones (Li et al., 2015). As in vitro and in vivo studies have shown, polyphenolic compounds provide beneficial properties for health. For instance, anthocyanins prevent the proliferation of cancer cells (Hsu et al., 2012). Soy isoflavones inhibited the transcription of genes responsible for cancer cell proliferation (Hillman and Singh-Gupta, 2011). As for curcumin, it can modulate several biochemical pathways relevant in carcinogenesis, exerting cytotoxic and apoptotic effects on cancer cells and in animal models (Mirzaei et al., 2017). Additionally, polyphenolic antioxidants act as sacrificial substrates to prevent the generation of reactive oxygen species (ROS) in many biological systems and have shown to possess good properties on the inhibition of oxidation processes and the repair of ROS-mediated tissue alterations. The main mechanisms include the donation of hydrogen, the deactivation of singlet and triplet oxygen, the capture of free radicals, and the chelation and the decomposition of peroxide (Zhu et al., 2018).

However, despite the bioactive properties of polyphenolic compounds, unwanted odors and flavors, low aqueous solubility, the instability of ultraviolet (UV) light, gastrointestinal conditions, and their extremely low absorption rate restrict their bioactivity in vivo and severely limit their application (Li et al., 2017; Qiu et al., 2016). The anthocyanins of berries and the resveratrol of grapes and wines had an extremely low percentage of absorption, lower than 1%; the proacyanidins of tea and chocolate had an absorption of <5%; and some flavanols (quercetin) and flavanones (hesperidin) have an absorption percentage <9% (Li et al., 2015). The development of nanocarriers has been a promising solution to counteract the disadvantages of low stability, poor absorption, and low solubility of different bioactive compounds. In the next sections, the scientific information regarding the nanoencapsulation of some polyphenolic compounds in starch nanoparticles is addressed.

5.3 NANOENCAPSULATION

A nanotechnology topic in food science and nutrition is the nanoencapsulation of bioactive molecules to provide greater stability and absorption (Khorasani et al., 2018). Nanotechnology exerted a high impact on several industries and scientific fields such as nutraceuticals. The term "nanotechnology" arises from the Greek word "nanos" meaning dwarf or small, and it is an interdisciplinary scientific area that involves the manipulation, preparation, and implementation of devices with at least one dimension in the submicron size range 10–1,000 nm (Khorasani et al., 2018; Santos et al., 2013). Submicron is the size range in which the properties of the material are different from what those in the macroscale, and for many materials and substances, it is mainly 1–300 nm (Khorasani et al., 2018).

In this size range, the features of the material change because they acquire a greater surface-to-volume ratio and greater reactivity. Therefore, nanovehicle systems are implemented as medical devices, as diagnostic imaging agents, and in the nanoencapsulation of medicines and bioactive molecules for therapies for different pathologies.

Encapsulation is the technology of encasing substances in solid, liquid, or gaseous states in matrices, namely, capsules. The concept emanates from the cell model in which the genetic information is protected in the nucleus in which a semipermeable membrane controls the transportation of several agents (Jafari, 2017). The ingredient inside the capsule is defined as the internal phase, while the wall is variously called the shell, coating, wall material, membrane shell, carrier material, or encapsulating agent. Based on this, the ingredient compounds can be adsorbed onto the surface or the interior of the nanovehicle by covalent or noncovalent binding (Khorasani et al., 2018). Several types of carriers that can be employed in the nutraceutical sector, including polymeric micelles, nanoliposomes, nanospheres, nanoemulsions, and biodegradable nanoparticles such as starch nanoparticles, are the focus of this chapter.

5.3.1 STARCH AS THE PRECURSOR OF NANOSTRUCTURES AS NANOCARRIERS

Nanoparticles derived from natural resources have become a topic of rising interest in academic research and biomedical and food industries due to their biodegradable nature, low cost, zero toxicity, sustainability, and their nonhazardous effect on health or the environment (Boufi et al., 2018). Starch is a very abundant, naturally renewable, and biodegradable carbohydrate-type polymer that has no toxicity and that is cataloged by the US Food and Drug Administration (FDA) as "Generally Recognized as Safe" (GRAS) (Zhu, 2017). Therefore, there is a rising interest in using the biopolymer starch for obtaining nanomaterials for various applications. Nanostructures from starch possess unique physicochemical properties that can be exploited to several applications (Ye et al., 2017). Starch nanoparticles can transport nutraceuticals and medicines to the organ objective and control the release, increasing the bioavailability and absorption of nutraceuticals with better therapeutic control and a reduction in dosing frequency, consequently obtaining significant benefits in nutrition and well-being (El-Naggar et al., 2015; El-Sheikh, 2017).

Starch is the main source of energy produced by photosynthesis in plants. It is found in nature as granular structures with dimensions ranging from 1 to 100 μm and with different morphologies. It is mainly composed of linear amylose and branched amylopectin molecules, polymers integrated into anhydrous glucose units (Kim et al., 2015). In the common native starch, amylose percentages vary between 72% and 82%, and amylopectin ranges between 18 and 28%. However, there are mutant starches such as high amylose (up to 70% and more for amilomaize) and low amylose (1% for waxy corn) contents (Bras and Dufresne, 2010). Amylose consists of linear molecules linked mainly by α-(1–4)-D-glycosidic bonds; however, it has now been established that some molecules are slightly branched by α-(1–6) bonds; amylopectin consists of branched chains formed by α-(1–6) bonds and with an average molar mass (Daltons) of up to hundreds of millions (Gul et al., 2016). The upgrowth of the microgranules starts in the hilum such as an onion-like structure with growth rings composed of crystalline and amorphous lamellae densely packed with a certain number of blocklets, with diameters of 20–500 nm (Kim et al., 2015). Concisely, considering a multiscale structure, there are the starch granules (1–100 μm) formed of growth rings (120–500 nm) composed of blocks (20–50 nm) made of amorphous and crystalline lamellae (9 nm) that contain amylopectin and amylose chains (0.1–1 nm) (Bras and Dufresne, 2010).

Different chemical, physical, and enzymatic methods, including the conventional procedures of acid hydrolysis (and/or enzymatic hydrolysis, or in combination with other techniques), regeneration/nanoprecipitation, and the use of some sophisticated technological tools, such as reactive extrusion, homogenization at high pressures, and the use of gamma radiation and ultrasound, have been employed for obtaining the starch nanoparticles. Depending on the procedure employed for obtaining nanoparticles, these have been classified as "top-down" processes when the destructuring of large molecules into nanoparticles occurs due to a structural refinement and "bottom-up" when the nanoparticles result from the accumulation of atoms and molecules under controlled conditions. Several authors have reviewed and reported in detail the respective information on the methods of obtaining nanoparticles (Ali Razavi and Amini, 2016; Bras and Dufresne, 2010; Gul et al., 2016; Kim et al., 2015; Sandhu and Nain, 2017; Sun, 2018) and the encapsulation of food ingredients in starch-based systems in a general manner (Zhu, 2017). However, information regarding the encapsulation in nanostructured systems obtained from starch is scarce.

Within the starch nanoparticles that have been evaluated as encapsulating materials for nutraceuticals, we can highlight at least six types of nanostructures as follows (Figure 5.1): nonesterified starch nanoparticles, esterified starch nanoparticles, hollow starch nanoreactors, adsorbent starch nanoparticles, hybrid starch–magnetic nanoparticles, and ternary nanoparticles; all of these nanodevices with unique and specific functionalities that will be addressed in the next sections. In this chapter, the encapsulation and protection of different nutraceutical compounds in starch nanoparticles are summarized (see Table 5.1), as is the bioactive properties of nanocapsules determined by *in vivo* and *in vitro* studies (see Table 5.2). As will be discussed later, the synthesis and use of starch nanoparticles as nanocarriers of nutraceuticals have been carried out using various "top-down" procedures, such as high-intensity ultrasound, as well as "bottom-up" processes. Likewise, in the main part of the research work, nanoencapsulation of nutraceuticals has been developed *in situ* during the methods for obtaining the nanoparticles, resulting in a versatile and recommendable process for future industrial applications.

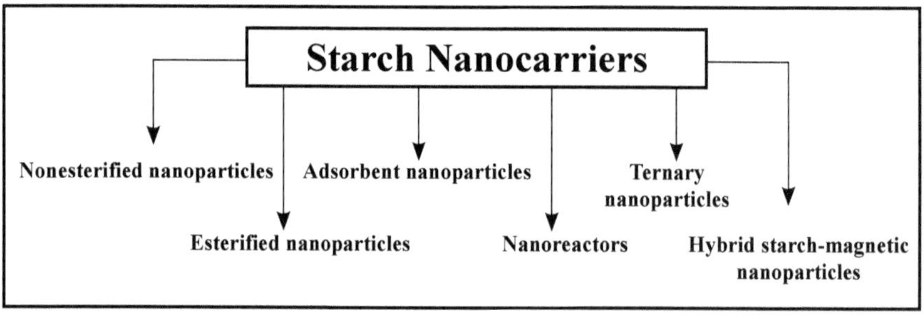

FIGURE 5.1 Classification of starch nanocarriers.

5.3.1.1 NONESTERIFIED STARCH NANOPARTICLES

Nonesterified starch nanoparticles comprise nanostructured systems that are mostly used and studied as nanovehicles of nutraceuticals. The term "nonesterified starch nanoparticles" refers to starch nanoparticles with a native chemical structure; that is, the hydroxyl groups (O–H) are not subject to some processes of esterification or chemical substitution and the nanostructures form hydrogen bonds with O–H groups of nutraceuticals (Figure 5.2). By means of these encapsulating systems, the main nutraceutical compounds have been encapsulated, including some vitamins, essential fatty acids, and polyphenolic compounds.

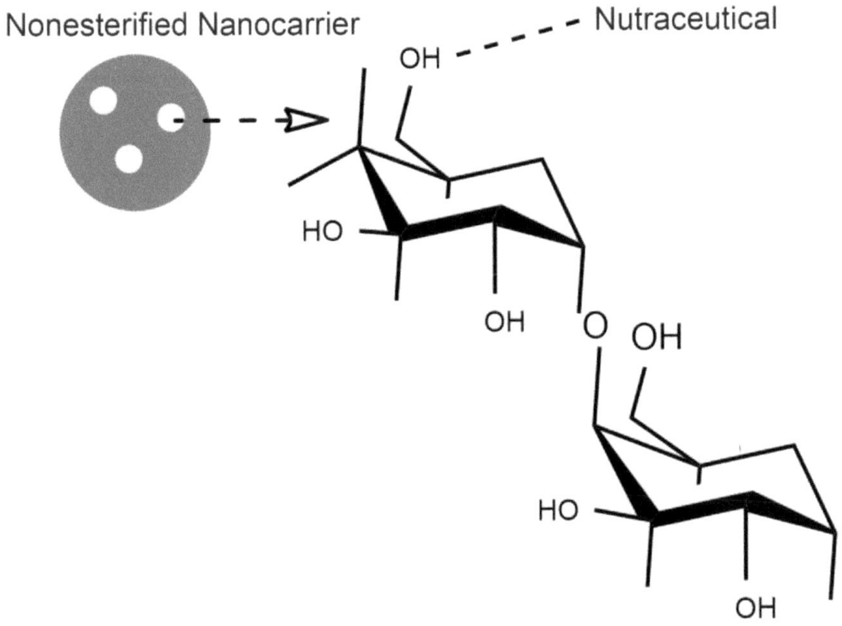

FIGURE 5.2 Chemical structure of nonesterified starch nanoparticles.

5.3.1.2 NANOENCAPSULATION OF VITAMINS AND ESSENTIAL FATTY ACIDS

The nutraceutical properties of essential fatty acids are mentioned above (see Section 5.2). However, the disadvantages of these highly sensitive molecules were also highlighted, whereby they have been nanoencapsulated into starch nanoparticles. This is the case of peppermint oil, which was laden into starch nanostructures (Liu et al., 2017), specifically debranched short glucans from high amylopectin corn starch, using a high-intensity ultrasonic energy method. The study showed that a decrease in the chain length of the glucans caused a decrease in the encapsulation efficiency (87%–74%) because the loading capacity and yield were higher (93%) for glucans with long chains. The introduction of peppermint oil did not exert an influence on the morphology of the nanoparticles but on their size, such that, after encapsulation, the nanoparticles exhibited good uniformity and spherical shape; nonetheless, the obtention of nanoparticles with diameters longer than 200 nm was observed. Additionally, starch glucans acted as effective encapsulating materials because, after encapsulation, peppermint oil demonstrated greater stability against heat treatment (80 °C).

Moreover, vitamins have also been encapsulated in nonesterified starch nanoparticles, such as vitamin E or tocopherols. Tocopherols and tocotrienols belong to a group of phenolic derivatives of chroma with a long chain of saturated hydrocarbons (in the case of tocopherols) or unsaturated hydrocarbons (tocotrienols). Vitamin E is the most efficient biological antioxidant (Delgado-Zamarreño et al., 2016). In this regard, D-α-tocopherol polyethylene glycol succinate is a water-soluble vitamin E derivative that has shown to have effective skin-care and tissue-regeneration functions due to its anti-inflammatory activities (Wu et al., 2011; Zhang et al., 2012). Kheradvar et al. (2018) nanoencapsulated D-α-tocopherol in starch nanoparticles through nanoprecipitation in a nonsolvent method, and subsequently, these nanocapsules were embedded to into an optimal formulation of silk fibroin–poly(vinyl alcohol)–Aloe vera (SF:PVA:AV; 40:50:10) nanofibers. The resulting nanocapsules had a controlled size (44.7 nm), adequate size distribution (0.403), and adequate encapsulation efficiency (91%). Transmission electron microscopy studies revealed that the spherical nanocapsules were satisfactorily entrapped within the SF:PVA:AV nanofibers (296 nm), which provided a core–shell structure. Moreover, the addition of nanocapsules significantly increased the metabolic activity of fibroblasts, as well as the secretion of collagen, suggesting that SF–PVA–AV nanofibers with a content of 5% of starch nanoparticles loaded with vitamin E can be a potential dressing to treat skin wounds.

Another vitamin encapsulated in starch nanoparticles is a fat-soluble sterol present in the nature in forms of ergocalciferol (D2) and cholecalciferol (D3), called vitamin D, which is essential for the development of teeth and bones. It is found in foods of animal origin and is synthesized in the skin when exposed to sunlight. A daily intake of 50 nmol/L is required for the skeletal health for all individuals; however, according to epidemiological studies, a deficiency prevails (<50 nmol/L) (Ložnjak and Jakobsen, 2018) due to a lack of exposure to sun, the excessive use of sunscreen, and dietary restrictions such as veganism or lactose intolerance, resulting in a public-health problem (Ložnjak and Jakobsen, 2018; Walia et al., 2017). Food supplements are a promising solution. In this regard, starch nanoparticles were evaluated to fortify dairy foods and to improve the bioavailability of vitamin D (Hasanvand et al., 2015). In another similar study, high-intensity ultrasound was evaluated in the synthesis of high amylose corn-starch nanoparticles and their use as an encapsulating material for vitamin D3 (Hasanvand et al., 2018). The impact of ultrasonication caused the exposition of starch particles to severe collisions by means of the explosion of cavities or bubbles within the medium (cavitation), allowing a reduction in size (32 nm) and an increase in chemical reactions, contributing to an increase in the interaction of vitamin D and starch nanoparticles, and resulting in a high encapsulation load of up to 94% (Hasanvand et al., 2018). These latter authors mentioned that the ultrasonic treatment led to an increase in the exposition of the starch chain, which resulted in hydrogen bonds and van der Waals interaction between vitamin D3 and potato starch, resulting in greater stability of vitamin D3 against high temperatures.

5.3.1.3 NANOENCAPSULATION OF POLYPHENOLIC COMPOUNDS

Polyphenolic antioxidants have been the most utilized nutraceuticals for their nanoencapsulation in nonesterified starch nanoparticles. For example, in a recent study, the ginko biloba leaf was used to obtain an antioxidant extract rich in polyphenolic compounds, which was encapsulated in nonesterified starch nanoparticles (Wang et al., 2016). The ginko biloba extract contains a high content of polyphenols (isorhamnetin, kaempferol, and quercetin) that possess anticancer activities, playing a role in the improvement of the symptoms of cerebral insufficiency and dementia, including Alzheimer's disease. Moreover, these can protect against the neuronal degeneration resulting from ischemic events (van Beek and Montoro, 2009). However, due to the passive diffusion, they lead to extremely low oral bioavailability and exhibit low absorption and penetration rates in human beings. In the study, the authors used different starches, among them are ginkgo biloba seeds. The starch nanoparticles were obtained by the antisolvent method, and encapsulation was carried out in situ using the said procedure. Starch nanoparticles had sizes from 201 to 250 nm and desirable polydispersity indexes from 0.428 to 0.478 with a uniform size distribution. However, the size range (255–396 nm) increased when the nanoparticles were charged with gingko biloba polyphenols. The content of polyphenols loaded in the starch nanoparticles was within a range of 0.560–1.045 mg/mg. According to in vitro studies, starch nanoparticles had a weakness in gastric digestion conditions (pH 2), in comparison with intestinal digestion environments (pH 7.4), in that during the first hour of digestion in the gastric phase, there was a rapid release of polyphenols because these were on the surface of nanoparticles. Later, in intestinal juices, the starch nanoparticles showed sustained release. The nanodevice improved the bioavailability through the improved absorption of the drug in the gastrointestinal tract by active endocytosis (Wang et al., 2018). This study suggests that ginkgo starch nanocapsules have the potential to be used as controlled-release nanovehicles for antioxidant extracts and may increase their application in the fields of pharmaceutical and nutraceutical industries.

Another group of polyphenolic compounds nanoencapsulated in nonesterified starch nanoparticles are catechins, flavonoid compounds found in several foods, mainly grapes, apples, pears, cherries, and green tea. Catechins have health properties for human beings by preventing various diseases, such as the inhibition of carcinogenesis of the skin, lungs, esophagus, stomach, liver, small intestine, colon, bladder, prostate, and mammary glands (Ahmad et al., 2019). However, according to in vitro studies, the main limitations for catechins lie in their low bioavailability, permeability, and degradation during digestion in humans due to the high pH resulting from digestive secretions during the intestinal phase, with (−)-epigallocatechingallate and (−)-epigallocatechin being the most sensitive (Neilson et al., 2007). Nanoencapsulation has been an efficient approach to increase its bioactivity, bioavailability, and stability under conditions of digestion. Ahmad et al. (2019) employed horse chestnut, lotus stem, and water chestnut as novel sources of starch to obtain nanoparticles. The authors synthesized nanocapsules by dissolving starch in 0.1 M NaOH, and subsequently adding a solution of catechin dissolved in ethanol into the hydrolyzed starch solution. Nanoparticle sizes were between 322 and 615 nm, with an encapsulation efficiency of up to 57%. According to the in vitro digestion studies, during the first 30 min of the gastric phase, the catechins trapped on the nanoparticle surface were rapidly released. Subsequently, in the following 30 min of gastric digestion, amounts of free catechins remained constant, which indicated their successful encapsulation. However, the catechins were progressively released from the nanocapsules during the next 2 h in the intestinal phase. It is worth noting that nanoencapsulation of catechins, in comparison to free catechin molecules, promoted a significant retention of the antiobesity nutraceutical properties, with an inhibition higher than 50% and 60% of the enzymes pancreatic lipase and cholesterol esterase, respectively. The authors suggested that future research should be conducted to verify the bioavailability and retention of the bioactivity of encapsulated catechin molecules of starch nanoparticles when fortified as a bioactive ingredient in nutraceutical foods.

Curcumin comprises one of the bioactive molecules that have had the greatest study. Curcumin is an orange crystalline powder obtained from the *Curcuma longa (turmeric)* herb. It is a hydrophobic polyphenol with promising medicinal properties, including antioxidant, anti-inflammatory, antimicrobial, antineoplastic, and anticancer properties. However, despite these healthy properties, their clinical applications are restricted because of their low bioavailability, low solubility (1.99×10^{-4} mg/mL), and degradation at an alkaline pH. For overcoming these limitations, curcumin entrapment within nanocarriers has been widely employed (Faridi et al., 2016; Sadeghi et al., 2017; Saikia et al., 2016). A detailed review of the encapsulation of curcumin in carbohydrate systems has been previously reported (Malaekeh-Nikouei and Salarbashi, 2018). With respect to nano-encapsulation in starch nanoparticles (Figure 5.3), studies published in the literature have shown that the nanoencapsulation of curcumin in these can significantly increase the stability, bioaccesibility, and solubility in aqueous solutions (6.0×10^{-2} mg/mL) up to 300 times more than those of pure curcumin (Pang et al., 2014; Sadeghi et al., 2017; Uzun and Kokini, 2014), which facilitates to formulate new food and nutraceutical products with a greater impact on the health.

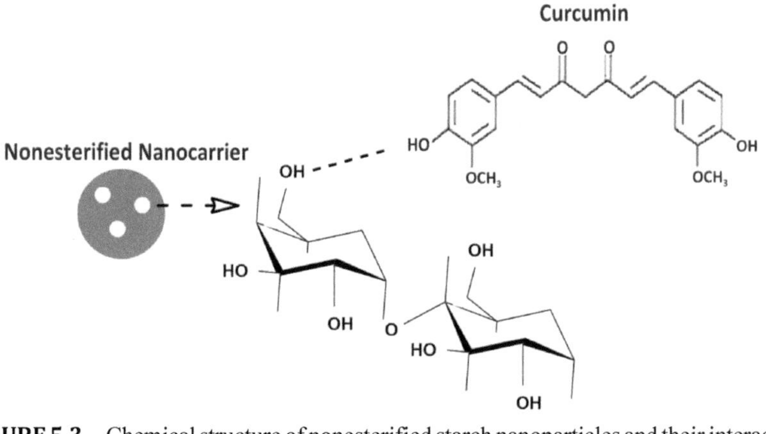

FIGURE 5.3 Chemical structure of nonesterified starch nanoparticles and their interaction with the nutraceutical curcumin.

TABLE 5.1 Starch Nanocarriers, Methods and Features of Nanoencapsulation of Several Nutraceuticals

Nanocapsule System	Nutraceutical	Size	Encapsulation Properties	Encapsulation Method	Reference
Hollow nanospheres/ nanoreactors					
Short linear glucans from waxy starch	Chlorogenic acid and ferulic acid	<123 nm	EE[a] of 52-95% and LE[b] of >20%	Ostwald ripening	Li et al. (2017)
Esterified starch nanoparticles					
- *Acetylated nanoparticles-based cassava and oxidized starch*	Gallic acid and butylatedhydroxytoluene	500-900 nm	EE: 10 y 20% for butylatedhydroxytoluene and gallic acid, respectively	Nanoprecipitation (*in situ*)	Oliveira et al. (2018)
- *Acetylated nanoparticles-based banana starch*	Curcumin	<190 nm	EE: 80% and LE of 3.205 mg/mg	Nanoprecipitation (*in situ*)	Acevedo-Guevara et al. (2018)
- *Nanoparticles from hydroxyethyl starch-deoxycholic acid*	Morin	197 nm	EE of 15.6% and drug loading of 92.5%	Solvent evaporation and sonication approaches (*in situ*)	Li et al. (2018)
Nanospheres from succynilated starch	Rutin	150-170 nm	EE < 85% and drug loading of 1.045 μg/mg	Dialysis	Li et al. (2018)
Adsorbent starch nanoparticles					
- *Worm-like amylopectin nanoparticles and spherical amylose nanoparticles*	Procyanidins, epicatechins and catechins	20-50 nm	Adsorption capacity of 1.2-4.3 g/g	Adsorption experiments	Qiu et al. (2016)
- *Waxy maize starch nanoparticle based nanocarriers*	(+)- catechin, (−)-epicatechin, (−)-epigallocatechin-3-gallate, and proanthocyanidins	100-400 nm	Adsorption capacity up to 235.57 μg/mg	Adsorption experiments	Liu et al. (2016)

TABLE 5.1 *(Continued)*

Nanocapsule System	Nutraceutical	Size	Encapsulation Properties	Encapsulation Method	Reference
			Nonesterified starch nanoparticles		
- *Soluble starch*	Curcumin	182-253 nm	EE of 97%	Nanoprecipitation and ultrasonication treatment	Jinglei Li et al. (2016)
- *Soluble starch*	Curcumin	61 nm	EE of 14.4%	Desolvation method by ethanol precipitation *(in situ)*	Maghsoudi et al. (2017)
- *Starch-based nanoparticles from horse chestnut, water chestnut and lotus stem.*	Catechin	322-615 nm	EE of 48-57%	Precipitation combined with ultrasonication	Ahmad et al. (2019).
- *Ginkgo starch nanospheres*	Ginkgo biloba extracts	201-250 nm	Loading charge of 0.661–1.045 mg/mg.	Anti-solvent precipitation (*in situ)*	Wang et al. (2018)
			Hybrid-magnetic nanoparticles		
- *Thiolated starch-coated iron oxide magnetic nanoparticles*	Curcumin	16 nm	EE of 70%	N/A	Saikia et al. (2017)
- *Aminated starch coated iron oxide magnetic nanoparticles*	Curcumin	16 nm	EE up to 84% and LE up to 74%	N/A	Saikia et al. (2016)
			Ternary nanoparticles		
Amylose, α-linoleic acid and β-lactoglobulin	Naringin	210-230 nm	EE of 74% and LE of 14%.	Inclusion complex	Feng et al. (2017)

TABLE 5.2 Previous Studies Concerning the in vitro and in vivo Findings of Nanoencapsulated Nutraceuticals in Starch Nanocarriers

Nutraceutical	Nanocarrier system	Author	Studies *in vitro and in vivo*
D-α-tocoferol	Nonesterified starch nanoparticles	Kheradvar *et al., 2018*	Studies in cell lines showed a marked increase in the metabolic activity of fibroblasts and collagen secretion.
Curcumi	Esterified starch nanoparticles	Acevedo-Guevara *et al., 2018.*	The *in vitro* release studies of curcumin showed that acetylated banana starch nanoparticles allowed a more controlled release, probably due to their stronger interaction of the hydrogen bond with curcumin.
Morin	Esterified starch nanoparticles	Li *et al., 2018*	Nanoencapsulation raised the plasma level of morin in rats and prolonged circulation time by up to 12 times more than free morin. The administration of the nanocapsules significantly reduced the level of serum uric acid at 6 h.
Ginkgo biloba extract	Nonesterified starch nanoparticles	Wang *et al., 2018*	The nanocapsules were stable under conditions of intestinal digestion. The administration of the nanocapsules improved the absorption of the extract by active endocytosis and improved its bioavailability.
(+)-catechin, (−)epicatechin, (−)epigallocatechin-3-gallate, and proanthocyanidins	Adsorbent starch nanoparticles	Liu *et al., 2016*	The starch nanoparticles, as well as the nanocapsules with the polyphenols, did not show toxicity in mouse embryonic fibroblast cells. The nanocapsules showed a sustained release under acidic and neutral conditions.
Curcumin	Nonesterified starch nanoparticles	Maghsoudi *et al., 2017.*	The starch nanoparticles were more efficient in the *in vitro* release of curcumin due to a resistance to water absorption and swelling. The nanoencapsulation improved the antimicrobial activity against *Streptococcus mutans* of curcumin and allowed greater absorption by the cells.
Chlorogenic acid and ferulic acid	Hollow starch nanoparticles as nanoreactors	Li *et al., 2017.*	The hollow nanoparticle shells were able to function as a protective barrier to stabilize the dynamic environment for the reaction of phenolic acids and free radicals. Hollow nanocapsules as nanoreactors enhanced the antioxidant activity of phenolic acids and improved the stability of their antioxidant activity in challenging environments with high salt concentrations and when exposed to UV radiation and high temperatures

TABLE 5.2 *(Continued)*

Nutraceutical	Nanocarrier system	Author	Studies *in vitro and in vivo*
Curcumin	Starch nanoparticles/ film	Athira and Jyothi, 2015	The nanoparticles showed no toxicity in L929 fibroblast cells. However, they had a toxicity in skin-cancer cells (SKMEL-28), which indicated high anticancer potential
Catechin	Nonesterified starch nanoparticles	Ahmad *et al.*, 2019	Catechins were effectively nanoencapsulated during the first 60 min of *in vitro* digestion. In addition, the nanoencapsulation of the catechins increased their nutraceutical properties antiobesity, and antidiabetic.
Naringina	Ternary nanoparticles	Feng *et al., 2017*	The encapsulating complexes were stable under simulated gastric conditions. The nanoparticle-naringin inclusion ternary complex showed strong release in the intestine.
Curcumin	Hybrid starch-magnetic nanoparticle	Saikia *et al., 2016*	Nanoencapsulation improved mucoadhesive properties in intestinal membrane, and enhanced cytotoxicity properties in MCF7 and HepG2 cancerous cells.

In another recent study, starch nanoparticles were evaluated as curcumin encapsulants. For this, starch was dissolved in water (5 mg/mL) and subjected to heating; afterward, an ethanolic solution with different concentrations of curcumin (1%–5%) was added to the solution of soluble starch previously mentioned to obtain charged nanocapsules. The size of nanoparticles was influenced by the introduction of curcumin molecules, resulting in sizes of 182, 203, 223, and 253 nm as the concentration of curcumin increased (1%, 2%, 3%, and 5%, respectively). However, the polydispersity indexes were lower than 0.2, indicating high homogeneity in all of the nanoparticulate systems. The 3% formulation of curcumin presented the highest encapsulation efficiency (97%) and antioxidant capacity; in addition, due to the high zeta potential value (−26) presented, it had the best colloidal stability for 30 days (Li et al., 2016).

Other authors have reported an increased solubility of curcumin through its encapsulation in starch nanoparticles due to the high hydrophilicity and the high surface-to-volume ratio, which permitted interaction with the aqueous solvent. The hydrodynamic size of nanocapsules was 61.1 nm, and the actual dry size within the range of 50–88 nm. The authors reported that starch nanocapsules were more resistant to water absorption up to 60 h, so they were more effective in releasing curcumin, compared to chitosan and alginate nanocapsules. In addition, the reduction in particle size on the nanoscale positively influenced the antimicrobial activity of curcumin. The reduced size of curcumin nanocapsules compared to that of pure curcumin provoked greater cell absorption and decreased the minimal inhibitory concentration (Maghsoudi et al., 2017). In another similar research work, curcumin was nanocapsulated into starch nanoparticles and these were added to films composed of different concentrations of cassava starch (ST) and PVA (4:2, 3:3, and 2:4), as well as to polymer films separately. Cytotoxicity studies demonstrated that both curcumin and ST–PVA films with curcumin were no toxic for L929 fibroblast cells; however, the viability (42.03%) of cancer skin cells treated with nanoencapsulated curcumin was significantly ($p<0.05$) lower than that of those treated with pure curcumin, which evidenced the anticancer potential of the former (Athira and Jyothi, 2015).

5.3.2 ESTERIFIED STARCH NANOPARTICLES

Hydroxyl (OH) groups of starch nanoparticles have been subjected to esterification processes employing several chemical reagents, such as acetic acid, 2-octenyl succinic acid, ethylene oxide, and deoxycholic acid, among others, to add surface functionality and promote the molecular interaction between starch nanoparticles and nutraceuticals (Figure 5.4). Esterified starch nanoparticles have been mostly used for the encapsulation of phenolic acids, vitamins, and some flavonoids such as morin and rutin.

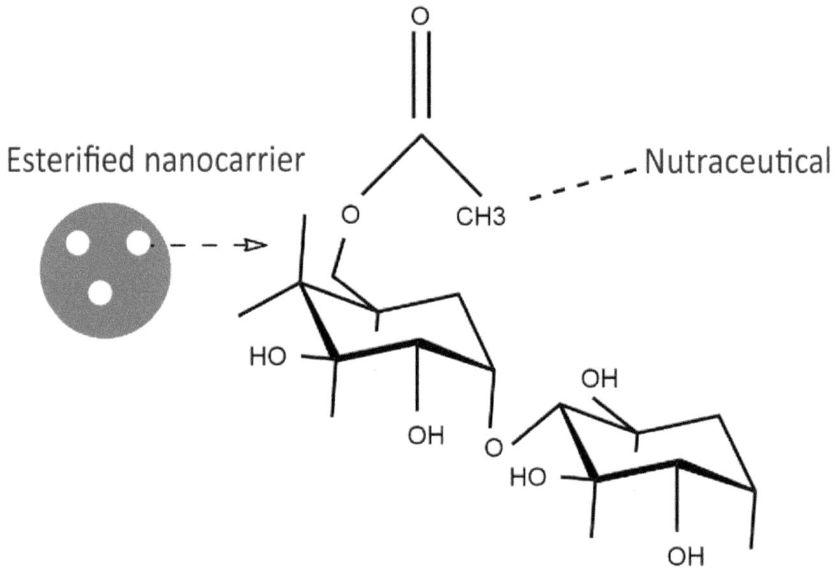

FIGURE 5.4 Chemical structure of acetylated starch nanoparticles.

Regarding the nanoencapsulation of vitamins, a recent research work reported the satisfactory encapsulation of coenzyme Q10 (CoQ10) in esterified starch nanoparticles. Here, it is important to highlight that the nanoencapsulation of vitamins in esterified starch nanoparticles is scarce. With great nutritional importance, CoQ10 forms an integral part of the electron transport chain of the mitochondrial membrane and, therefore, is essential for aerobic metabolism; moreover, it is currently the third most consumed nutritional supplement after fish oil. Despite the fact that human beings can synthesize CoQ10, internal (genetic) or external factors (statins) can reduce the concentrations of CoQ10, and the lack of CoQ10 is symptomatic in cells and tissues rich in mitochondria, such as muscle and nerve tissues, resulting in several types of myopathy and neuropathy. Supplementation with CoQ10 has been shown to decrease the symptoms of these patients (Boitier et al., 1998). Recently, the high-pressure homogenization method (170 kPa) was utilized to treat succinate starch and CoQ10 emulsions. The authors reported obtaining nanoparticles with sizes of about 230 nm. Efficiency and encapsulation capacity values were not reported; however, CoQ10 nanoformulation is ideal, not only to add nutritional value to common beverages such as fruit juices but also useful for increasing CoQ10 levels in baked products due to the thermostability and high retention (98%) presented when the nanoparticles were exposed to supercritical fluid treatments at high temperatures (Cheuk et al., 2015).

Succinylated starch as amphiphilic molecules has been evaluated for the nanoencapsulation of the bioflavonoid rutin. Rutin (also known as vitamin P) is found in *Fagopyrum esculentum*, *Rutagraveolens*, and *Sophora japonica* plants and in other sources such as passionflower, buckwheat, tea, and apple, from which it has been demonstrated through in vitro and in vivo studies to have antidiabetic, anti-inflammatory, antioxidant, neuroprotective, nephroprotective, hepatoprotective, cytoprotective, vasoprotective, and anticancer activities (Ganeshpurkar and Saluja, 2017; Ghorbani, 2017). However, its clinical application is limited due to its poor solubility in water and lipids phases. When rutin molecules were nanoencapsulated in starch nanospheres, a considerable increase in the particle size of up to 150 nm was evident. On the other hand, an increase in the degree of substitution resulted in a decrease in the encapsulation efficiency (85%–75%), the amount of loaded rutin (0.36–0.30 μg/mg), and the particle size (70–150 nm). The latter was explained because the greater the degree of substitution, the greater the hydrophobic interaction among the micellar nuclei, resulting in more compact nuclei with a reduced particle size and a smaller amount of rutin embedded in the nucleus. According to in vitro release studies, the authors indicated that nanocellular systems were capable of exerting a slow and controlled release (Li et al., 2018).

In a recent research paper, acetylated cassava starch nanoparticles with large sizes (500–900 nm), spherical morphology, and reduced crystallinity resulted in an encapsulation efficiency of gallic acid (hydrophilic) and butylated hydroxyl toluene (hydrophobic) of 10% and 90%, respectively (de Oliveira et al., 2018). Similarly, in another research study, nanoparticles of acetylated and native banana starch with narrow sizes of <190 nm and polydispersity indexes of less than 0.3 revealed high efficacy of curcumin encapsulation of up to 80%, which is higher for acetylated nanoparticles. Additionally, when the encapsulates were exposed to gastrointestinal fluids in vitro, both systems showed similar release profiles, indicating that both nanovehicles entertain a similar interaction with the intestinal fluid (Acevedo-Guevara et al., 2018).

Morin flavonoid (3,5,7,2′,4′-pentahydroxyflavone) is other polyphenol encapsulated in esterified starch nanoparticles. Morin has been reported as a potent antihyperuricemic because it markedly inhibits the synthesis of uric acid and reduces the absorption of renal urate in vitro. However, its application in the treatment of hyperuricemia is limited due to its poor solubility in water and its rapid degradation in vivo. A murine nanoencapsulant from hydroxyethyl starch (HES) esterified with deoxycholic acid (DOCA) was developed. In the first instance, the hydrophobically modified HES polymer was prepared by covalently bonding DOCA with HES through an ester linkage, producing a new type of amphiphilic polymer. Subsequently, morin was easily encapsulated in HES-DOCA by utilizing a simple combination of solvent evaporation and sonication. The nanocapsules had a particle size of 197 nm, a polydispersity index of 0.05, a zeta potential of −14 mV, and a loading capacity and encapsulation efficiency of 92.5% and 15.6%, respectively. Regarding pharmacokinetic results in rats, morin nanocapsules exhibited a remarkably higher plasma level and a prolonged plasma-elimination time after injection, compared to those of free morin. In addition, nanocapsules with morin showed a significantly longer blood circulation time of up to 12 h after injection. Surprisingly, the half-life of morin (0.37 h) was significantly improved by more than four times by its nanoencapsulation (1.52 h). Treatment with free morin and nanoencapsulated morin reduced the level of serum uric acid at 4 h, compared to that of the hyperuricemia (control) group, although only the nanoencapsulated morin further reduced the serum level of uric acid (55 μmol/L) at 6 h, indicating its superior antihyperuricemia efficacy. These encouraging results suggest a possible greater therapeutic effect of nanoencapsulated morin compared to that of the free morin solution (Li et al., 2018).

5.3.3 HOLLOW STARCH NANOPARTICLES AS NANOREACTORS

In recent years, hollow biopolymer nanospheres have gained considerable recognition and have been postulated as drug nanocarriers due to their large surface volume, low density, and good surface permeability (Ge et al., 2016; Hajfathalian et al., 2016; Li et al., 2017). The synthesis of hard-core templates comprises the most direct and efficient procedure for obtaining hollow nanoparticles. Starch nanoparticles obtained by the retrogradation of short linear glucans can have a considerable application as a hard-core template due to their null toxicity, biocompatibility, and biodegradability (Li et al., 2017). The assembly of phenolic acids in hollow nanoparticles synthesized from short glucans of waxy starch was conjugated with arabica gum to increase the antioxidant capacity and stability against a high condensation of salt, UV light, and thermal treatments. Sweeping activity occurred in an accelerated manner within the nanospheres, which were cataloged as nanoreactors due to which the phenolic acids, specifically ferulic and chlorogenic acids, were immobilized in the inner cavity of the nanospheres by means of hydrogen bridges and hydrophobic interactions, while the free radicals permeated the outer shell of the nanospheres of gum arabica (Figure 5.5). The encapsulation efficiency and load capacity of the nanoreactors were 51–95% and 20%, respectively, and the colloidal properties comprise a hydrodynamic size of 123 nm, a desirable polydispersity index of 0.454, and a high zeta potential value of (−43 to −51 mV) (Li et al., 2017).

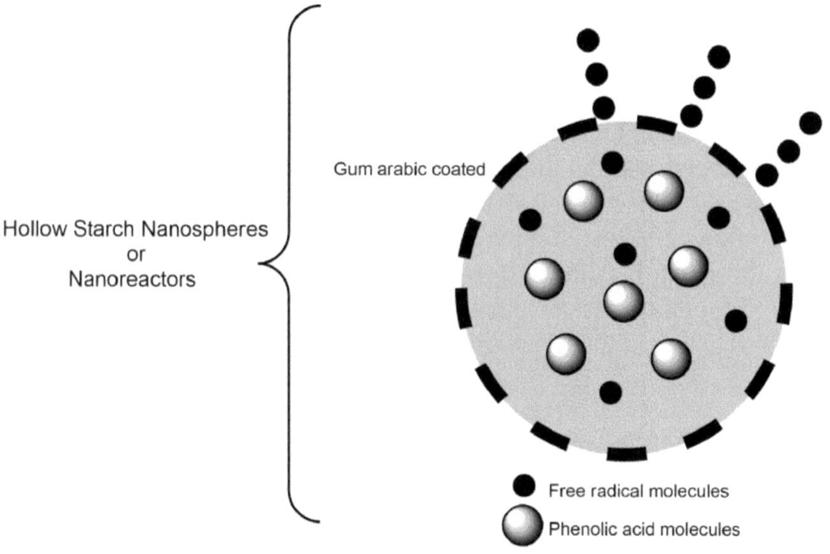

FIGURE 5.5 Representative scheme of hollow nanoparticles and their functionalization as nanoreactors.

5.3.4 ADSORBENT STARCH NANOPARTICLES

Starch nanoparticles have been used effectively in the adsorption of nutraceuticals due to the high affinity and binding capacity of amylose and amylopectin molecules to polyphenolics compounds. Descriptions of various adsorption mechanisms and kinetic models are reported in the literature (Dehghanian et al., 2016). Qiu et al. (2016) predicted through the Langmuir isotherm model that some polyphenols such as procyanidins, epicatechins, and catechins have high binding capacity (1.2–4.3 g/g) to starch nanoparticles. The authors reported that amylopectin molecules presented a new worm-like morphology with a size of up to 20 nm and a length of less than 500 nm. Otherwise, the amylose molecules showed sizes from 20 to 50 nm and a spherical morphology. Both nanostructures showed polydispersity indexes lower than 0.185 and negative zeta potential values. The adsorption of polyphenol molecules on the surface of the nanoparticles caused an increase in size; however, it did not surpass 200 nm.

Similarly, others investigated the mechanisms involved and capacity adsorption of epigallocatechin-3-gallate, catechin, epicatechin, and proanthocyanins in starch nanoparticles (Liu et al., 2016). Predictably, up to 21.95 μg/mg of polyphenols could be adsorbed onto starch nanoparticles; however, the experimental results showed that when the concentration of polyphenols in the reaction was 20 mM, up to 235.57 μg/mg was adsorbed onto the nanostructures. Adsorption of polyphenols onto starch nanoparticles increased the particle size within a range of 100–400 nm, and according to in vitro cytotoxicity studies, both the nanoparticles and the nanoparticles with adsorbed polyphenols did not show cellular toxicity. In addition, starch nanoparticles showed a sustained release of polyphenols under acid and neutral digestion conditions.

5.3.5 HYBRID STARCH MAGNETIC NANOPARTICLES

The delivery of the anticancer agent in the target tissues without affecting normal cells is of maximal interest in the treatment of cancer. Nanoparticles with magnetic properties have attracted a special recognition in the oncological field due to their ability to localize the therapeutic agent within cells under the help of magnetism (Sensenig et al., 2012). However, the biggest challenge for obtaining magnetic nanoparticles is to control the size and supermagnetism. In response to these challenges, the biopolymer starch has been used to coat and modify the surface of nanoparticles, for instance, of iron oxide nanoparticles, which increased their biocompatibility and biodegradability. Thiolated starch is a good candidate as a coating material for magnetic nanoparticles (Figure 5.6), in that it was possible to decrease the particle size from 75 to 16 nm and increase the colloidal stability of the coated magnetic nanoparticles in different media (phosphate-buffered saline and Dulbecco's modified Eagle's minimal essential medium). Additionally, the coated magnetic nanoparticles showed a curcumin-encapsulation efficiency of less than 70%, with high compatibility with lymphocyte cells, and anticancer potential in MCF7 and HepG2 cells (Saikia et al., 2017). In another investigation (Saikia et al., 2016), magnetic iron oxide nanoparticles coated with cross-linked starch decreased the particle size from 100 to 16 nm and potentiated the stability of the nanoparticles due to a more compact structure, which also facilitated the encapsulation of curcumin molecules with a charge capacity and encapsulation efficiency of 74% and 84%, respectively. Among other results, the coated nanoparticles showed mucoadhesive properties, thus improving the cytotoxicity in MCF7 and HepG2 cancer cells and a significant uptake in the HepG cell line.

5.3.6 TERNARY NANOPARTICLES

The design and application of ternary complexes formed from organic phases, among these by starch, have engendered rising attention in the encapsulation of polyphenolic nutraceuticals, such as naringin (4′,5,7-trihydroxyflavanone-7-β-l-rhamnoglucoside-(1,2)-α-D-glucopyranoside), a flavonoid that possesses enormous health benefits, as it can substantially strengthen the immune system and have antiatherogenic, anticancer, and antimicrobial properties. However, its low bioavailability has reduced its application in the nutraceutical field (Feng et al., 2017). Therefore, these authors nanoencapsulated naringin in ternary nanoparticles synthesized from high-amylose corn starch in combination with β-lactoglobulin and α-linoleic acid. The encapsulation efficiency and the encapsulation capacity of these nanoparticles were up to 74%. The ternary nanoparticles and the inclusion complexes between these and naringin showed a narrow size distribution and sizes smaller than 230 nm and a stability under simulated gastric conditions, which allowed more naringin molecules to reach the colon for absorption. When the nanoencapsulates were exposed to simulated intestinal conditions, the release of naringin was

43% at 2 h; therefore, the authors concluded that the nanoparticle–naringin inclusion ternary complex demonstrated strong release in the intestine, permitting more molecules of naringin to be absorbed in this site, which could increase its absorption and bioavailability (Feng et al., 2017).

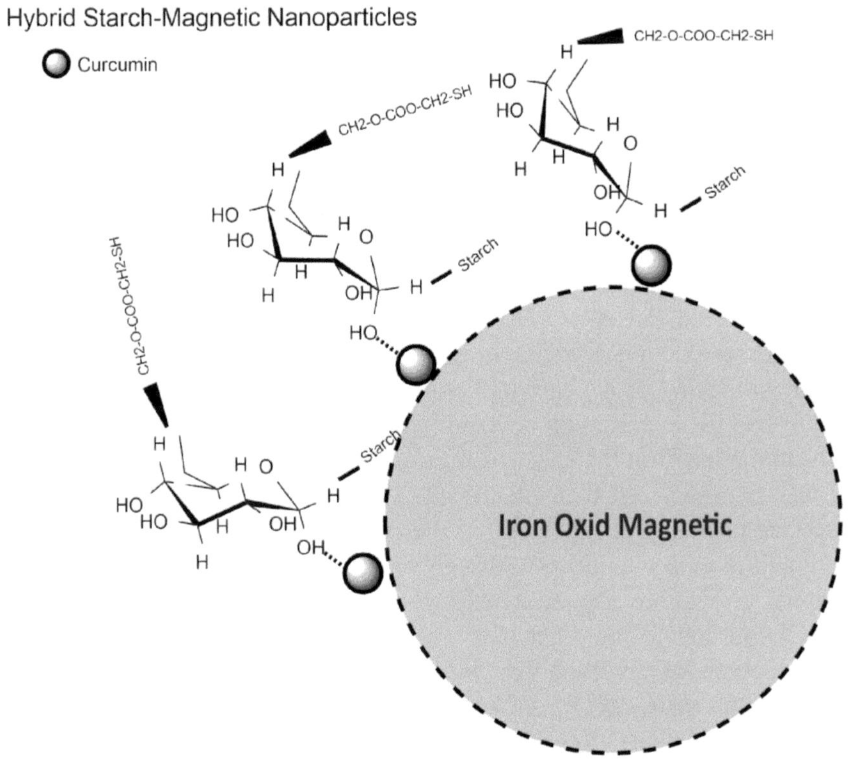

FIGURE 5.6 Hybrid starch–magnetic nanoparticles formed by iron oxide nanoparticles coated with thiolated starch and their functionalization as encapsulating materials of the anticancer curcumin.

5.4 INTERACTION OF NUTRACEUTICAL-STARCH NANOCARRIERS

The molecular interaction of nutraceutical compounds with starch nanoparticles has been widely studied. The molecular weight, the polarity of the nutraceuticals, and the available reactive sites, as well as the molecular structures of the nanoparticles, exert a great influence on their properties as nanoencapsulating materials. To study this interaction, various methods such as Fourier transform infrared (FTIR) spectroscopy, X-ray diffraction (XRD), scanning electron microscopy (SEM), and differential scanning calorimetry (DSC) have been employed.

Through FTIR spectroscopy, the chemical interactions involved between polyphenol curcumin and starch nanoparticles have been elucidated. The IR spectra of acetylated nanoparticles, as well as of native nanoparticles showed a strong spectroscopic signal at 3520 cm^{-1}, commonly related to the narrowing of the O–H group in the benzene ring of curcumin. The authors added that, due to an increase in oxygen atoms by the esterifying agent anhydride acetic, more curcumin molecules interacted by means of hydrogen bonds (Acevedo-Guevara et al., 2018). Similarly, through FTIR studies, the occurrence of a possible formation of hydrogen bonds between the O–H groups of curcumin and the O–H groups of the glucose molecules was suggested (Li et al., 2016). In other studies, for vitamin E, FTIR spectra revealed a new signal in the band at 1750 cm^{-1} broadly attributed to the stretching of the carbonyl ester groups (C=O) of vitamin E, which confirmed the loading of the vitamin into the starch nanoparticles.

On the basis on the spectroscopic signals of XRD and FTIR analyses, it was hypothesized that vitamin D3 was loaded in greater amounts in the side chains of amylopectin in the starch, forming a helical inclusion interacting with vitamin D3 through van der Waals bonds, the methylene groups of vitamin D3, and hydrogens of the five carbons of glucose. It is important to highlight that scarce information has reported on the introduction of hydrophobic molecules into amylopectin molecules (Hasanvand et al., 2018), with the main data involving the inclusion in amylose molecules. Regarding the crystallinity pattern determined by XRD, the authors have shown that nanoparticles obtained from waxy starch had the property to form a helical type-V structure with essential oils under ultrasonic treatments, and they suggested that the interaction between the starch nanoparticles and the oil of mint was through hydrogen bridges (Liu et al., 2017).

The reactive sites play an important role in the interaction and adsorption time of the nutraceuticals in the starch nanoparticles. It is previously suggested that the molecular interaction of starch nanoparticles with antioxidant compounds was not limited by the available reactive sites of the nanoparticles but rather by that of the polyphenol content in the reaction medium during nanoprecipitation because an increase in the content of ginkgo biloba extract led to an increase in the amount loaded in the nanoparticles (Wang et al., 2018). Based on this, the adsorption would depend on the number of free polyphenol molecules in the medium. However, other studies have shown otherwise. It has been reported that the highest adsorption of polyphenols occurred during the first 30 min, which could be due to the availability of many binding sites on the surface of the nanoparticles. However, the allocation of polyphenols decreased significantly after 30 min, and this was explained by the saturation of the reactive sites, coupled with the forces of repulsion between loaded and free polyphenols (Liu et al., 2016).

The molecular weight of the nutraceuticals also exerts a positive influence on their adsorption onto starch nanoparticles, as it has been shown, for example, that higher molecular weight polyphenols such as procyanidins (Mw = 594.52) had higher adsorption compared to those of epicatechins (Mw = 290.27) and catechins (Mw = 332.43) (Qiu et al., 2016). Moreover, the polarity of nutraceuticals has also been key in the interaction and encapsulation in starch nanoparticles. For example, the high polarity of the galic acid antioxidant caused a greater interaction with the aqueous medium and decreased the efficiency of encapsulation in acetylated starch nanoparticles. Also, the chemical interaction of gallic acid with acetylated nanoparticles was sterically prevented by the esterifying groups in the nanoparticles, occurring only in unesterified O–H groups (de Oliveira et al., 2018). The structure and crystallinity of nanoparticles also play an important role in the starch nanoparticle–polyphenol interaction. For example, amylopectin nanoparticles had the highest binding capacity with polyphenols, which could be due to the worm-like structure with a greater superficial area. In contrast, amylose nanoparticles had a low adsorption capacity, which could be due to higher crystallinity, with a more compact structure (Qiu et al., 2016). Similarly, some authors have suggested that curcumin was encapsulated in the amorphous regions of the starch nanoparticles instead of in the crystalline ones (Acevedo-Guevara et al., 2018).

5.5 CONCLUSIONS

The application of nutraceutical compounds, specifically phenolic acids, polyphenols, vitamins, and essential fatty acids, has been reduced due to their high sensitivity. However, the encapsulation of various nutraceuticals in starch nanoparticles offers a wide range of applications in the fields of both biomedical and food industries.

Starch-based nanoparticles do not present toxicity against normal cell lines and increased the bioactive properties of different nutraceuticals such as polyphenols; for example, the nanoencapsulation catechin promoted a significant retention of the antiobesity nutraceutical properties. Likewise, when some polyphenols were nanoencapsulated, they exhibited greater stability against simulated gastrointestinal conditions and greater solubility and penetration in cancer cells. Moreover, in vivo studies have shown that nanoencapsulation has improved the serum half-life of certain polyphenols such as murin, increasing its antiuricemic properties. Also, starch nanoparticles can function as nanoreactors, where they render it possible to carry out the antioxidant reactions of some polyphenols inside them. With respect to the nanoencapsulation of essential fatty acids in starch nanoparticles, information is scarce; some successful studies have suggested an increase in the stability of peppermint oil, for example, under high-temperature conditions.

Regarding the encapsulation of vitamins, different types of starch-based nanostructures have been employed for their nanoencapsulation and the results have been successful. For instance, it has been possible to increase the solubility of vitamin D in a food system with the aim of enriching it without impairing sensory properties. Concerning the biomedical area, it was demonstrated that vitamin E nanocapsules embedded in nanofibers demonstrated a sustained release in fibroblast culture media, resulting in an improvement in tissue production and opening up numerous biomedical applications in skincare. Regarding nanochemotherapy or nanochemo-prevention, it has been shown that the conjugation of magnetic nanoparticles with starch utilized in the treatment of cancer has improved their physical properties, greatly reducing their size, increasing their biocompatibility, and enhancing anticancer-encapsulation efficiency.

At present, the study on and application of starch nanoparticles in various areas have been growing. The importance of expanding the in vivo and in vitro studies of these biodegradable materials as efficient tools in the protection of different sensitive bioactive compounds and their transport and release is clear.

KEYWORDS

- **bioactivities**
- **based-starch nanoparticles**
- **nanoencapsulation**
- **nutraceuticals**
- **phytochemicals**

REFERENCES

Acevedo-Guevara, L.; Nieto-Suaza, L.; Sanchez, L. T.; Pinzon, M. I.; Villa, C. C. Development of Native and Modified Banana Starch Nanoparticles as Vehicles for Curcumin. *Int. J. Biol. Macromol.* **2018**, *111*, 498–504.

Ahmad, M.; Mudgil, P.; Gani, A.; Hamed, F.; Masoodi, F. A.; Maqsood, S. Nano Encapsulation of Catechin in Starch Nanoparticles: Characterization Release Behavior and Bioactivity Retention during In-Vitro Digestion. *Food Chem.* **2019**, *270*(1), 95–104.

Ali Razavi, S. M.; Amini, A. M. Starch Nanomaterials: A State-of-the-Art Review and Future Trends. In *Novel Approaches of Nanotechnology in Food*; Elsevier Inc.: the Netherlands, 2016; Vol. 49, pp 237–269.

Asbahani, A. El; Miladi, K.; Badri, W.; Sala, M.; Addi, E. H. A.; Casabianca, H.; Mousadik, A. El; Hartmann, D.; Jilale, A.; Renaud, F. N. R.; et al. Essential Oils: From Extraction to Encapsulation. *Int. J. Pharm.* **2015**, *483*(1–2), 220–243.

Athira, G. K.; Jyothi, A. N. Cassava Starch-Poly(Vinyl Alcohol) Nanocomposites for the Controlled Delivery of Curcumin in Cancer Prevention and Treatment. *Starch Stärke* **2015**, *67*(5–6), 549–558.

Boitier, E.; Degoul, F.; Desguerre, I.; Charpentier, C.; François, D.; Ponsot, G.; Diry, M.; Rustin, P.; Marsac, C. A Case of Mitochondrial Encephalomyopathy Associated with a Muscle Coenzyme Q10 deficiency. *J. Neurol. Sci.* **1998**, *156*(1), 41–46.

Boufi, S.; Bel Haaj, S.; Magnin, A.; Pignon, F.; Impéror-Clerc, M.; Mortha, G. Ultrasonic Assisted Production of Starch Nanoparticles: Structural Characterization and Mechanism of Disintegration. *Ultrason. Sonochem.* **2018**, *41*, 327–336.

Bras, J.; Dufresne, A. Starch Nanoparticles: A Review. *Biomacromolecules* **2010**, *11*, 1139–1153.

Cheuk, S. Y.; Shih, F. F.; Champagne, E. T.; Daigle, K. W.; Patindol, J. A.; Mattison, C. P.; Boue, S. M. Nano-Encapsulation of Coenzyme Q10 Using Octenyl Succinic Anhydride Modified Starch. *Food Chem.* **2015**, *174*, 585–590.

de Oliveira, N. R.; Fornaciari, B.; Mali, S.; Carvalho, G. M. Acetylated Starch-Based Nanoparticles: Synthesis, Characterization, and Studies of Interaction With Antioxidants. *Starch* **2018**, *70*(3–4), 1700170.

Dehghanian, N.; Ghaedi, M.; Ansari, A.; Ghaedi, A.; Vafaei, A.; Asif, M.; Agarwal, S.; Tyagi, I.; Gupta, V. K. A Random Forest Approach for Predicting the Removal of Congo Red from Aqueous Solutions by Adsorption onto Tin Sulfide Nanoparticles Loaded on Activated Carbon. *Desalin. Water Treat.* **2016**, *57*(20), 9272–9285.

Delgado-Zamarreño, M. M.; Fernández-Prieto, C.; Bustamante-Rangel, M.; Pérez-Martín, L. Determination of Tocopherols and Sitosterols in Seeds and Nuts by QuEChERS-Liquid Chromatography. *Food Chem.* **2016**, *192*, 825–830.

El-Naggar, M. E.; El-Rafie, M. H.; El-sheikh, M. A.; El-Feky, G. S.; Hebeish, A. Synthesis, Characterization, Release Kinetics and Toxicity Profile of Drug-Loaded Starch Nanoparticles. *Int. J. Biol. Macromol.* **2015**, *81*, 718–729.

El-Sheikh, M. A. New Technique in Starch Nanoparticles Synthesis. *Carbohydr. Polym.* **2017**, *176*, 214–219.

Faridi, A.; Seid, E.; Jafari, M. Biopolymer Nano-Particles and Natural Nano-Carriers for Nano-Encapsulation of Phenolic Compounds. *Colloids Surfaces B: Biointerfaces* **2016**, *146*, 532–543.

Feng, T.; Wang, K.; Liu, F.; Ye, R.; Zhu, X.; Zhuang, H.; Xu, Z. Structural Characterization and Bioavailability of Ternary Nanoparticles Consisting of Amylose, α-Linoleic Acid and β-Lactoglobulin Complexed with Naringin. *Int. J. Biol. Macromol.* **2017**, *99*, 365–374.

Ganeshpurkar, A.; Saluja, A. K. The Pharmacological Potential of Rutin. *Saudi Pharm. J.* **2017**, *25*(2), 149–164.

Ge, K.; Zhang, C.; Sun, W.; Liu, H.; Jin, Y.; Li, Z.; Liang, X.; Jia, G.; Zhang, J. Up-Conversion Y_2O_3:Yb^{3+},Er^{3+} Hollow Spherical Drug Carrier with Improved Degradability for Cancer Treatment. *ACS Appl. Mater. Interfaces* **2016**, *8*(38), 25078–250860.

Ghorbani, A. Mechanisms of Antidiabetic Effects of Flavonoid Rutin. *Biomed. Pharmacother.* **2017**, *96*, 305–312.

Grumezescu, A. M.; El-Sherbiny, I. M.; El-Baz, N. M.; Hefnawy, A. Potential of Nanotechnology in Nutraceuticals Delivery for the Prevention and Treatment of Cancer. In *Nutraceuticals*; Elsevier Inc., 2016; pp 117–152.

Gul, F.; Efeoglu, H.; Efeoglu, H. Cereal Starch Nanoparticles—A Prospective Food Additive: A Review. *Crit. Rev. Food Sci. Nutr.* **2016**, *58*(7), 1–46.

Hajfathalian, M.; Gilroy, K. D.; Golze, S. D.; Yaghoubzade, A.; Menumerov, E.; Hughes, R. A.; Neretina, S. A Wulff in a Cage: The Confinement of Substrate-Based Structures in Plasmonic Nanoshells, Nanocages, and Nanoframes Using Galvanic Replacement. *ACS Nano* **2016**, *10*(6), 6354–6362.

Hasanvand, E.; Fathi, M.; Bassiri, A. Production and Characterization of Vitamin D3 Loaded Starch Nanoparticles: Effect of Amylose to Amylopectin Ratio and Sonication Parameters. *J. Food Sci. Technol.* **2018**, *55*(4), 1314–1324.

Hasanvand, E.; Fathi, M.; Bassiri, A.; Javanmard, M.; Abbaszadeh, R. Novel Starch Based Nanocarrier for Vitamin D Fortification of Milk: Production and Characterization. *Food Bioprod. Process.* **2015**, *96*, 264–277.

Hillman, G. G.; Singh-Gupta, V. Soy Isoflavones Sensitize Cancer Cells to Radiotherapy. *Free Radic. Biol. Med.* **2011**, *51*(2), 289–298.

Hsu, C. P.; Shih, Y. T.; Lin, B. R.; Chiu, C. F.; Lin, C. C. Inhibitory Effect and Mechanisms of an Anthocyanins- and Anthocyanidins-Rich Extract from Purple-Shoot Tea on Colorectal Carcinoma Cell Proliferation. *J. Agric. Food Chem.* **2012**, *60*(14), 3686–3692.

Jafari, S. M. An Overview of Nanoencapsulation Techniques and Their Classification. In *Nanoencapsulation Technologies for the Food and Nutraceutical Industries*; Elsevier Inc., 2017; pp 1–34.
Katouzian, I.; Jafari, S. M. Nano-Encapsulation as a Promising Approach for Targeted Delivery and Controlled Release of Vitamins. *Trends Food Sci. Technol.* **2016**, *53*, 34–48.
Kheradvar, S. A.; Nourmohammadi, J.; Tabesh, H.; Bagheri, B. Starch Nanoparticle as a Vitamin E-TPGS Carrier Loaded in Silk Fibroin-Poly(Vinyl Alcohol)-Aloe Vera Nanofibrous Dressing. *Colloids Surfaces B: Biointerfaces* **2018**, *166*, 9–16.
Khorasani, S.; Danaei, M.; Mozafari, M. R. Nanoliposome Technology for the Food and Nutraceutical Industries. *Trends Food Sci. Technol.* **2018**, *79*, 106–115.
Kim, H.; June, D.; Kim, J.; Lim, S. Preparation of Crystalline Starch Nanoparticles Using Cold Acid Hydrolysis and Ultrasonication. *Carbohydr. Polym.* **2013**, *98*(1), 295–301.
Kim, H.; Park, S. S.; Lim, S. Preparation, Characterization and Utilization of Starch Nanoparticles. *Colloids Surfaces B Biointerfaces* **2015**, *126*, 607–620.
Kozlowska, M.; Gruczyńska, E.; Ścibisz, I.; Rudzińska, M. Fatty Acids and Sterols Composition, and Antioxidant Activity of Oils Extracted from Plant Seeds. *Food Chem.* **2016**, *213*, 450–456.
Li, J.; Shin, G. H.; Lee, I. W.; Chen, X.; Park, H. J. Soluble Starch Formulated Nanocomposite Increases Water Solubility and Stability of Curcumin. *Food Hydrocoll.* **2016**, *56*, 41–49.
Li, J.; Yang, Y.; Lu, L.; Zhang, J. Preparation, Characterization and Systemic Application of Self-Assembled Hydroxyethyl Starch Nanoparticles-Loaded Flavonoid Morin for Hyperuricemia Therapy. *Int. J. Nanomed.* **2018**, *13*, 2129–2141.
Li, S.; Gu, F.; Gao, Q. Preparation of Rutin-Loaded Starch Nanospheres. *Starch* **2018**, *70*(3–4), 1–7.
Li, Z.; Jiang, H.; Xu, C.; Gu, L. A Review: Using Nanoparticles to Enhance Absorption and Bioavailability of Phenolic Phytochemicals. *Food Hydrocoll.* **2015**, *43*, 153–164.
Li, X.; Li, M.; Liu, J.; Ji, N.; Liang, C.; Sun, Q.; Xiong, L. Preparation of Hollow Biopolymer Nanospheres Employing Starch Nanoparticle Templates for Enhancement of Phenolic Acid Antioxidant Activities. *J. Agric. Food Chem.* **2017**, *65*(19), 3868–3882.
Liu, C.; Ge, S.; Yang, J.; Xu, Y.; Zhao, M.; Xiong, L.; Sun, Q. Adsorption Mechanism of Polyphenols onto Starch Nanoparticles and Enhanced Antioxidant Activity under Adverse Conditions. *J. Funct. Foods* **2016**, *26*, 632–644.
Liu, C.; Li, M.; Ji, N.; Liu, J.; Xiong, L.; Sun, Q. Morphology and Characteristics of Starch Nanoparticles Self-Assembled via a Rapid Ultrasonication Method for Peppermint Oil Encapsulation. *J. Agric. Food Chem.* **2017**, *65*(38), 8363–8373.
Ložnjak, P.; Jakobsen, J. Stability of Vitamin D3 and Vitamin D2 in Oil, Fish and Mushrooms after Household Cooking. *Food Chem.* **2018**, *254*, 144–149.
Maghsoudi, A.; Yazdian, F.; Shahmoradi, S.; Ghaderi, L.; Hemati, M.; Amoabediny, G. Curcumin-Loaded Polysaccharide Nanoparticles: Optimization and Anticariogenic Activity Against Streptococcus Mutans. *Mater. Sci. Eng. C* **2017**, *75*, 1259–1267.
Malaekeh-Nikouei, B.; Salarbashi, D. Nanoencapsulation and Delivery of Curcumin Using Some Carbohydrate Based Systems: A Review. *Nanomed J.* **2018**, *5*(2), 57–61.
Mirzaei, H.; Shakeri, A.; Rashidi, B.; Jalili, A.; Banikazemi, Z.; Sahebkar, A. Phytosomal Curcumin: A Review of Pharmacokinetic, Experimental and Clinical Studies. *Biomed. Pharmacother.* **2017**, *85*, 102–112.

Neilson, A. P.; Hopf, A. S.; Cooper, B. R.; Pereira, M. A.; Bomser, J. A.; Ferruzzi, M. G. Catechin Degradation with Concurrent Formation of Homo- and Heterocatechin Dimers during in Vitro Digestion. *J. Agric. Food Chem.* **2007**, *55*, 8941–8949.

Oliveira, A.; Amaro, A. L.; Pintado, M. Impact of Food Matrix Components on Nutritional and Functional Properties of Fruit-Based Products. *Curr. Opin. Food Sci.* **2018**, *22*, 153–159.

Pang, S. C.; Tay, S. H.; Chin, S. F. Facile Synthesis of Curcumin-Loaded Starch-Maleate Nanoparticles. *J. Nanomater.* **2014**, 1–7.

Patel, C. M.; Chakraborty, M.; Murthy, Z. V. P. Fast and Scalable Preparation of Starch Nanoparticles by Stirred Media Milling. *Adv. Powder Technol.* **2016**, *27*(4), 1287–1294.

Qin, Y.; Liu, C.; Jiang, S.; Xiong, L.; Sun, Q. Characterization of Starch Nanoparticles Prepared by Nanoprecipitation: Influence of Amylose Content and Starch Type. *Ind. Crop. Prod.* **2016**, *87*, 182–190.

Qiu, C.; Qin, Y.; Zhang, S.; Xiong, L.; Sun, Q. A Comparative Study of Size-Controlled Worm-like Amylopectin Nanoparticles and Spherical Amylose Nanoparticles: Their Characteristics and the Adsorption Properties of Polyphenols. *Food Chem.* **2016**, *213*, 579–587.

Quintin, D.; Garcia-Gomez, P.; Ayuso, M.; Sanmartin, A. M. Active Biocompounds to Improve Food Nutritional Value. *Trends Food Sci. Technol.* **2019**, *84*, 19–21.

Sadeghi, R.; Daniella, Z.; Uzun, S.; Kokini, J. Effects of Starch Composition and Type of Non-Solvent on the Formation of Starch Nanoparticles and Improvement of Curcuminy Stability in Aqueous Media. *J. Cereal Sci.* **2017**, *76*, 122–130.

Saikia, C.; Das, M. K.; Ramteke, A.; Maji, T. K. Controlled Release of Curcumin from Thiolated Starch-Coated Iron Oxide Magnetic Nanoparticles: An in Vitro Evaluation. *Int. J. Polym. Mater. Polym. Biomater.* **2017**, *66* (7), 349–358.

Saikia, C.; Das, M. K.; Ramteke, A.; Maji, T. K. Effect of Crosslinker on Drug Delivery Properties of Curcumin Loaded Starch Coated Iron Oxide Nanoparticles. *Int. J. Biol. Macromol.* **2016**, *93*, 1121–1132.

Sandhu, K. S.; Nain, V. Starch Nanoparticles: Their Preparation and Applications. In *Plant Biotechnology: Recent Advancements and Developments*; Springer: Singapore, 2017, pp 213–232.

Santos, I. S.; Ponte, B. M.; Boonme, P.; Silva, A. M.; Souto, E. B. Nanoencapsulation of Polyphenols for Protective Effect against Colon Rectal Cancer. *Biotechnol. Adv.* **2013**, *31*(5), 514–523.

Sensenig, R.; Sapir, Y.; MacDonald, C.; Cohen, S.; Polyak, B. Magnetic Nanoparticle-Based Approaches to Locally Target Therapy and Enhance Tissue Regeneration in Vivo. *Nanomedicine* **2012**, *7*(9), 1425–1442.

Simón-Vázquez, R.; Peleteiro, M.; Lozano, T.; González-Fernández, A.; Casal, A. Nanotoxicology. In *Frontiers of Nanoscience*; Elsevier Inc.: New York, 2012; Vol. 4, pp. 443–485.

Song, D.; Thio, Y. S.; Deng, Y. Starch Nanoparticle Formation via Reactive Extrusion and Related Mechanism Study. *Carbohydr. Polym.* **2011**, *85*(1), 208–214.

Sun, Q. Starch Nanoparticles. In *Starch in Food*; Elsevier Ltd.: New York, 2018; pp 691–745.

Uzun, S.; Kokini, J. L. Preparation and Characterization of Starch Nanoparticles by Desolvation Method. *Int. J. Polym. Sci.* **2014**, *Technical Proceedings of the 2014 Clean Technology Conference and Trade Show*, 292–295.

van Beek, T. A.; Montoro, P. Chemical Analysis and Quality Control of Ginkgo Biloba Leaves, Extracts, and Phytopharmaceuticals. *J. Chromatogr. A* **2009**, *1216*(11), 2002–2032.
Walia, N.; Dasgupta, N.; Ranjan, S.; Chen, L.; Ramalingam, C. Fish Oil Based Vitamin D Nanoencapsulation by Ultrasonication and Bioaccessibility Analysis in Simulated Gastro-Intestinal Tract. *Ultrason. Sonochem.* **2017**, *39*, 623–635.
Wang, L.; Zhao, X.; Zu, Y.; Wu, W.; Li, Y.; Zu, C.; Zhang, Y. Enhanced Dissolution Rate and Oral Bioavailability of Ginkgo Biloba Extract by Preparing Nanoparticles via Emulsion Solvent Evaporation Combined with Freeze Drying (ESE-FR). *RSC Adv.* **2016**, *6*(81), 77346–77357.
Wang, S.; Su, R.; Nie, S.; Sun, M.; Zhang, J.; Wu, D.; Moustaid-Moussa, N. Application of Nanotechnology in Improving Bioavailability and Bioactivity of Diet-Derived Phytochemicals. *J. Nutr. Biochem.* **2014**, *25*(4), 363–376.
Wang, T.; Wu, C.; Fan, G.; Li, T.; Gong, H.; Cao, F. Ginkgo Biloba Extracts-Loaded Starch Nano-Spheres: Preparation, Characterization, and in Vitro Release Kinetics. *Int. J. Biol. Macromol.* **2018**, *106*, 148–157.
Wu, X. M.; Branford-White, C. J.; Yu, D. G.; Chatterton, N. P.; Zhu, L. M. Preparation of Core-Shell PAN Nanofibers Encapsulated? Tocopherol Acetate and Ascorbic Acid 2-Phosphate for Photoprotection. *Colloids Surf. B: Biointerfaces* **2011**, *82*(1), 247–252.
Ye, F.; Miao, M.; Lu, K.; Jiang, B.; Li, X.; Cui, S. W. Structure and Physicochemical Properties for Modified Starch-Based Nanoparticle from Different Maize Varieties. *Food Hydrocoll.* **2017**, *67*, 37–44.
Young, E.; Kim, M.; Cho, M.; Hun, J.; Kim, J. Production of Starch Nanoparticles Using Normal Maize Starch via Heat-Moisture Treatment under Mildly Acidic Conditions and Homogenization. *Carbohydr. Polym.* **2016**, *151*, 274–282.
Zhang, Z.; Tan, S.; Feng, S. S. Vitamin E TPGS as a Molecular Biomaterial for Drug Delivery. *Biomaterials* **2012**, *33*(19), 4889–4906.
Zhu, F. Encapsulation and Delivery of Food Ingredients Using Starch Based Systems. *Food Chem.* **2017**, *229*, 542–552.
Zhu, M.; Huang, Y.; Wang, Y.; Shi, T.; Zhang, L.; Chen, Y.; Xie, M. Comparison of (Poly) Phenolic Compounds and Antioxidant Properties of Pomace Extracts from Kiwi and Grape Juice. *Food Chem.* **2018**, *271*, 425–432.

PART II

Role of Nutraceuticals in Cancer Treatment

CHAPTER 6

Study of Nutraceuticals in Cancer Treatment: An *In Silico* Approach

ANAMIKA BASU

Department of Biochemistry, Gurudas College, Kolkata 700054, West Bengal, India

**Corresponding author. E-mail: basuanamikaami@gmail.com*

ABSTRACT

Cancer is a disorder caused by unregulated cell growth and is present throughout the world. Uncontrolled cell growth forms tumors in the human body. According to the type of cells that originates the tumor, cancers can be broadly classified into seven types, for example, carcinoma, sarcoma, leukemia, lymphoma, myeloma, blastoma, and central nervous system cancers. A majority of these disorders are generated by carcinogens, radiation, and infectious agents (viruses/bacteria). In cancer, six hallmark changes occur in cell physiology, which are different from the normal cells. Several cell signaling pathways, transcription factors, proteoglycans, microRNAs, carcinogenic molecules, and viruses are responsible for human cancerous cell growth. So, cancers can be cured by using surgery, chemotherapy, radiation therapy, and immunotherapy depending on various aspects of cancerous cells. Nutraceuticals for cancer are mostly phytochemicals that are derived from plants. These chemical compounds can be derived from dietary or medicinal plants. The active constituents of these nutraceuticals have chemopreventive activities as well as chemotherapeutics activities. Synergistic effects on cancer treatment by using two or more than two nutraceutical compounds with microRNAs are also reviewed. Identification of targets at the molecular level for the active constituents present in nutraceuticals, gene network construction with oncogenes, and documentation of synergistic effect can be executed with the help of computational techniques. Almost all types of cancers can be barred from spreading in different parts of the body and treated by nutraceuticals, and their molecular mechanism of action can be explained and visualized by using computational studies.

6.1 INTRODUCTION

6.1.1 CANCER

In medical science, cancer is known as growth of malignant cells, which is a collection of diseases, relating cell growth without normal regulation. Cancer, which is a genetic disorder, is the main cause of death in economically advanced countries and the second one in developing countries (Jemal et al., 2011). The human body is composed of many types of cells that are structural and functional units of life, and these cells grow and divide in a controlled manner to maintain the normal functions of the body. When living cells become aged or injured, they expire and are substituted with new cells; this process is known as apoptosis or programmed cell death. However, sometimes, this programmed process changes abnormally. The genetic material (DNA) of a living cell regulates the cell growth and cell division in normal conditions. During mutation, cells containing altered DNA becozme immortal and keep dividing in an uncontrolled manner. Due to unwanted cell growth, these extra cells form a mass of tissue that is called tumor. Tumors are classified into two groups: the first one is benign tumors and the second one is malignant tumors. Not all tumor cells are cancerous in nature. Benign tumors do not grow under uncontrolled conditions, do not invade neighboring tissues of tumor cells, and do not spread all over the body. However, malignant tumors are cancerous and cells of these tumors can spread to the unaffected parts of the body (metastasis) and penetrate nearby parts of the body. The cancer cells can even spread to some distant parts of the body by using the route of the lymphatic system or bloodstream. Till now, more than 200 different cancers are known to affect humans.

6.1.2 CLASSIFICATION OF CANCERS

Cancers can be classified according to the type of cells that produces the tumorous growth, and hence, this tissue is supposed to be the origin of that tumor. Various types of cancers are enlisted in Table 6.1.

TABLE 6.1 Types of Cancer Along With Their Origination Position

Types of Cancer	Type of Cells that Originates Tumor
Carcinoma	Cancer that affects the epithelial tissues that line internal organs, for example, breast, prostate, lung and colon cancer
Sarcoma	Cancer that begins in connective or supportive tissues (e.g., bone, cartilage, fat, muscle, and blood vessels)
Leukemia	Cancer related to blood-forming tissues
Lymphoma	Cancers that affects the lymphatic tissues
Myeloma	Cancer that initiates in bone marrow
Blastoma	Cancer that begins in embryonic tissues
Central nervous system cancer	Cancers that begin in the tissues of the brain and spinal cord

6.1.3 CAUSES OF CANCERS

Cancers are caused by irregularities in the genetic material (DNA) of the transformed cells. The common cancers are generated in the lungs, stomach, liver, colon, and breast (Chen et al., 2010). These irregular growths may be due to the effects of carcinogens (cancer-causing agents, such as the smoke of tobacco), chemicals (carbon tetrachloride, benzene, vinyl chloride, asbestos, polychlorinated biphenyls, etc.), UV radiation, and several infectious agents (viruses/bacteria). Different classes of cancers with their causative agents are discussed in Table 6.2.

Specifically, about 90% of cancer in the lungs is caused due to tobacco smoking. The smoke of tobacco is considered a vital reason for cancer because it contains more than 50 known carcinogenic compounds, for example, nitrosamines and polycyclic aromatic hydrocarbons. Radiation such as gamma and X-ray can cause cancerous growth in the human body. It is assessed that 2% of future cancers will be due to current CT scans. Radiation from mobile phones and mobile towers has also been anticipated as a reason for cancer, but this is not proven experimentally. Microorganisms like bacteria, viruses, and parasites are accountable for up to 20% of human cancers. The risk of cancer formation can be related to lifestyle, for example, consumption of alcohol and tobacco and environmental factors.

TABLE 6.2 Types of Cancer According to Their Causative Agents

Cause of Cancer	Types of Cancer
Tobacco smoking and chewing	Lung, mouth, larynx, head, neck, stomach, bladder, kidney, esophagus, and pancreas cancers
Prolonged exposure to asbestos fibers	Mesothelioma (cancer of the protective tissues covering the lungs, abdomen, and heart)
Ultraviolet radiation	Melanoma and other skin cancers
Human immune deficiency virus (HIV)	Kaposi's sarcoma and non-Hodgkin's lymphoma
Hepatitis B virus (HBV) and hepatitis C virus (HCV)	Hepatocellular carcinoma (liver cancer)
Human papillomavirus (HPV-16 and -18)	Cervical, vaginal, vulvar, oropharyngeal, and penile cancers
Epstein–Barr virus	Burkitt lymphoma (cancer starts in immune cells called B-cells), non-Hodgkin lymphoma, Hodgkin lymphoma, and nasopharyngeal carcinoma
Kaposi's sarcoma herpes virus (KSHV)	Kaposi sarcoma (a type of cancer that can form masses in the skin, lymph nodes, or other organs)
Human T-cell lymphotropic virus type-1 (HTLV1)	Adult T-cell leukemia
Helicobacter pylori	Stomach cancer
Schistosomes *(Schistosoma hematobium)* parasite	Bladder cancer
Liver flukes (*Opisthorchis viverrin*) parasite	Liver cancer

6.1.4 CHARACTERISTICS OF CANCER CELLS

Six hallmarks of cancer recommend that tumor progression occurs due to DNA mutation in genes involved in controlling cell signaling and metastatic pathways.

In cancer, six common changes occur in cell physiology, which is different from the normal cells. According to the heterotypic model of cancer, proposed by Douglas Hanahan and Robert Weinberg, the changes are as follows:

1. Normal cells have a finite life span, but cancer cells can survive indefinitely by producing several proteins.
2. Normal cell depends on the cell signaling system from their surroundings for their division, whereas cancer cells can produce their own chemical messenger and autonomous signaling system for their division.
3. Cancer cells can override the signal from their surrounding cells and divide continuously.
4. At the time of stress, self-destruction occurs in normal cells, whereas cancer cells can grow continuously avoiding this situation.
5. By stimulating the new blood vessel construction processes, cancerous cells can divide continuously with a constant supply of nutrients and oxygen.
6. In the last stage of tumor progression, migration of tumor cells occurs through blood and lymphatic system (metastasis) from the site of origination to different parts of the human body.

Several cell signaling pathways, transcription factors, proteoglycans, microRNAs, carcinogens, and various viruses are involved in human cancerous cell growth. Various signaling pathways such as MAPK (ERK) signaling, RAS signaling, PI3K signaling, WNT signaling, NOTCH signaling, HH signaling, TGFB signaling, JAK-STAT signaling, calcium signaling, HIF-1 signaling, KEAP1-NRF2 signaling, NR signaling, cell cycle G1/S, apoptosis and telomerase activity are altered in cancerous cells. Various cell signaling pathways involved in cell lines of breast cancer are displayed in Figure 6.1 KEGG pathway related with breast cancer in Homo sapiens (human).

6.1.5 CANCER TREATMENT

Cancer can be treated by using different methods, for example, surgery, chemotherapy, radiation therapy, and immunotherapy. The way of treatment depends upon the various factors, for example, the position of the tumor, stage of the cancer, and health condition of cancer patients. Nutraceuticals (pharmaceutical grade and standardized nutrient) for cancer are mostly phytochemicals derived from different dietary or medicinal plants such as soyabean, tomato, garlic, turmeric, mango, ginger, tea, propolis, and honey. The active constituents of these nutraceuticals may have chemopreventive

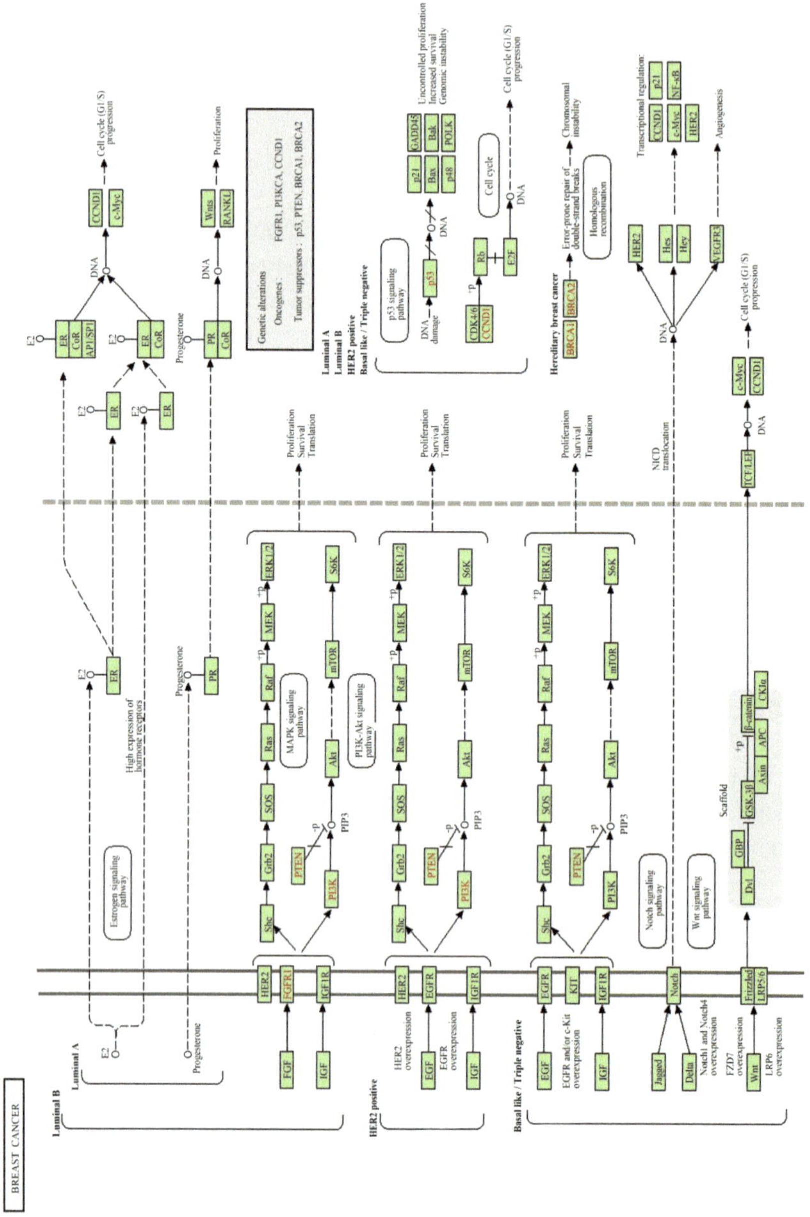

FIGURE 6.1 Cell signaling pathways in cancer.

activities as well as chemotherapeutics activities. Their ability to reduce cancer occurrence is commonly related to programmed cell death, that is, apoptosis. The potential of nutraceuticals as chemopreventive reagents has been confirmed by analyzing their biological effects on cultured human cells, both normal and cancerous cells. Chemoprevention can be defined as the use of small chemical molecules, including dietary or herbal compounds, to avoid cancers, whereas chemotherapeutics, which are mostly synthetic chemicals, are used to eliminate or ease cancer symptoms. A study on the biological roles of these phytochemical compounds at the cellular level delivers the molecular basis for their antitumor action and helps researchers to launch a platform for predicting more powerful chemopreventive and even chemotherapeutic agents. Some selective phytochemicals that can be used as nutraceuticals in different cancers are shortlisted in Table 6.3.

TABLE 6.3 Phytochemicals in Different Cancers

Phytochemicals	Nutraceuticals in cancers
Some flavonoids	Lung, colon, breast, prostate, stomach, brain, cervical cancer, and head, neck squamous carcinoma
Carotenoids, for example, Lycopene and β-carotene	Prostate cancer and lymphoblastoma
Caffeic acid phenethyl ester, an active phenolic component extracted from honey bee propolis	Skin cancer
Curcumin	Colon carcinoma, leukemia, prostate cancer, melanoma, and breast cancer
Allyl sulfur compounds derived from garlic	Non-small-cell lung cancer, prostate cancer, and breast cancer

With increasing scientific knowledge on the chemistry and biology of nutraceuticals, the research on nutraceuticals will spread more into the region of chemoprevention and chemotherapeutics. It has been recommended that the following approaches should be considered for the future use of nutraceuticals for cancer prevention and treatment:

1. To increase the effectiveness of promising nutraceuticals in various types of cancers, analysis of their analogs is considered as the first step. In computational studies, this can be done with a quantitative

structure activity relation study of phytochemicals with anticancer activity.

2. Identification of molecular targets for the active constituents present in nutraceuticals is the second step. By identifying the molecular targets of nutraceuticals, it can be possible to develop more advanced chemicals that precisely target one or more targets at the molecular level. Here, molecular modeling and molecular docking techniques can be executed to identify the molecular target of active constituents of nutraceuticals.
3. Documentation of the synergistic effect is the last but not least step. With a detailed understanding of the molecular mechanism of each nutraceutical, probable synergistic effects on cancer treatment by using two or more nutraceuticals or their derivatives can be tested in the near future.

6.2 ANTITUMOR ACTIVITIES AND MOLECULAR TARGETS OF PHYTOCHEMICALS

Natural herbs have been used for thousands of years for the prevention and/or treatment of diverse diseases (Lai and Roy, 2004). The presence of bioactive components in plants makes them appropriate choices to be used due to their little or no toxicity, high efficacy in multiple target sites, capability of oral consumption, and low cost (Meybodi et al., 2017). Pieces of evidence confirmed the anticancer activities of natural plant-derived bioactive components (Pezzuto, 1997). The bulk of research has been devoted to dietary phytochemicals that resulted in an increase in the comprehension of these compounds as a chemical and biological functional agent that has a constructive effect on human health.

6.3 ANTICANCER ACTIVITIES OF PHYTOCHEMICALS DURING DIFFERENT STAGES OF CANCER

Different molecular targets, for example, oncoproteins, enzymes, transcription factors, growth factor proteins, and expression of oncogenes can be regulated in the presence of different phytochemicals that are present in different dietary sources. According to anticancer activities, these phytochemicals are listed in Table 6.4.

TABLE 6.4 In Different Stages of Cancer Dietary Phytochemicals With Their Mode of Action

Activity	Molecular Targets	Phytochemicals	Dietary Source
Inhibition of infectious agent	Suppression of human papillomavirus oncoproteins	Curcumin, luteoline	Turmeric
Carcinogen metabolism	hSULT1A1, hSULT2A1	Genistein (flavonoids)	Soybean
Cell proliferation	G2/M arrest	B-sitosterol	Nuts, grains, seeds
Replicative immortality	Telomerase inhibition	(−)- Epigallocatechin-3-gallate (EGCG)	Green tea
Cell death evasion	Bax: Bcl-2 ratio increased	Lycopene	Tomato, watermelon, and red grapes
Angiogenesis	Inhibition of vascular endothelial growth factor (VEGF) induced angiogenesis	Luteolin	Broccoli, pepper, thyme, celery
Invasion and metastasis	MMP-9 suppression	Carnosol	Rosemary (*Rosmarinus officinalis*), Mountain desert sage (*Salvia pachyphylla*)
Immune evasion	IDO suppression	(−)-Epigallocatechin-3-gallate (EGCG)	Tea
Inflammation	COX2 gene expression decreased	Luteolin	Broccoli, pepper, thyme, and celery
Cell metabolism alteration	HIF-1α expression decreased	Sulforaphane	Cruciferous vegetables, for example, broccoli, brussels sprouts, and cabbages
Genome instability	Protein p53 increased	Curcumin	Turmeric

6.3.1 SUPPRESSION OF HUMAN PAPILLOMAVIRUS ONCOPROTEINS BY NUTRACEUTICALS

Cervical cancer is the most common cancer in Indian females and the second foremost basis of cancer death for women in the world. This cancer is connected with infection with high-risk human papillomaviruses (HPVs). Curcumin (diferuloyl methane), a phytochemical, is a natural chemical compound extracted from *Curcuma longa* that allows suppression of carcinogenesis by preventing this virus (Figure 6.2). Curcumin is responsible for distinct inhibition of human telomerase reverse transcriptase (hTERT), the catalytic core of telomerase, thereby preventing the proliferation of cancer cells (Singh and Singh, 2009). Experimental exposure to benzo[a]pyrene (BaP), a tobacco carcinogen, increases the expression of esophageal cancer oncoprotein, signifying a molecular link between smoking and cervical cancer. Curcumin decreases the BaP-induced growth in the expression of HPV E7 oncoprotein in a three-dimensional raft culture system (Mather et al., 2011). Curcumin and one of its analogs 1,5-bis(2-hydroxyphenyl)-1,4-pentadiene-3-one cause downregulation of E6 and E7 oncogenes and tissue-specific gene expression of HPVs (Paulraj et al., 2015). Curcumin, that is, diferuloyl methane and its naturally occurring analogs, namely, bisdemethoxy, demethoxy, and cyclocurcumin, present in rhizomes of curcuma species turmeric have been reported to inhibit the propagation of a wide variety of tumor cells, for example, bladder cancer, breast cancer, cervical cancer, colorectal cancer, and esophageal cancer. Target nuclear proteins HPV-16 E6 (PDB ID: 4GIZ) and E7 are the major proteins actively contributing to oral and cervical cancers. In silico studies elaborates that curcumin and its several natural analogs have effective binding capacity with different active sites on HPV-16 E6 protein, which is an ideal target for restoring the tumor suppressor function of p53 and thus allowing the programmed cell death of infected cells. E7 protein has been found to be associated with tumor formation, whereas E6 plays a key role in tumor progression (Song et al., 2000). Flavonoid luteolin can bind directly with HPV-16 E6 (Table 6.4). Docking analysis of HPV-16 E6 proposes that this compound binds in a hydrophobic pocket at the point of interface between E6 and E6AP (Cherry et al., 2013).

6.3.2 NUTRACEUTICALS IN CARCINOGEN METABOLISM

Cruciferous vegetables play a main and unique role in the widely documented protective functions of vegetables against the risk of cancers of breast, colorectal, lung, and prostate. Hydrolysis products of glucosinolates and their isothiocyanates, for example, sulforaphane [1-isothiocyanato-(4*R*)-methylsulfinyl)-butane; $CH_3S(CH_2)4NCS$] are renowned protectors against tumorigenesis (Farnham et al., 2004). Several isothiocyanates can prevent the action of tobacco-specific nitrosamine carcinogen

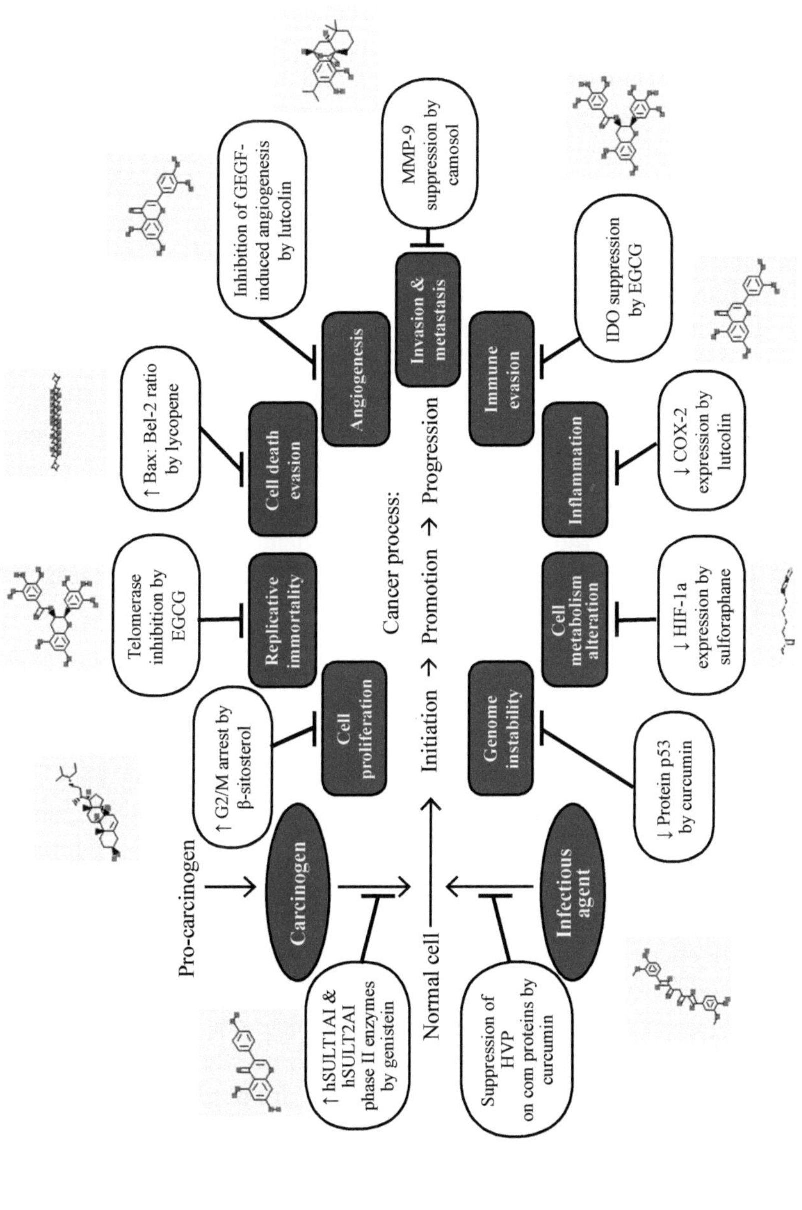

FIGURE 6.2 Role of nutraceuticals along with their structures in different stages of cancer.

4-(methylnitrosamino)-1-(3-pyridyl)-1-butanone, which is perhaps the most prominent carcinogenic compounds derived from cigarettes, by inhibiting its metabolism. Two kinds of DNA-damaging compounds can induce neoplastic transformations, that is, electrophilic compounds and reactive oxygen species, produced mostly from normal cellular oxidations. Most electrophiles need activation by metabolism, usually by enzyme cytochromes P450. They convert usually harmless procarcinogen compounds to extremely reactive electrophilic carcinogens that can cause damage to susceptible DNA bases and thus can initiate carcinogenesis (Figure 6.2). DNA and other similar macromolecules are chiefly protected against damage by electrophilic compounds and reactive oxygen compounds by a family of enzymes via a variety of mechanisms, for example, conjugation with endogenous ligands (e.g., glutathione, glucuronic acid) (Talalay and Fahay, 2001). Several enzymes, for example, glutathione *S*-transferases (GST), g-glutamylcysteine synthase, quinone reductase, UDP-glucuronosyl transferases, and aflatoxin B1 aldehyde reductase, present in our body can inactivate these agents and endorse their excretion. Dietary broccoli sprouts and their component glucoraphanin, a glucosinolate, and 4-methylsulphinylbutyl glucosinolate), its isothiocyanate derivative, can induce above-mentioned enzymes and can impart protection against chemically induced tumors in the early stage of cancer (Shapiro et al., 2001; Farnham et al., 2004).

6.3.3 NUTRACEUTICALS AS TELOMERASE ENZYME INHIBITORS

Eukaryotic telomerase consists of a catalytic protein subunit known as telomerase reverse transcriptase component (hTERT) and an RNA component (hTR) (Ganesan and Xu, 2018). Numerous telomerase inhibitors have been acknowledged that can inhibit the catalytic activity of enzyme through targeting its catalytic components or RNA. Telomerase inhibitors are generally diverse chemical compounds such as natural, synthetic products and modified oligonucleotides. Telomerase is generally overexpressed in most cancerous cells and is essential for cancer cell development. So, the deactivation or inhibition of telomerase is critical in cancer therapy.

Anticancer potentials of natural products from plants on targeting telomerase are shown in Table 6.5.

Epigallocatechin gallate (EGCG) can act as a competitive enzyme inhibitor to a telomerase substrate primer. This observation indicates that EGCG competitively interacts with the substrate-binding site of the telomerase enzyme (Figure 6.2). Similarly, curcumin interrupts the interaction of p23, which is a co-chaperon, with the hTERT enzyme. This activity of curcumin weakens the nuclear translocation through a nuclear pore complex of the hTERT protein and thereby prevents telomerase activity. Genistein inhibits DNA methyltransferase expression and causes hypomethylation of the E2F-1 recognition sequence in the hTERT promoter, thus leading to increased binding of E2F-1 to its promoter andrepressing hTERT and telomerase activities.

TABLE 6.5 Anticancer Compounds as Telomerase Inhibitors

Plant Source	Compounds	Mechanism of Action
Targeting hTERT—Inhibition of the Catalytic Function		
Brassica oleracea (wild cabbage)	Indole-3-carbinol	Inhibition of telomerase and downregulated expression of the catalytic subunit of hTERT
Camellia sinensis (tea plant)	Epigallocatechin gallate	Binding competitively at the active site of hTERT
Trigonella foenum-graecum (fenugreek)	Diosgenin	Prevention of telomerase activity by downregulation of the hTERT gene expression
Zingiber officinale Roscoe (ginger)	Gingerol	Reduction of hTERT expression and decrease of c-Myc (myelocytomatosis viral oncogene)
Suppression of Transcriptional and Post-Transcriptional Regulation		
Papaveraceae	Papaverine	Downregulation of the telomerase activity and hTERT expression
Blueberries	Resveratrol	
Camellia sinensis	Epigallocatechin gallate	Inhibition of hTERT expression and decreased the expression of both mRNA and protein
Curcuma longa (turmeric)	Curcumin	
Vitis vinifera (grapes)	Resveratrol and pterostilbene	Reduction of hTERT mRNA levels

6.4 SYNERGISTIC EFFECTS OF NUTRACEUTICALS WITH MICRORNAS IN CANCER TREATMENT

MicroRNAs (miRNAs), one type of noncoding RNAs, are about 23 nucleotides in length (Bartel, 2009). Their mode of action is that they downregulate or degrade their target mRNAs after transcription by binding to the recognition sites in the 30 untranslated region (30UTR) of their precise mRNA. Thus, they repress the gene expression of oncogenes in cancerous cells. Various types of miRNAs have been recognized as the target of hTERT mRNA. This leads to a reduction in cell proliferation in different types of cancers such as neuroblastoma by miR-138 (Chakraborti et al., 2013), anaplastic thyroid carcinoma by miR-138 (Mitomo et al., 2008), gastric cancer by miR-1207-5p, miR-1266, and miR-1182 (Chen et al., 2014, Zhang et al., 2015), cervical cancer by miR-491-5p (Zhao et al., 2015), head and neck squamous cell carcinoma by miR-512-5p (Li et al., 2015), ovarian cancer by miR-532, miR-3064 (Bai et al., 2017), and breast cancer by miR-296 and miR-512 (Dinami et al., 2017).

6.4.1 IN SILICO BINDING STUDY OF MICRORNA WITH MRNA OF HTERT

Here, miRNA–mRNA binding study has been performed with hTERT mRNA and miR-138 by RNAhybrid 2.2 software (Rehmsmeier et al., 2004).

miR-138 binds with mRNA of the human telomerase reverse transcriptase enzyme with a minimum free energy of 103.7 kcal/mol (Figure 6.3). This proves that miR-138 can be used to degrade the mRNA of this enzyme and hTERT gene expression.

MiRNAs have been concerned at all stages of cancer, from initiation, promotion, and progression of the tumor, by influencing cell proliferation, cell differentiation, apoptosis, new blood cell formation and metastasis of tumor cells (Calin and Croce, 2006). miRNAs can mimic oncogenes or tumor suppressor genes due to their respective up- or downregulated expression patterns in different cancers, for example, lymphomas, T cell acute lymphoblastic leukemia, glioblastoma, and B cell lymphoblastic leukemia. Specifically, miR-21 has been found to be steadily upregulated in cancers of the breast, colon, lung, pancreas, and stomach, as well as in chronic lymphocytic leukemia, acute myeloid leukemia, glioblastoma, and myeloma (Pan et al., 2011). The most significant upregulation of clusters of miRNAs, for example, miR-196a, miR-190, miR-186, miR-221, miR-222, miR-200b, miR-15b, and miR-95, is observed in most pancreatic cancer tissues by real-time PCR study (Zhang et al., 2009). Several bioactive agents, present in our diet, are now considered as modulators of miRNA profiles in cancer, and this is a promising method of chemoprevention in cancer. Effects of phytochemicals on temporary changes in miRNA profiles

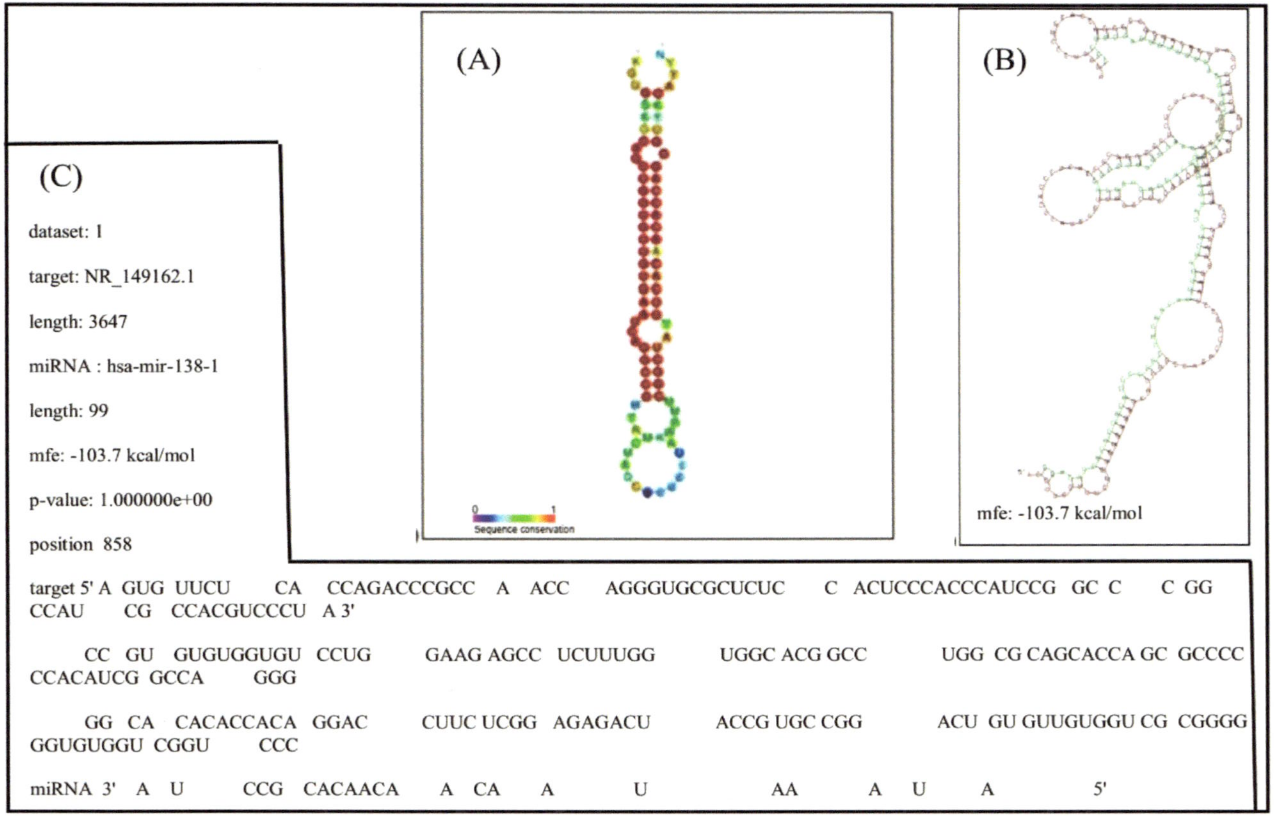

FIGURE 6.3 (A) Structure of miR-138, (B) bound structure of miR-138 with hTERT mRNA, (C) target binding of miR-138.

during cancer initiation and progression must provide new insights into early biomarkers for cancer prevention. However, various issues, for example, in vivo bioavailability, target selection, and the mechanism of action of such bioactive metabolites require further in silico study. Several computational techniques, for example, secondary and tertiary structure prediction of miRNAs and molecular docking study of miRNAs with target gene, can be applied to explore the role of nutraceuticals in cancer treatment with the help of miRNAs. Some phytochemicals with their predicted targets along with their epigenetics in cancer cells are listed in Table 6.6.

6.5 ACTIVE CONSTITUENTS IN NUTRACEUTICALS WITH ANTICANCER ACTIVITY

Different nutrients, for example, phytochemicals in fruits, vegetables, edible oils, are proven to be anticancer therapeutics. Nutraceuticals with their active constituents and their mode of action in various types of cancers are shown in Table 6.7.

There are several methods to identify the molecular target in cancer treatment with active constituents of natural product binding interaction, including equilibrium dialysis, pH-metric and spectrophotometric methods, affinity chromatography, surface plasmon resonance, nuclear magnetic resonance, X-ray crystallographic analysis, and computational molecular docking analysis (Partouche et al., 2018). Computational techniques have several advantages, that is, it is fast, more accurate, and offers visualization at the molecular level.

6.6 COMPUTATIONAL TECHNIQUES FOR MOLECULAR-LEVEL VISUALIZATION OF TARGET MOLECULES

Molecular docking study shows that phytochemical curcumin can act as a binding inhibitor for HPV E6 protein. Target nuclear protein HPV-16 E6 (PDB ID: 4GIZ) is the major protein actively participating in oral and cervical cancers. This in silico study reveals that curcumin has an effective binding affinity with the substrate-binding site on HPV-16 E6 protein (Figure 6.4). This protein can be considered as an ideal target for re-establishing the tumor suppressor function of p53, and thus, curcumin binding ultimately results in the programmed cell death of infected cells.

TABLE 6.6 Synergistic Effect of Nutraceuticals With miRNAs in Cancer Treatment

Sr. No.	Name of Bioactive Ingredients Present in Diet	Affected Cancer Cells	Epigenetics With miRNAs	Predicted Targets
1.	Curcumin	Gastric cancer	Downregulation of miR-196, miR-200, and miR-21	Tumor suppressor PTEN (phosphatase and tensin homolog deleted on chromosome 10) (Bao et al., 2011)
		Pancreatic cancer	Upregulation of miR-22	ERα (Estrogen receptor alpha (ERα), also known as NR3A1 (nuclear receptor subfamily 3, group A, member 1) (Sun et al., 2008)
2.	Resveratrol	Colon cancer	Downregulation of miR-21, miR-196a, miR-25, miR-17, and miR-92a-2,	
			Upregulation of miR-155	Activator protein (AP-1) through the Jun signaling pathway (Tili et al., 2010), (Tili et al., 2010)
3.	Catechins, for example, epigallocatechin-3-gallate (EGCG)	Hepatocellular carcinoma	Upregulation of miR-16	Bcl-2 (B-cell lymphoma 2) (Tsang and Kwak, 2010)
4.	Ellagitannins, for example, 1,3-di-O-galloyl-4,6-(s)-HHDP-b-D-glucopyranose	Liver cancer	Downregulation of miR-370, miR-373, and miR-526b (Wen et al., 2009)	
5.	Isoflavones, for examine, genistein	Breast, prostate, bladder, and kidney cancers (Taylor et al., 2009)		
		Prostate cancer	Upregulation of miR-1296	MCM (mini chromosome maintenance) (Majid et al., 2010)

TABLE 6.6 *(Continued)*

Sr. No.	Name of Bioactive Ingredients Present in Diet	Affected Cancer Cells	Epigenetics With miRNAs	Predicted Targets
		Uveal cancer	Downregulation of miR-27a	Upregulation of *ZBTB10* (zinc finger and BTB domain-containing 10) gene (Sun et al., 2009)
6.	Indole, for example, indole-3-carbinol (Izzotti et al., 2009)		Regulation of miR-34b	p53 (also known as TP53 or tumor protein (EC: 2.7.1.37)
			miR-26a	TGF-β (transforming growth factor beta1 or TGF-β1)
			miR-125a	ERBB2 (human epidermal growth factor receptor 2) or tyrosine kinase-type cell surface receptor HER2
			miR-10a	Angiogenesis
7.	Isothiocyanates, for example, phenethyl isothiocyanate (PEITC) (Izzotti et al., 2010)		Regulation of miR-192	Ras (rat sarcoma activation
			let-7a, let-7c	Cell proliferation, angiogenesis, and Ras activation
			miR-146 miR-123	NFκB (nuclear factor-κB (NF-κB)/Rel) proteins activation
			miR-222	Angiogenesis and cell proliferation
			miR-99b	Apoptosis

TABLE 6.7 Detailed List of Nutraceuticals With Their Active Constituents in Cancer Therapeutics

Nutraceuticals	Active Constituents in Cancer Treatment	Mode of Action of Phytochemicals
Cranberry	Proanthocyanidins 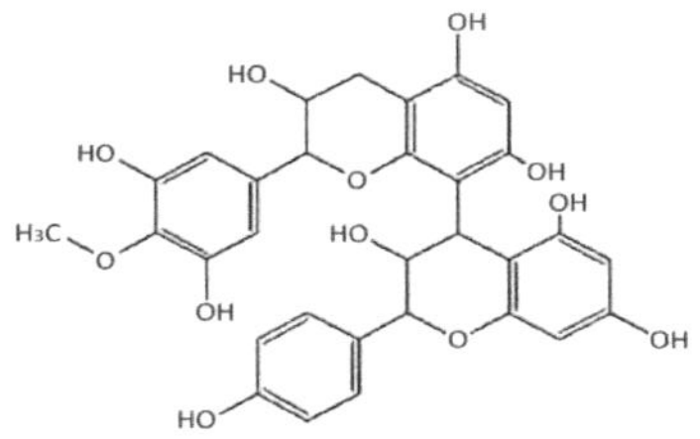	Reduced tumor cell growth in breast, prostate, colon, and lung cancers (Sun and Hai, 2006)
	Flavonoids	Inhibit oxidative DNA damage in the promotion of cancer and also cell proliferation is inhibited (Ferguson et al., 2006)
	Quercetin 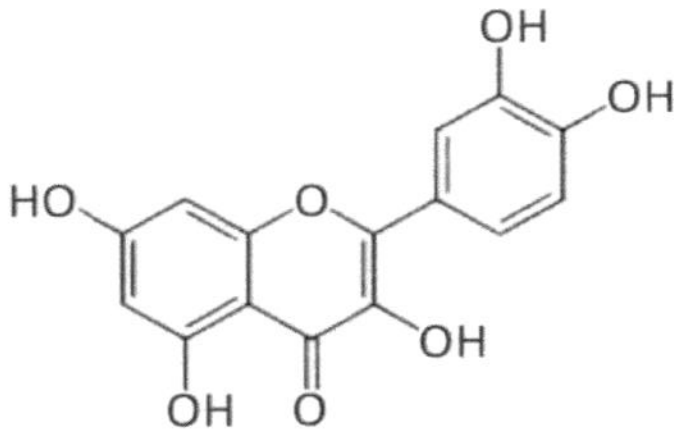	Inhibit growth in tumor cells in breast, colon cancers, adenocarcinoma, and chronic myelogenous leukemia
Papaya	Lycopene	Antitumor activity by free radical scavenging effect in liver cancer (Rahmat et al., 2002)
Cherries	Polyphenols	Inhibits tumor cell (Serra et al., 2010)
	Perillyl alcohol	Proliferation, apoptosis of cancerous cells in colon, skin, lung, and prostate cancers and neuroblastoma

TABLE 6.7 *(Continued)*

Nutraceuticals	Active Constituents in Cancer Treatment	Mode of Action of Phytochemicals
Pineapple	Bromelain	Protease activity in malignant peritoneal mesothelioma (MPM)
		Increase expression of p53 and Bax activator of apoptosis (Chobotova et al., 2010)
Guava	Lupeol, a triterpene	Suppresses AKT/mTOR/S6K kinase signaling in prostate cancer (Ryu et al., 2012)
Apple	Flavonoids, for example, quercetin-3-O-β-D-glucopyranoside,	Inhibits cell proliferation in liver, colon and breast cancer
	quercetin-3-O-β-D-galactopyranoside, quercetin, catechin, epicatechin, quercetin-3-O-α-L-arabinofuranoside, 2-α-hydroxyursolic acid	Apple extract increases the expression of five established or candidate tumor suppressor genes, namely,BRCA2 (breast cancer 2, early onset), PTPRT (protein tyrosine phosphatase, receptor type, T), PTPN6 (protein tyrosine phosphatase, nonreceptor type 6), PTPRN (protein tyrosine phosphatase, receptor type, N) and PTPRJ (protein tyrosine phosphatase, receptor type, J) (Dep-1)
	Triterpenoids including ursolic acids, 3 β-trans-p coumaroyloxy-2-alpha hydroxyolean-12-en-28-oic acid	
	Procyanidin B2	
	Polyphenols, for example, chlorogenic acid and phlorizin found exclusively in apples	
Grape	Proanthocyanidins	Induce apoptosis in colorectal and non-small-cell lung cancers (Zhou and Raffoul, 2012)
Mango	Lupeol, a triterpene	Antioxidant in leukemia, lung, breast, prostate, and colon cancers (Prasad et al., 2008)
Peach	Phenolic compounds	Inhibit the proliferation of breast cancer cells (Noratto et al., 2009)

TABLE 6.7 *(Continued)*

Nutraceuticals	Active Constituents in Cancer Treatment	Mode of Action of Phytochemicals
Plum	Neochlorogenic and chlorogenic acid	Cytotoxic effect on cervical and breast cancers (Prajapati et al., 2012)
Date	Water-soluble glucan containing (1,3) β-D glucan linkage	Antitumor activity in sarcoma (Ishurd et al., 2007)
Avocado	Lutein, a carotenoid	Antiproliferative effects on cancer cells (D'Ambrosio et al., 2011)
	Two aliphatic acetogenins, for example, (2S,4S)-2,4 dihydroxyheptadec-16-enyl acetate, (2S,4S)-2,4 dihydroxyheptadec-16-ynyl acetate	Blocks phosphorylation of EGFR (at Tyr 1173), C-RAF (Ser 338), ERK1/2 (Thr 202/Tyr 204)
Sapodilla	Polyphenols	Scavenger of NO and inhibits growth of breast cancer cells (Lu et al., 2005)
Watermelon	Lycopene	Inhibits cell proliferation in mammary cancerous cells (Jayakumar and Kanthinathi, 2011)
Pomegranate	Urolithin glucuronides and dimethyl ellagic acid	Inhibits the capacity of the chemokine SDFα to chemoattract prostate cancer cells (SDFα and its receptor CXCR4 play vital role in metastasis of cancer cells to bone (Bhandari, 2012)
	Fruit extract with tamoxifen	Improve and sensitive the TAM's effect in estrogen receptor-alpha (ER-α) positive breast cancer (Jeune et al., 2005)
	Punicic acid, an ω-5 long-chain polyunsaturated fatty acid	Inhibits cell proliferation in breast cancer due to its effect on lipid peroxidation and PKC pathway
	Ellagitannins and their intestinal urolithins	Inhibits TCDD induced CYP1-mediated EROD activity in colon cancer by inhibiting cell proliferation and induction of apoptosis
	Pomegranate emulsion	Increases mRNA and protein of hepatic nuclear factor E2-related factor (Nrf2) and prevents hepatocellular carcinoma

TABLE 6.7 *(Continued)*

Nutraceuticals	Active Constituents in Cancer Treatment	Mode of Action of Phytochemicals
	Pomegranate juice extract (flavonoid-rich)	Stimulates cell cycle arrest, induction of apoptosis, inhibition of leukemia
	Pomegranate extracts and genistein	Inhibits the growth of breast cancer cells by the induction of apoptosis (Motaal and Shaker, 2011)
Berry	Anthocyanins, a class of flavonoids	Inhibition of cell proliferation in oral, breast, colon, and prostate cancers and stimulates apoptosis of colon cancer cells (Faria et al., 2010)
Citrus fruits	Limonin, nomilin, types of limonoids	Inhibits tumor formation by stimulating the enzyme GST (Okwu, 2008)
	Flavonoids, for example, quercetin, myricetin, rutin, taneritin, narigenin, and hesperidin	Antiproliferative effect in myeloid and lymphoid leukemia, prostate, and ovary cancers and squamous cell carcinoma
Apricot	Cyclic triterpenes, for example, oleanolic and ursolic acids	Through a reactive oxygen-species-dependent mechanism in pancreatic cancer (Hattori, 2013)
Musk melon	Cucurbitacin B, cucurbitacin A, and cucurbitacin E	Inhibits STAT3 activation and the RAF/MEK/RRK pathway in leukemia (Milind, 2011)
Litchi	Tannins (Polymeric Proanthocyanidins), epicatechin, procyanidin A2	Inhibits cell proliferation and apoptosis in breast cancer and hepatocellular carcinoma (Wang, 2006)
Blackberry	Anthocyanin	Inhibits lipid peroxidation, COX inhibiting activity in colon, lung, and breast cancers (Bowen-Forbes et al., 2010)
Blueberry	Anthocyanin and anthocyaninpyruvic acid adduct	Antiproliferation, induction of apoptosis, act as cell anti-invasive factors in breast and lung cancers (Madhuri and Pandey, 2009)
Olive	Maslinic acid	Inhibits the growth of colon cancer (Reyes et al., 2009)

TABLE 6.7 *(Continued)*

Nutraceuticals	Active Constituents in Cancer Treatment	Mode of Action of Phytochemicals
Pumpkin	Proteins and polysaccharides	Cytotoxic, antioxidant property in melanoma and leukemia (Kawase et al., 2003)
Barberry	Berberine alkaloid	Apoptotic effect on prostate cancer and oral epithelioid carcinoma (Sun et al., 2009)
		Suppresses the androgen signaling in prostate cancer (Li et al., 2011)
Tomato	Lycopene, carotenoid, and A-tomatine	Complex interactions among multiple components may responsible for anticancer activity on prostate cancer
Eggplant	Solasodine glycosides	Useful in skin cancer treatment (Cham, 2007)
Spinach	Glycolipids	Antiproliferative effect on hepatocarcinoma (Sani et al., 2004)
Mushroom	Polysaccharides, for example, six-branched 1,3 beta-D-glucan	Prevents tumor metastasis (Wasser, 2002)
Cabbage	Glucosinolates	Inhibits growth of cancer cells in breast, colon, and lung (Devi and Thangam, 2012)
	Sulforaphane, an isothiocyanate	Antiproliferative and induces apoptosis in larynx cancer (Devi and Thangam, 2010)
Beetroot	Betanin, a betacyanin constituent	Very high free radical scavenging activity and multi-organ antitumor activity (Kapadia and Rao, 2012)
Shallot	Allyl sulfides	Suppress tumor proliferation (Moradi et al., 2013)
	Organosulfur compounds, allicin	Inhibit cell proliferation in breast cancer, cervical cancer, and adenocarcinoma

TABLE 6.7 *(Continued)*

Nutraceuticals	Active Constituents in Cancer Treatment	Mode of Action of Phytochemicals
Broccoli	Sulforaphane (SFN), an isothiocyanate	Anticancer cytotoxic antioxidant effect in barrette adenocarcinoma (Quazi et al., 2010)
Garlic	Allicin	
	Lipid-soluble allyl sulfur compounds, for example, diallyl disulfide (DADS), diallyl trisulfide (DATS)	Arrests cell cycle, apoptosis with caspase-3 in lung and breast cancers (Seki et al., 2008)
	Water soluble S-allyl cysteine,	
	S-allyl mercapto cysteine	
Sweet potato	Sweet potato protein	Inhibits Kunitz-type trypsin and causes apoptosis in leukemia (Li et al., 2013)
Rhubarb	Anthraquinones, for example, aloe-emodin	Through p-53 and downstream p-21 pathway (Huang et al., 2007)
		Inhibits the uptake of glucose in tumor cells and results in cell death
	Rhein	
Black tea	Polyphenols, theaflavins	Reduce tumor growth, protection of the immune system from cancer attack, reduce tumor-induced hepatotoxicity, and protection against oxidative damage in colon, lung, and skin cancers (Bhattacharyya et al., 2003)
Green tea	Catechins	Antioxidant, antiangiogenetic, and antiproliferative (Lin et al., 1999)
	Caffeine	Anticancer effect increases p53 and p21

TABLE 6.7 *(Continued)*

Nutraceuticals	Active Constituents in Cancer Treatment	Mode of Action of Phytochemicals
	EGCG and theaflavin-3,3′-digallate	Inhibits ERK phosphorylation, inhibits AP1 DNA binding activity, and inhibits NF-κB transcription factors (Bode and Dong, 2003)
Olive oil	Oleic acid	Blocks HER2/neu, a cancer-causing oncogene in breast cancer (Timsina et al., 2012), (Win, 2005)
Palm oil	Tocotrienols	Inhibits the growth of different cancerous cells (Alegre et al., 2013)
Clove oil	Eugenol	Cytotoxic effect in hepatoma (Dwivedi et al., 2011)
Soybean oil	Three isoflavones, For example, aglycone daidzein, genistein, and glycitein	Increases concentration of TGF-β, a potent inhibitor of angiogenesis, and reduces DNA synthesis
Turmeric	Curcumin	Inhibits cancerous cell growth in leukemia and breast and colon cancerd (Rani and Saikishore et al., 2013)
Saffron	Crocetin, a carotenoid compound	Inhibits nucleic acid synthesis (Gutheil et al., 2010)
Red chili pepper	Capsaicin	TRAIL (tumor necrosis factor-related apoptosis-induced ligand) mediated apoptosis via DR5 upregulation and survinin downregulation (Kim et al., 2010)
Black pepper	Capsaicin	Immunomodulatory role in colon cancer (Lu et al., 2010)

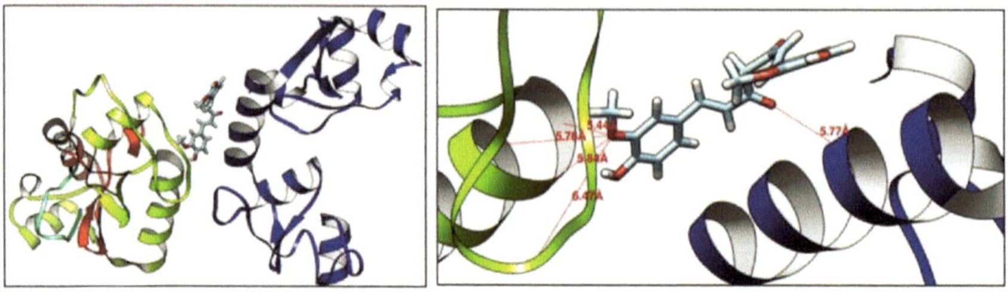

FIGURE 6.4 Binding of curcumin with HPV E6 protein.

6.7 GENE NETWORK IN BREAST INVASIVE CARCINOMA

Co-expression network figures are drawn using R package igraph. Only the top 20 genes with the highest correlations are shown in Figure 6.5. Co-expressed genes are C1QBP, COX10, CTDNEP1, EIF4A1, EIF5AL1, ELAC2, GEMIN4, METTL16, MYBBP1A, NDEL1, NUP88, PFAS, PITPNA, PLD2, RPA1, RPL26, SCO1, SENP3, TP53, TSR1, andYWHAE.

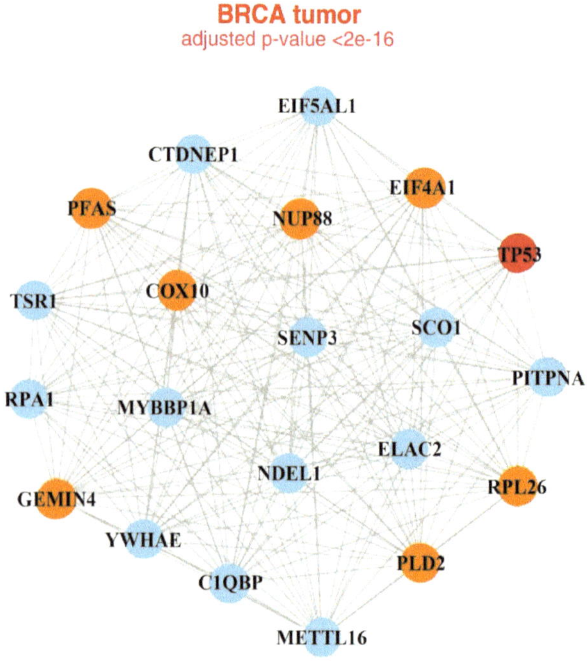

FIGURE 6.5 Co-expressed gene network in breast invasive carcinoma.

6.8 ROLE OF PYRIDOXAL PHOSPHATE AS A NUTRACEUTICAL IN CANCER BY GENE NETWORK STUDY

Pyridoxal phosphate or vitamin B_6 is found most abundantly in bread, cereals, grains, chicken, and fruits such as orange, tomato juice, banana, and avocado. Dietary vitamin B_6 reduces colon tumorigenesis by reducing cell proliferation, oxidative stress, NO production, and angiogenesis (Komatsu et al., 2003). Structures of pyridoxal phosphate and its drug-centered gene interaction network that target genes in cancer are shown in Figure 6.6. Pyridoxal phosphate (Drug bank ID DB00114) interacted with several genes, as shown in Table 6.8.

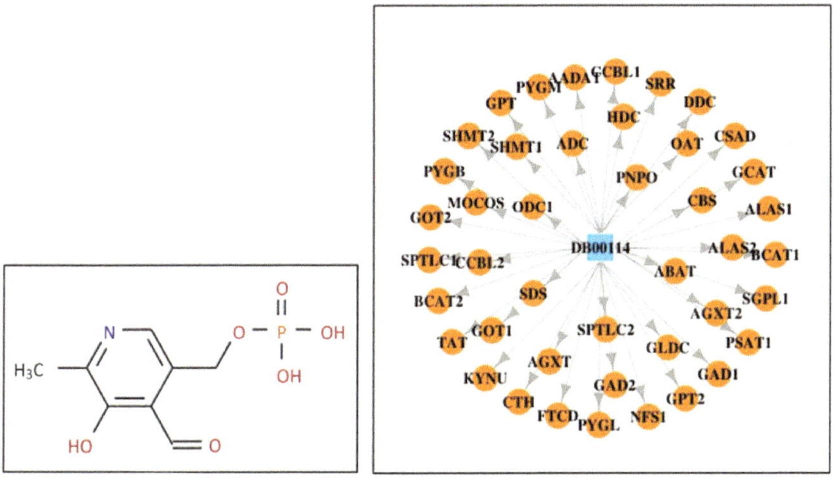

FIGURE 6.6 Structure and gene interaction of pyridoxal phosphate.

6.9 CONCLUSIONS

Structure-based, ligand-based, hybrid protein–ligand pharmacophore methods, homology modeling, molecular docking, and molecular simulation studies help in different steps of drug discovery pipeline that can save both the time and expenditure. In silico tools also find applications in the domain of cancer drug development in natural products. Phytochemicals are potential sources of medicines in different diseases. In cancer therapeutics, along with chemotherapy, radiotherapy, and immunotherapy, nutrients can take part in the prevention and healing of certain types of cancer. Different stages of cancers can be checked by nutraceuticals, and their mechanism of action at the molecular level can be explained and visualized by using computational studies.

TABLE 6.8 List of Target Genes Interacted with Pridoxal Phosphate

Symbol	Name of Target Gene	Name of Nutraceutical
AADAT	Aminoadipate aminotransferase	Pyridoxal phosphate
ABAT	4-Aminobutyrate aminotransferase	Pyridoxal phosphate
AZIN2	Arginine decarboxylase	Pyridoxal phosphate
AGXT	Alanine-glyoxylate aminotransferase	Pyridoxal phosphate
AGXT2	Alanine-glyoxylate aminotransferase 2	Pyridoxal phosphate
ALAS1	Aminolevulinate, delta-, synthase 1	Pyridoxal phosphate
ALAS2	Aminolevulinate, delta-, synthase 2	Pyridoxal phosphate
ALOX5	Arachidonate 5-lipoxygenase	Hyperforin
BCAT1	Branched chain amino-acid transaminase 1, cytosolic	Pyridoxal phosphate
BCAT2	Branched chain amino-acid transaminase 2, mitochondrial	Pyridoxal phosphate
CBS	Cystathionine-beta-synthase	Pyridoxal phosphate
CCBL1	Cysteine conjugate-beta lyase, cytoplasmic	Pyridoxal phosphate
CCBL2	Cysteine conjugate-beta lyase 2	Pyridoxal phosphate
CSAD	Cysteine sulfinic acid decarboxylase	Pyridoxal phosphate
CTH	Cystathionase (cystathionine gamma-lyase)	Pyridoxal phosphate
DDC	Dopa decarboxylase (aromatic L-amino acid decarboxylase)	Pyridoxal phosphate
FTCD	Formimidoyltransferase cyclodeaminase	Pyridoxal phosphate
GAD1	Glutamate decarboxylase 1 (brain, 67 kDa)	Pyridoxal phosphate
GAD2	Glutamate decarboxylase 2 (pancreatic islets and brain, 65 kDa)	Pyridoxal phosphate
GCAT	Glycine C-acetyltransferase	Pyridoxal phosphate
GLDC	Glycine dehydrogenase (decarboxylating)	Pyridoxal phosphate
GOT1	Glutamic-oxaloacetic transaminase 1, soluble	Pyridoxal phosphate

TABLE 6.8 *(Continued)*

Symbol	Name of Target Gene	Name of Nutraceutical
GOT2	Glutamic-oxaloacetic transaminase 2, mitochondrial	Pyridoxal phosphate
GPT	Glutamic-pyruvate transaminase (alanine aminotransferase)	Pyridoxal phosphate
GPT2	Glutamic pyruvate transaminase (alanine aminotransferase) 2	Pyridoxal phosphate
HDC	Histidine decarboxylase	Pyridoxal phosphate
HSD17B1	Hhydroxysteroid (17-beta) dehydrogenase 1	Dehydroepiandrosterone
KYNU	Kynureninase	Pyridoxal phosphate
MOCOS	Molybdenum cofactor sulfurase	Pyridoxal phosphate
NFS1	NFS1 nitrogen fixation 1 homolog (*S. cerevisiae*)	Pyridoxal phosphate
OAT	Ornithine aminotransferase	Pyridoxal phosphate
ODC1	Ornithine decarboxylase 1	Pyridoxal phosphate
PNPO	Pyridoxamine 5′-phosphate oxidase	Pyridoxal phosphate
PSAT1	Phosphoserine aminotransferase 1	Pyridoxal phosphate
PTGS1	Prostaglandin-endoperoxide synthase 1 (prostaglandin G/H synthase and cyclooxygenase)	Hyperforin
PYGB	Phosphorylase, glycogen; brain	Pyridoxal phosphate
PYGL	Phosphorylase, glycogen, liver	Pyridoxal phosphate
PYGM	Phosphorylase, glycogen, muscle	Pyridoxal phosphate
SDS	Serine dehydratase	Pyridoxal phosphate
SGPL1	Sphingosine-1-phosphate lyase 1	Pyridoxal phosphate
SHMT1	Serine hydroxymethyltransferase 1, soluble	Pyridoxal phosphate
SHMT2	Serine hydroxymethyltransferase 2, mitochondrial	Pyridoxal phosphate
SPTLC1	Serine palmitoyltransferase, long-chain base subunit 1	Pyridoxal phosphate
SPTLC2	Serine palmitoyltransferase, long-chain base subunit 2	Pyridoxal phosphate
SRR	Serine racemase	Pyridoxal phosphate
TAT	Tyrosine aminotransferase	Pyridoxal phosphate

KEYWORDS

- **nutraceuticals**
- **cancer**
- **phytochemicals**
- **miRNA**
- **molecular docking**
- **gene network**

REFERENCES

Alegre, M. M.; Knowles, M. H.; Robison, R. A.; O'Neill, K. L. Mechanics Behind Breast Cancer Prevention—Focus on Obesity, Exercise and Dietary Fat. *Asian Pac. J. Cancer Prev.* **2013**, *14*(4), 2207–2212.

Bai, L.; Wang, H.; Wang, A. H.; Zhang, L. Y.; Bai, J. MicroRNA-532 and MicroRNA-3064 Inhibit Cell Proliferation and Invasion by Acting as Direct Regulators of Human Telomerase Reverse Transcriptase in Ovarian Cancer. *PLoS One* **2017**, *12*, e0173912.

Bao, B.; Ali, S.; Kong, D.; Sarkar, S. H.; Wang, Z.; Banerjee, S.; Aboukameel, A.; Padhye, S.; Philip, P. A.; Sarkar, F. H. Anti-Tumor Activity of a Novel Compound-CDF is Mediated by Regulating miR-21, miR-200, and PTEN in Pancreatic Cancer. *PLoS One* **2011**, *6*(3), e17850.

Bartel, D. P. MicroRNAs: Target Recognition and Regulatory Functions. *Cell* **2009**, *36*, 215–233.

Bhandari, P. R. Pomegranate (*Punica granatum* L). Ancient Seeds for Modern Cure? Review of Potential Therapeutic Applications. *Int. J. Nutr., Pharmacol., Neurol. Dis.* **2012**, *2*, 171–184.

Bhattacharyya, A.; Choudhuri, T.; Pal, S.; Chattopadhyay, S.; Datta, G. K.; Sa, G.; Das, T. Apoptogenic Effects of Black Tea on Ehrlich's Ascites Carcinoma Cell. *Carcinogenesis* **2003**, *24*(1), 75–80.

Bode, A. M.; Dong, Z. Signal Transduction Pathways: Targets for Green and Black Tea Polyphenols. *J. Bioch. Mol. Biol.* **2003**, *36*(1), 66–77.

Bowen-Forbes, C. S.; Zhang, Y.; Nai, M. G. Anthocyanin Content, Antioxidant, Anti-inflammatory and Anticancer Properties of Blackberry and Raspberry Fruits. *J. Food Compos. Anal.* **2010**, *23*(6), 554–560.

Calin, G. A.; Croce, C. M. MicroRNA-Cancer Connection: The Beginning of a New Tale. *Cancer Res.* **2006**, *66*(15), 7390–7394.

Chakrabarti, M.; Banik, N. L.; Ray, S. K. miR-138 Overexpression Is More Powerful Than hTERT Knockdown to Potentiate Apigenin for Apoptosis in Neuroblastoma in Vitro and in Vivo. *Exp. Cell Res.* **2013**, *319*, 1575–1585.

Cham, B. E. *The Eggplant Cancer Cure: A Treatment for Skin Cancer and New Hope for Other Cancers from Nature's Pharmacy*. Smart Publications: Vadacode, India, 2007; p 122.

Chen, M. S.; Chen, D.; Dou, Q. P. Inhibition of Proteasome Activity by Various Fruits and Vegetables Is Associated With Cancer Cell Death. *In Vivo*. 2004, *18*(1), 73–80.

Chen, Y.; Huang, C.; Zhou, T.; Chen, G. Genistein Induction of Human Sulfotransferases in HepG2 and Caco-2 Cells. *Basic Clin. Pharmacol. Toxicol.* **2008**, *103*(6), 553–559.

Chen, L; Lü, M. H.; Zhang, D.; Hao, N. B.; Fan, Y. H.; Wu, Y. Y.; Wang, S. M.; Xie, R.; Fang, D. C.; Zhang, H.; Hu, C. J.; et al. miR-1207-5p and miR-1266 Suppress Gastric Cancer Growth and Invasion by Targeting Telomerase Reverse Transcriptase. *Cell Death Dis*. **2014**, *5*, e1034.

Chen, W.; Zheng, R.; Zhang, S.; Zhao, P.; Zeng, H.; Zou, X.; et al. Annual Report on Status of Cancer in China, 2010. *Chin. J. Cancer Res.* **2014**.

Cherry, J. J.; Rietz, A.; Malinkevich, A.; Liu, Y.; Xie, M.; et al. Structure Based Identification and Characterization of Flavonoids That Disrupt Human Papillomavirus-16 E6 Function. *PLoS One* **2013**, *8*, e84506.

Chobotova, K.; Vernallis, A. B.; Majid, F. A. Bromelain's Activity and Potential as an Anti-Cancer Agent: Current Evidence and Perspectives. *Cancer Lett.* **2010**, 290, 148–156.

D'Ambrosio, S. M.; Han, C.; Pan, L.; Kinghorn, A. D.; Ding, H. Aliphatic Acetogenin Constituents of Avocado Fruits Inhibit Human Oral Cancer Cell Proliferation by Targeting the EGFR/RAS/RAF/MEK/ERK1/2 Pathway. *Biochem. Biophys. Res. Commun.* **2011**, *409*, 465–469.

Devi, J. R.; Thangam, E. B. Extraction and Separation of Glucosinolates from *Brassica oleraceae* var Rubra. *Adv. Biol. Res.* **2010**, 4(6), 309–313.

Devi, J. R.; Thangam, E. B. Mechanisms of Anticancer Activity of Sulforaphane From Brassica Oleracea in Hep-2 Human Epithelial Carcinoma Cell Line. *Asian Pac. J. Cancer Prev*. **2012**, 13, 2095–2100.

Dinami, R.; Buemi, V.; Sestito, R.; Zappone, A.; Ciani, Y.; Mano, M.; Petti, E.; Sacconi, A.; Blandino, G.; Giacca, M.; et al. Epigenetic Silencing of miR-296 and miR-512 Ensures hTERT Dependent Apoptosis Protection and Telomere Maintenance in Basal-Type Breast Cancer Cells. *Oncotarget* **2017**, *8*, 95674–95691.

Dwivedi, V.; Shrivastava, R.; Hussain, S.; Ganguly, C.; Bharadwaj, M. Comparative Anticancer Potential of Clove (*Syzygium aromaticum*)—An Indian Spice—Against Cancer Cell Lines of Various Anatomical Origin. *Asian Pac. J. Cancer Prev*. **2011**, *12*(8), 1989–1993.

Faria, A.; Pestana, D.; Teixeira, D.; de Freitas, V.; Mateus, N.; Calhau, C. Blueberry Anthocyanins and Pyruvic Acid Adducts: Anticancer Properties in Breast Cancer Cell Lines. *Phytother. Res*. **2010**, *24*(12), 1862–1869.

Farnham, M. W.; Wilson, P. E.; Stephenson, K. K.; Fahey, J. W. Genetic and Environmental Effects on Glucosinolate Content and Chemoprotective Potency of Broccoli. *Plant Breed.* **2004**, *123*(1), 60–65.

Ferguson, P. J.; Kurowska, E. M.; Freeman, D. J.; Chambers, A. F.; Koropatnick, J. In Vivo Inhibition of Growth of Human Tumor Lines by Flavonoid Fractions From Cranberry Extract. *Nutr. Cancer* **2006**, *56*, 86–94.

Ganesan, K.; Xu, B. Telomerase Inhibitors From Natural Products and Their Anticancer Potential. *Int. J. Mol. Sci*. **2018**, *19*(1), 13.

Gutheil, W. G.; Reed, G.; Ray, A.; Anant, S.; Dhar, A. Crocetin: An Agent Derived From Saffron for Prevention and Therapy for Cancer. *Current Pharm. Biotechnol.* **2010**, *13*(1), 173–179.

Hattori, M.; Kawakami, K.; Akimoto, M.; Takenaga, K.; Suzumiya, J.; Honma, Y. Antitumor Effect of Japanese Apricot Extract (MK615) on Human Cancer Cells in Vitro and in Vivo Through a Reactive Oxygen Species-Dependent Mechanism. *Tumori.* **2013**; 99: 239248.

Huang, Q.; Lu, G.; Shen, H. M.; Chung, M. C.; Ong, C. N. Anti-Cancer Properties of Anthraquinones From Rhubarb. *Med. Res. Rev.* **2007**, *27*(5), 609–630.

Ishurd, O.; Zgheela, F.; Kermagia, A.; Flefleaa, N.; Elmabruka, M.; Keneedy, J. F.; Ashour, A. (1ã 3)-β-D-Glucans from Libyan Dates (*Phoenix dactylifera* L.) and Their Anticancer Activities. *J. Biol. Sci.* **2007**, *7*(3), 554–557.

Izzotti, A.; Calin, G. A.; Arrigo, P.; Steele, V. E.; Croce, C. M.; De Flora, S. Downregulation of microRNA Expression in the Lungs of Rats Exposed to Cigarette Smoke. *FASEB J.* **2009**, *23*(3), 806–812.

Izzotti, A.; Larghero, P.; Cartiglia, C.; Longobardi, M.; Pfeffer, U.; Steele, V. E.; De Flora, S. Modulation of microRNA Expression by Budesonide, Phenethyl Isothiocyanate and Cigarette Smoke in Mouse Liver and Lung. *Carcinogenesis* **2010**, *31*(5), 894–901.

Jayakumar, R.; Kanthimathi, M. S. Inhibitory Effects of Fruit Extracts on Nitric Oxide-Induced Proliferation in MCF-7 Cells. *Food Chem.* **2011**, *126*(3), 956–960.

Jemal, A.; Bray, F.; Center, M. M.; Ferlay, J.; Ward, E.; Forman, D. Global Cancer Statistics. *CA Cancer J. Clin.* **2011**, *61*(2), 69–90.

Jeune, M. A.; Kumi-Diaka, J.; Brown, J. Anticancer Activities of Pomegranate Extracts and Genistein in Human Breast Cancer Cells. *J. Med. Food* **2005**, *8*(4), 469–475.

Kapadia, G. J.; Rao, G. S. Anticancer Effects of Red Beet Pigments. In: *Red Beet Biotechnology*; Neelwarne, B., Ed. Springer: USA, 2012; pp 125–154 (chapter 7).

Kawase, M.; Motohashi, N.; Satoh, K.; Sakagami, H.; Nakashima, H.; Tani, S.; Shirataki, Y.; Kurihara, T.; Spengler, G.; Wolfard, K.; Molnár, J. Biological Activity of Persimmon (*Diospyros kaki*) Peel Extracts. *Phytother. Res.* **2003**, *17*, 495–500.

Kim, J. Y.; Kim, E. H.; Kim, S. U.; Kwon, T. K.; Choi, K. S. Capsaicin Sensitizes Malignant Glioma Cells to TRAIL-Mediated Apoptosis via DR5 Upregulation and Survivin Downregulation. *Carcinogenesis* **2010**, *31*(3), 367–375.

Komatsu, S.; Yanaka, N.; Matsubara, K.; Kato, N. Antitumor Effect of Vitamin B6 and Its Mechanisms. *Biochim. Biophys. Acta Proteins Proteom.* **2003**, *1647*(2), 127–130.

Lai, P. K.; Roy, J. Antimicrobial and Chemopreventive Properties of Herbs and Spices. *Curr. Med. Chem.* **2004**, *11*(11), 1451–1460.

Li, J.; Cao, B.; Liu, X.; Fu, X.; Xiong, Z.; Chen, L.; Sartor, O.; Dong, Y.; Zhang, H. Berberine Suppresses Androgen Receptor Signaling in Prostate Cancer. *Mol. Cancer Ther.* **2011**; *10*(8), 1346–1356.

Li, J.; Lei, H.; Xu, Y.; Tao, Z. Z. miR-512-5p Suppresses Tumor Growth by Targeting hTERT in Telomerase Positive Head and Neck Squamous Cell Carcinoma in Vitro and in Vivo. *PLoS One* **2015**, *10*, e0135265.

Li, P. G.; Mu, T. H.; Deng, L. Anticancer Effects of Sweet Potato Protein on Human Colorectal Cancer Cells. *World J. Gastroenterol.* **2013**, *19*(21), 3300–3308.

Lin, J. K.; Liang, Y. C.; Lin-Shiau, S. Y. Cancer Chemoprevention by Tea Polyphenols Through Mitotic Signal Transduction Blockade. *Biochem. Pharmacol.* **1999**, *58*, 911–915.

Lu, Q. Y.; Arteaga, J. R.; Zhang, Q.; Huerta, S.; Go, V. L.; Heber, D. Inhibition of Prostate Cancer Cell Growth by an Avocado Extract: Role of Lipid-Soluble Bioactive Substances. *J. Nutr. Biochem.* **2005**, *16*, 23–30.

Lu, H. F.; Chen, Y. L.; Yang, J. S.; Yang, Y. Y.; Liu, J. Y.; Hsu, S. C.; Lai, K. C.; Chung, J. G. Antitumor Activity of Capsaicin on Human Colon Cancer Cells in Vitro and colo 205 Tumor Xenografts in Vivo. *J. Agric. Food Chem.* **2010**, *58*, 12999–13005.

Madhuri, S; Pandey, G. Some Anticancer Medicinal Plants of Foreign Origin. *Curr. Sci.* **2009**, *96*(6), 779–783.

Maher, D. M.; Bell, M. C.; O'donnell, E. A.; Gupta, B. K.; Jaggi, M.; Chauhan, S. C. Curcumin Suppresses Human Papillomavirus Oncoproteins, Restores p53, rb, and ptpn13 Proteins and Inhibits Benzo [A] Pyrene-Induced Upregulation of HPV E7. *Mol. Carcinogen.* **2011**, *50*(1), 47–57.

Majid, S.; Dar, A. A.; Saini, S.; Chen, Y.; Shahryari, V.; Liu, J.; Zaman, M. S.; Hirata, H.; Yamamura, S.; Ueno, K.; et al. Regulation of Minichromosome Maintenance Gene Family by microRNA-1296 and Genistein in Prostate Cancer. *Cancer Res.* **2010**, *70*(7), 2809–2818.

Meybodi, N. M.; Mortazavian, A. M.; Monfared, A. B.; Sohrabvandi, S.; Meybodi, F. A. Phytochemicals in Cancer Prevention: A Review of the Evidence. *Iran. J. Cancer Prev.* **2017**, *10*(1).

Milind, P.; Kulwant, S. Musk Melon Is Eatmusk Melon. *Int. Res. J. Pharmacy* **2011**, *2*(8), 52–57.

Mitomo, S.; Maesawa, C.; Ogasawara, S.; Iwaya, T.; Shibazaki, M.; Yashima-Abo, A.; Kotani, K.; Oikawa, H.; Sakurai, E.; Izutsu, N.; et al. Downregulation of miR-138 Is Associated With Overexpression of Human Telomerase Reverse Transcriptase Protein in Human Anaplastic Thyroid Carcinoma Cell Lines. *Cancer Sci.* **2008**, *99*, 280–286.

Moradi, Y.; Moradi-Sardareh, H.; Ghasemi, H.; Mohamadi, N.; Moradi, M. N.; Hosseini-Zijoud, S. M. Medicinal Properties of Persian Shallot. *Eur. J. Exp. Biol.* **2013**, *3*(1), 371–379.

Motaal, A. A.; Shaker, S. Anticancer and Antioxidant Activities of Standardized Whole Fruit, Pulp, and Peel Extracts of Egyptian Pomegranate. *Open Conf. Proc. J.* **2011**, *2*, 41–45.

Noratto, G.; Porter, W.; Byrne, D.; Cisneros-Zevallos, L. Identifying Peach and Plum Polyphenols With Chemopreventive Potential Against Estrogen-Independent Breast Cancer Cells. *J. Agric. Food Chem.* **2009**, *57*(12), 5219–5226.

Okwu, D. E. Citrus Fruits: A Rich Source of Phytochemicals and Their Roles in Human Health. *Int. J. Chem. Sci.* **2008**, *6*(2), 451–471.

Pan, X.; Wang, Z. X.; Wang, R. MicroRNA-21: A Novel Therapeutic Target in Human Cancer. *Cancer Biol. Ther.* **2011**, *10*(12), 1224–1232.

Parasramka, M. A.; Ho, E.; Williams, D. E.; Dashwood, R. H., MicroRNAs, Diet, and Cancer: New Mechanistic Insights on the Epigenetic Actions of Phytochemicals. *Mol. Carcinogen.* **2012**, *51*(3), 213–230.

Partouche, D.; Turbant, F.; El Hamoui, O.; Campidelli, C.; Bombled, M.; Trépout, S.; Wien, F.; Arluison, V. Epigallocatechin Gallate Remodelling of Hfq Amyloid-Like Region Affects *Escherichia coli* Survival. *Pathogens* **2018**, *7*(4), 95.

Paulraj, F.; Abas, F.; Lajis, N. H.; Othman, I.; Hassan, S. S.; Naidu, R. The Curcumin Analogue 1,5-Bis(2-Hydroxyphenyl)-1,4-Pentadiene-3-One Induces Apoptosis and Downregulates E6 and E7 Oncogene Expression in HPV16 and HPV18-Infected Cervical Cancer Cells. *Molecules* **2015**, *20*(7), 11830–11860.

Pezzuto, J. M. Plant-Derived Anticancer Agents. *Biochem. Pharmacol.* **1997**, *53*(2), 121–133.

Prajapati, P. M.; Solanki, A. S.; Sen, D. J. Nutrition Value of Plum Tree for Health. *Int. Res. J. Pharmacy* **2012**, *3*(5), 54–56.

Prasad, S.; Kalra, N.; Shukla, Y. Induction of Apoptosis by Lupeol and Mango Extract in Mouse Prostate and LNCaP Cells. *Nutr. Cancer* **2008**, *60*(1), 120–130.

Qazi, A.; Pal, J.; Maitah, M.; Fulciniti, M.; Pelluru, D.; Nanjappa, P.; Lee, S.; Batchu, R. B.; Prasad, M.; Bryant, C. S.; Rajput, S.; Gryaznov, S.; Beer, D. G.; Weaver, D. W.; Munshi, N. C.; Goyal, R. K.; Shammas, M. A. Anticancer Activity of a Broccoli Derivative, Sulforaphane, in Barrett Adenocarcinoma: Potential Use in Chemoprevention and as Adjuvant in Chemotherapy. *Transl. Oncol.* **2010**, *3*(6), 389–399.

Rahmat, A.; Rosli, R.; Zain, W. N. I. Z. W. M.; Endrini, S.; Sani, H. A. Antiproliferative Activity of Pure Lycopene Compared to Both Extracted Lycopene and Juice From Watermelon (*Citruis vulgaris*) and Papaya (*Carica papaya*) on Human Breast and Liver Cancer Cell Lines. *J. Med. Sci.* **2002**, *2*(2), 55–58.

Rani, A. P.; Saikishore, V. The Potential Role of Curcumin in Cancer Prevention and Treatment. *J. Pharmacy Pharm. Sci.* **2013**, *2*(3), 1–4.

Rehmsmeier, M.; Steffen, P.; Hoechsmann, M.; Giegerich, R. Fast and Effective Prediction of microRNA/Target Duplexes RNA, *RNA* **2004,** *10*(10), 1507–1517.

Reyes-Zurita, F. J.; Rufino-Palomares, E. E.; Lupiáñez, J. A.; Cascante, M. Maslinic Acid, A Natural Triterpene from *Olea europaea L.*, Induces Apoptosis in HT29 Human Colon-Cancer Cells via the Mitochondrial Apoptotic Pathway. *Cancer Lett.* **2009**, *273*, 44–54.

Ryu, N. H.; Park, K. R.; Kim, S. M.; Yun, H. M.; Nam, D.; Lee, S. G.; Jang, H. J.; Ahn, K. S.; Kim, S. H.; Shim, B. S.; Choi, S. H.; Mosaddik, A.; Cho, S. K.; Ahn, K. S. A Hexane Fraction of Guava Leaves (*Psidium guajava* L.) Induces Anticancer Activity by Suppressing AKT/Mammalian Target of Rapamycin/Ribosomal p70 S6 Kinase in Human Prostate Cancer Cells. *J. Med. Food* **2012**, *15*(3), 231–241.

Sani, H. A.; Rahmat, A.; Ismail, M.; Rosli, R.; Endrini, S. Potential Anticancer Effect of Red Spinach (*Amaranthus gangeticus*) Extract. *Asia Pac. J. Clin. Nutr.* **2004**, *13*(4), 396–400.

Seki, T.; Hosono, T.; Hosono-Fukao, T.; Inada, K.; Tanaka, R.; Ogihara, J.; Ariga, T. Anticancer Effects of Diallyl Trisulfide Derived From Garlic. *Asia Pac. J. Clin. Nutr.* **2008**, *17*(1), 249–252.

Serra, A. T.; Seabra, I. J.; Braga, M. E.; Bronze, M. R.; de Sousa, H. C.; Duarte, C. M. Processing Cherries (*Prunus avium*) Using Supercritical Fluid Technology. Part 1: Recovery of Extract Fractions Rich in Bioactive Compounds. *J. Supercritic. Fluids* **2010**, *55*, 184–191.

Shapiro, T. A.; Fahey, J. W.; Wade, K. L.; Stephenson, K. K.; Talalay, P. Chemoprotective Glucosinolates and Isothiocyanates of Broccoli Sprouts: Metabolism and Excretion in Humans. *Cancer Epidemiol. Prev. Biomark.* **2001**, *10*(5), 501–508.

Singh, M.; Singh, N. Molecular Mechanism of Curcumin Induced Cytotoxicity in Human Cervical Carcinoma Cells. *Mol. Cell. Biochem.* **2009**, *325*(1–2), 107–119.

Song, S.; Liem, A.; Miller, J. A.; Lambert, P. F. Human Papillomavirus Types 16 E6 and E7 Contribute Differently to Carcinogenesis. *Virology* **2000**, *267*, 141–150.

Sun, Q.; Cong, R.; Yan, H.; Gu, H.; Zeng, Y.; Liu, N.; Chen, J.; Wang B. Genistein Inhibits Growth of Human Uveal Melanoma Cells and Affects microRNA-27a and Target Gene Expression. *Oncol Rep.* **2009**, *22*(3), 563–567.

Sun, M.; Estrov, Z.; Ji, Y.; Coombes, K. R.; Harris, D. H.; Kurzrock, R. Curcumin (Diferuloylmethane) Alters the Expression Profiles of microRNAs in Human Pancreatic Cancer Cells. *Mol. Cancer Ther.* **2008**, *7*(3), 464–473.

Sun, J.; Hai Liu, R. Cranberry Phytochemical Extracts Induce Cell Cycle Arrest and Apoptosis in Human MCF-7 Breast Cancer Cells. *Cancer Lett.* **2006**, *241*, 124–134.

Sun, Y.; Xun, K.; Wang, Y.; Chen, X. A Systematic Review of the Anticancer Properties of Berberine, A Natural Product From Chinese Herbs. *Anticancer Drugs* **2009**, *20*(9), 757–769.

Talalay, P.; Fahey, J. W. Phytochemicals From Cruciferous Plants Protect Against Cancer by Modulating Carcinogen Metabolism. *J. Nutr.* **2001**, *131*(11), 3027S–3033S.

Taylor, C. K.; Levy, R. M.; Elliott, J. C.; Burnett, B. P. The Effect of Genistein Aglycone on Cancer and Cancer Risk: A Review of in Vitro, Preclinical, and Clinical Studies. *Nutr. Rev.* **2009**, *67*(7), 398–415.

Tili, E.; Michaille, J. J.; Adair, B.; Alder, H.; Limagne, E.; Taccioli, C.; Ferracin, M.; Delmas, D.; Latruffe, N.; Croce, C. M. Resveratrol Decreases the Levels of miR-155 by Upregulating miR-663, a microRNA Targeting JunB and JunD. *Carcinogenesis.* **2010**, *31*(9), 1561–1566.

Tili, E.; Michaille, J. J.; Alder, H.; Volinia, S.; Delmas, D.; Latruffe, N.; Croce, C. M. Resveratrol Modulates the Levels of microRNAs Targeting Genes Encoding Tumor-Suppressors and Effectors of TGFbeta Signaling Pathway in SW480 Cells. *Biochem. Pharmacol.* **2010**, *80*(12), 2057–2065.

Timsina, B.; Shukla, M.; Nadumane, V. K. A Review of Few Essential Oils and Their Anticancer Property. *J. Nat. Pharm.* **2012**, *3*(1), 1–8.

Tsang, W. P.; Kwok, T. T. Epigallocatechin Gallate Up-Regulation of Mir-16 and Induction of Apoptosis in Human Cancer Cells. *J. Nutr. Biochem.* **2010**, *21*(2), 140–146.

Wang, X.; Wei, Y.; Yuan, S.; Liu, G.; Zhang, Y. L.; Wang, W. Potential Anticancer Activity of Litchi Fruit Pericarp Extract Against Hepatocellular Carcinoma in Vitro and in Vivo. *Cancer Lett.* **2006**, *239*, 144–150.

Wasser, S. P. Medicinal Mushrooms as a Source of Antitumor and Immunomodulating Polysaccharides. *Appl. Microbiol. Biotechnol.* **2002**, *60*, 258–274.

Wen, X. Y.; Wu, S. Y.; Li, Z. Q.; Liu, Z. Q.; Zhang, J. J.; Wang, G. F.; Jiang, Z. H.; Wu, S. G. Ellagitannin (BJA3121), an Anti-Proliferative Natural Polyphenol Compound, Can Regulate the Expression of MiRNAs in HepG2 Cancer Cells. *Phytother Res.* **2009**, *23*(6), 778–784.

Win, D. T. Oleic acid—The Anti-Breast Cancer Component in Olive Oil. *J, Technol.* **2005**, *9*(2), 75–78.

Zhang, Y.; Li, M.; Wang, H.; Fisher, W. E.; Lin, P. H.; Yao, Q.; Chen, C. Profiling of 95 microRNAs in Pancreatic Cancer Cell Lines and Surgical Specimens by Real-Time PCR Analysis. *World J. Surg.* **2009**, *33*(4), 698–709.

Zhang, D.; Xiao, Y. F.; Zhang, J. W.; Xie, R.; Hu, C. J.; Tang, B.; Wang, S. M.; Wu, Y. Y.; Hao, N. B.; Yang, S. M. miR-1182 Attenuates Gastric Cancer Proliferation and Metastasis by Targeting the Open Reading Frame of hTERT. *Cancer Lett.* **2015**, *360*, 151–159.

Zhao, Q.; Zhai, Y. X.; Liu, H. Q.; Shi, Y. A.; Li, X. B. MicroRNA-491-5p Suppresses Cervical Cancer Cell Growth by Targeting hTERT. *Oncol. Rep.* **2015**, *34*, 979–986.

Zhou, K.; Raffoul, J. J. Potential Anticancer Properties of Grape Antioxidants. *J. Oncol.* **2012**, *2012*, 1–8.

CHAPTER 7

Nutraceuticals in Cancer Therapy

ANDREW G. MTEWA[1,2*], LUCRECE AHOVEGBE[2,3], EMANUEL L. PETER[2,4], HEDMON OKELLA[2], CASSIM U. TOLO[2], and FANUEL LAMPIAO[5]

[1]*Department of Chemistry, Malawi Institute of Technology, Malawi University of Science and Technology, Limbe, Malawi*

[2]*Pharmbiotechnology and Traditional Medicine Center of Excellence, School of Medicine, Mbarara University of Science and Technology, Mbarara, Uganda*

[3]*Unite de Formation et de Recherché en Pharmacie, Faculté des Sciences de la Santé, Université d'abomey-Calavi, Cotonou, Benin*

[4]*Department of Innovation, Technology Transfer & Commercialization, National Institute for Medical Research, Dar-Es-Salaam, Tanzania*

[5]*Africa Centre of Excellency in Public health and Herbal Medicine, College of Medicine, University of Malawi, Chichiri, Blantyre 3, Malawi*

**Corresponding author. E-mail: amtewa@must.ac.mw; andrewmtewa@yahoo.com*

ABSTRACT

Cancers remain a serious public health concern in the whole world. Most hit parts of the world are developing countries that struggle with both risk actor mitigation to prevent cancers and availability of treatment options. However, plant sources have presented a good alternative source of therapy against cancers from which the whole world will and is benefiting but, more so, a better opportunity for most of the developing world countries that still have forests rich in flora that provide functional foods and pharmaceutical plant molecules. Studies have been done, are still underway to discover more of these plants, and have been deposited in various libraries for further actions. More studies are warranted on the chemistry of the properties of nutraceutical active molecules to understand their mechanisms of action, to know how their structures can be modified to modulate their activities, and for product development purposes.

7.1 BACKGROUND

Cancer remains one of the most threatening diseases in the world. Due to the chronic development of the disease, awareness occurs most of the time at an advanced stage. Conventional treatment of cancer include chemotherapy and invasive means (American Cancer Society, 2017). The success rate of these treatment approaches is very low, and the overall goal of the treatment is more about improving patient's life quality as they wait to pass on. Cancer care guidelines emphasize more on the prevention of the disease by avoiding risk factors or counteract them by good lifestyle. About 30%–50% of the occurring cancers can be prevented through the eviction of risk factors and implementation of evidence-based prevention strategies that include behavioral and dietary habits (Deane, 2013; Lancet, 2011). Traditional practitioners across Africa and outside Africa are using hundreds of medicinal plants to manage the disease partly because they work and also due to the low accessibility to high-cost and scarce conventional treatment. The success of cancer treatment in some parts of the developing world is claimed to be due to the use of herbs along with the consumption of "anticancer food supplements." Today, popular press usually reports articles focused on prevention strategies with the promise of therapeutic success using functional foods and nutraceuticals. Having in view this pressing demand and relevance of complementary and alternative treatment derived from daily nutrition and medicinal plants, the World Health Organization and the National Institutes of Health of the United States decided to take action toward scientific validation and development of those food supplements (Colic and Pavelic, 2002). This has led to a number of food supplements developed by pharmaceutical companies as well as food industries. This chapter aims at reviewing nutritional components, emphasizing the role of nutraceuticals in cancer therapy by highlighting some known compounds as well as sources of nutraceutical value.

7.2 INTRODUCTION

Natural botanical products have played a significant role in the treatment, inhibition, and prevention of cancers for many years. By the year 2013, at least 50% of marketed drugs, with anticancer therapies inclusive, were made from natural herbal products and/or their derivatives (Pan et al., 2013). Plant phytochemicals are largely responsible for bioactivities against cancers that characterize these botanical therapies possess. For plant-based molecules in herbal products to be effective, they need to have properties that enable them to reach biological targets such as cancers on which they are to work. These bioactive drug molecules could be nutritious as well as drug molecules with pharmaceutical properties. These are a category of bioactive molecules that are collectively called nutraceuticals. Because cancers are basically associated with prolonged excessive inflammation and oxidative radicals and other species that work in a similar mechanism, nutraceuticals that have antioxidant and, to a lower extent, anti-inflammatory activities are the first-line products as therapies against cancers. Antioxidants work by manipulating pathways to do with reduction–oxidation-mediated transcription factors. Sometimes, they modulate inflammation-related enzymes and immunological and endocrine systems directly.

7.3 PHYSIOPATHOLOGY OF CANCER

Cancer is characterized by uncontrolled cell proliferation in the body. Cancer patients usually show cancer cachexia that is a multifactorial syndrome characterized by a loss of skeletal muscle mass. Biologically, cancer cachexia is a significant alteration in protein, carbohydrate, and fat metabolism (Shukla et al., 2014).

Alterations in carbohydrates: Glucose molecules are the first metabolized molecules by tumors to produce energy, resulting in lactic acid as an end product. Cancer cells are thus a source of an increased rate of anaerobic glycolysis and lactic acid. Due to the skeletal muscle protein breakdown, liver protein synthesis and the whole-body synthesis increase to support tumor growth. The overexpression of the ubiquitin–proteasome proteolytic pathway is a major cause of skeletal muscle loss in cancer patients. Fat synthesis decrease is the second major characteristic in cancer cachexia, and it has been associated with a lipid-mobilizing factor that induces the release of free fatty acids and glycerol.

7.4 CHEMOPREVENTIVE MECHANISMS OF NUTRACEUTICALS

Mechanisms of chemopreventive activities of nutraceuticals involve the inhibition of cell proliferation of mediated transcription factors, modification of signaling pathways, and induction of apoptosis. Nutraceuticals directly affect the endocrine, immunological, and enzymatic systems involved in inflammation. Nutraceuticals are able to promote DNA cleavage and repair processes. Intervention of nutraceuticals in cancers can prevent, suppress, or reverse carcinogenesis. They are thus classified as antipromotion or antiprogression for clinical purposes. Selected nutraceuticals have been investigated and have been shown to modulate the activity of transcription factors such as sterol response element-binding proteins (SREBPs), activator protein, nuclear factor-kB, peroxisome proliferator-activated receptor-gamma and change the expression of antioxidant genes like in the signal transduction cascade and gene regulation (Orzechowski et al., 2002).

7.5 NUTRACEUTICAL USE IN CANCER THERAPY

A third of all cancer deaths were estimated in 2005 to be preventable through appropriate changes in lifestyle including nutrition (Danaei et al., 2005). Research findings (Weisburger, 1999; Ranzato et al., 2014) reported that with appropriate modulation of nutrition, cancer therapeutics may be found. Further, simple foods, rich in fiber, proteins, and omega 3 fatty acids and low in carbohydrates, have proven to have positive effects against cancers (Gimble, 2003).

7.5.1 SOME OF THE KNOWN NUTRACEUTICAL ACTIVITIES AGAINST CANCERS

Generally, photochemical compounds found in nutraceuticals possess anti-cancer proliferation properties and also inhibit cancer mutagenic activities (Surh, 2003). They are also able to modulate the inflammatory response and protect the body from the peroxidation of lipids (Issa et al., 2006). Bioactive substances from foods have the potential to influence the gene expression regulation, even when they are in low concentrations (Ranzato et al., 2014). Table 7.1 shows some of the proven nutraceutical compounds and groups of compounds as well as their known activities in cancer therapy.

TABLE 7.1 Some Proven Nutraceuticals in Cancer Therapy

Sr. No.	Antioxidant/Anticancer Compounds From Nutraceuticals	Some Known Sources	Reference
1	Tannins: Tannic acids, ellagitannins, catechin, proanthocyanidins, and epicatechin polymers	Walnuts, pomegranate, peach, plum, olives, chickpea, peas, grape skin and seeds, berries (raspberries, strawberries, and blackberry), apple juice, green fruits, haricot bean, red wine, cocoa, tea, chocolate, and coffee	Ranzato et al., 2012
2	Epigallocatechin-3-gallate	Major catechin richly found in green tea (*Camellia sinensi*); Shown to work against various cell lines including anaplastic thyroid carcinoma and malignant mesothelioma	DeAmicis et al., 2013; Ranzato et al., 2012
3	Dietary fiber	Dietary fiber from both fruits and vegetables; soluble and insoluble were reported to have anticancer properties against colon cancer	Khan et al., 2014
4	Lutein	One of the bioactive carotenoids against cancers, with proven antibreast and lung cancer properties	Radhika et al., 2011
5	Quercetin	A flavonoid, found in fruits and vegetables, was reported to induce cancer cell apoptosis; it inhibits colon cancer and melanoma invasion, growth, and metastasis	Zhang et al., 2012
6	B-Glucan	A bioactive nutraceutical compound reported to effectively work against colon cancer	Radhika et al., 2011
7	Resveratrol	A stilbene whose properties relate to cancers; it is reported to have antiproliferative potential, anticarcinogenesis, and anti-inflammatory activities	Singh et al., 2013
8	**Daidzein**	An isoflavone that was isolated from soybeans and reported to have weak anticancer activities	Colli and Amling, 2009

TABLE 7.1 (*Continued*)

Sr. No.	Antioxidant/Anticancer Compounds From Nutraceuticals	Some Known Sources	Reference
9	Tocotrienol	An antioxidant bioactive nutraceutical compound reported to have anticancer properties against various cancer cell lines	Radhika et al., 2011
10	Vitamins	Ascorbate and vitamins A and E have been proven to prevent several cancers	Ranzato et al., 2014
11	Equol	A metabolic derivative of isoflavone daidzein was reported to be an effective inhibitor of prostate cancer	Colli and Amling, 2009
12	Indole-3-carbinol	An isolate from cabbage of the Brassica species was reported to be able to inhibit invasion of breast carcinoma cells	Kapusta-Duch et al., 2012; Morley et al., 2007
13	Minerals	Trace elements (e.g., selenium) have been reported to have preventive properties against cancers	Ranzato et al., 2014
14	Phytoestrogens	A flavone that has been shown to be effective against prostate, colorectal, breast, lung, and stomach cancers	Bicikova et al., 2012

7.5.1.1 NOTES TO TABLE 7.1

It was reported that there is a significant biotransformation difference in rats, mice, and humans for green tea polyphenols that may affect the generalization of results in one species (Clifford et al., 2013). As such, there is a need to undertake more clinical trials on epigallocatechin-3-gallate as most of the anticancer activity data available on the same is from preclinical trials.

Research shows that data on resveratrol's anticancer mechanisms is very limited; however, sirtuin protein activation is understood to be the underlying molecular mechanism it uses (Nakata et al., 2012). This provides a good area of research to have enough pool of findings for validation purposes.

In redox reactions, ascorbate is understood to be acting as an electron donor and oxidative stress as a mechanism of action against cancers is supported by several research findings (Ranzato et al., 2011). For a cancer cell to live, one of the factors is an effective uptake of iron. Ascorbate is also reported to reduce iron uptake, activate apoptosis-causing factors, and cause cell cycle arrest to induce cell apoptosis in cancerous cells. Table 7.2 shows some sources of anticancer bioactive components essential in nutraceutical formulations as therapies against cancers.

7.5.1.2 NOTES TO TABLE 7.2

Some research studies report that eugenol interferes with some standard conventional anticancer therapies (Wargovich et al., 2010), which means robust research on the modulation of the nutraceuticals in lieu with the drug needs to be given much focus to design an effective agonistic therapy. Besides diosgenin being proven to inhibit 5-LOX as stated in Table 7.2, data on the suppression mechanism is still limited.

Piperine increases the bioavailability and uptake of tea polyphenols and curcumin in the liver through its drug metabolism modulation properties in the liver (Aggarwal, 2010). However, more research is required to determine whether piperine would also modulate the stem cell biology of breast cancer.

Genistein is an isoflavone contained in soybeans, as stated in Table 7.2. They have shown a great therapeutic potential against cancers but have a serious safety concern that requires more research, beyond the basic level.

TABLE 7.2 Some Sources of Active Anticancer Nutraceuticals

Plant Species	English Name	Known Bioactivity	References
Capsicum sp.	Chilli	Capsaicin is an active principle in the plant. In vitro, it was shown to inhibit the proliferation of cancer cells and inflammatory response. By influencing the GSKβ3 section of the β-catenin pathway, it has the potential to mediate apoptosis. In mice, capsaicin inhibited pancreatic cancer xenografts. Apart from cancer treatment, it also showed the potential of treating radiotherapy and chemotherapy, overcoming major adverse effects, and treating oral mucositis	Lee et al., 2010 ; Yang et al., 2009; Zhang et al., 2008
Vitis rotundiflora	Muscadine	Contain proanthocyanadins that have antiantiproliferative and anticancer properties. Their constituents inhibit COX-1 and COX-2 and have proven to be one of the best inhibitors of late-stage cancers	Clinical trials.gov, 2010; Guha et al., 2009
Zingiber officinale	Ginger	One of the active principles in ginger, gingerol (6-gingerol) showed anti-inflammatory activities and is able to inhibit invasion, motility, and adhesion in human hepatocarcinoma, breast cancer, and other cell lines. It is also able to offset nausea and vomiting as side effects of chemotherapy	Dugasani et al., 2010 ; Kundu et al., 2009 ; Lee et al., 2008; Nigam et al., 2009; Rhode et al.,2007
Trigonella foenum-graecum	Fenugreek	A steroid saponin, diosgenin, abundant in the seeds of the plant has anti inflammatory properties and suppresses human myelogenous leukemia proliferation, breast, and colon cancers cell lines. It also inhibits COX enzymes, NFκβ activities and other potential effects on 5-LOX. The saponin is reported to be safe in rats at the highest test doses	Raju and Bird., 2007; Shishodia et al., 2006; Srinivasan et al., 2009; Taylor et al., 2000
Curcuma longa	Turmeric	A diferuloylmethane, curcumin, from the tubers has known potential to heal unresolved inflammation, modulation of TNF-α, and inhibition of the nuclear transcription factor, NFκβ, which has an impact on cancerous cells. COX-2, interleukins IL-1 and IL-6, IGF, and chemokines MCP-1 and MCP-4 are some of the proteins modulated by curcumin. Curcuminoids are helpful in cancer therapy by cancer cells chemosensitization making cancer cells more amenable to therapy	Kamat et al., 2009; Kang et al., 2009; Kunnumakkara et al., 2009; Lin et al., 2009; Sandur et al., 2009

TABLE 7.2 (*Continued*)

Plant Species	English Name	Known Bioactivity	References
Camellia sinensis	Green tea	One of the most studied active compounds against cancers is epigallocatechin-3-gallate (EGCG) that has been reported to inhibit pathways involving MMPs, NFκβ, and COX-2, all implicated in cancer cell survival and expansion	Park et al., 2009; Roomi et al., 2010
Vitis vinafera	Grapes	Contain proanthocyanidins, the most studied of which is resveratrol. Proanthocyanidins have anticancer and antiproliferative properties. Resveratrol inhibits COX-1 and COX-2 and has proven to be one of the best inhibitors of late-stage cancers	Liu et al., 2010
Piper nigrum	Black pepper	Piperine is one of the main biologically active principles in black pepper was reported to inhibit lung cancer metastases and mammosphere formation in animal models	Kakarala et al., 2010; Pradeep et al., 2002
Glycine max	Soy	Said to be rich in anticancer agents, they contain isoflavones that modulate breast cancer signature estrogen receptors (ERs)	Barnes et al., 2010; Wargovich et al., 2010
Syzygium aromaticum	Cloves	The oils of the plant contain eugenol (4-allyl-1hydroxy-2-methoxybenzene), also in cinnamon and other aromatics which inhibit the growth of human and rat cancer cell lines and suppresses proliferation. The compound is also reported to have anti-inflammatory effects	Aggarwal et al., 2008; Manikandan et al., 2009; Slamenova et al., 2009

For the effective potential of these nutraceuticals, their bioactive molecules are supposed to be able to reach appropriate biological targets, thus having the right "drug-like" properties. The term "dug-like" properties refer to properties that molecules from promising bioactive extracts need to have to be considered as clinical leads and then qualify as approved marketable drugs later (Schneider, 2013). They are highly predictive of the bioavailability of drug molecules to biological targets. These properties are basically the description of the chemistry of the molecules to see to it that they can ably pass through various membranes, survive different pH environments without unnecessary alterations, be able to be transported to the right affected areas and interact only with the right cell targets, be able to partition into different phases in the body, be able to have optimal solubility, not too much to have everything gone before reaching the target and also not too little to have the whole molecule come out as it got into the body without acting on the right spots, and, finally, be able to be ejectable from the body after its rightful work is done.

This is essential as all of these known nutraceuticals can be there but their mechanisms of action not known, which can make it hard to properly modulate them by structural modifications where necessary for further studies or product development. Studying the chemistry of active molecules for drug-like properties, though of such essence, is lacking a lot in the field of drug discovery and nutraceutical product development. Most people are only satisfied with bioactivities proven in the lab but less care is put on ensuring the bioavailability of the molecules once taken by consumers or patients.

7.6 CONCLUSIONS

Cancers are still hitting hard despite various treatment options being available in conventional medicine. Among other reasons, this is due to the high costs the treatments attract from patients. Functional foods and pharmaceutical plants are now providing a promise of cheaper readily available preventive options against cancers, which are already in use today but with the potential of further designing into better products. The authors envision cheaper preventive medicines using locally available materials (PreMULAM) to help evade the hard impact of cancers. The same products are also vital in providing therapies against cancers, with the advantage of having no serious, if at all present, adverse effects to consumers. Nutraceuticals discovery is just a springboard for a better future against cancers and other diseases.

KEYWORDS

- **nutraceuticals**
- **nutraceutical therapy**
- **cancer therapy**
- **nuclear factor-kβ**
- **antitumor**
- **nutraceuptides**

REFERENCES

Aggarwal, B. B.; Kunnumakkara, A. B.; Harikumar, K. B.; Tharakan, S. T.; Sung, B.; Anand, P. Potential of Spice-Derived Phytochemicals for Cancer Prevention. *Planta Medica* **2008**, *74*:1560–1569.

Aggarwal, B. B. Targeting Inflammation-Induced Obesity and Metabolic Diseases by Curcumin and Other Nutraceuticals. *Annu Rev Nutri.* **2010**, *30*:173–199.

American Cancer Society. Cancer Facts and Figures. *Genes Dev.* **2017,** *21*: 2525–2538.

Barnes, S. The Biochemistry, Chemistry and Physiology of the Isoflavones in Soybeans and Their Food Products. *Lymphat. Res. Bio.* **2010**, *8*: 89–98.

Berger, A.; Henderson, M.; Nadoolman, W.; Duffy, V.; Cooper, D.; Saberski, L.; Bartoshuk, L. Oral Capsaicin Provides Temporary Relief for Oral Mucositis Pain Secondary to Chemotherapy/Radiation Therapy. *J. Pain Symptom Manag.* **1995**, *10*: 243–248.

Clifford, M. N.; Van-der-Hooft, J. J.; Crozier, A. Human Studies on the Absorption, Distribution, Metabolism, and Excretion of Tea Polyphenols. *Am J Clin Nutr.* **2013**, *98*: 1619S–1630S.

ClinicalTrials.gov. *A Clinical Study to Assess the Safety and Activity of srt501 alone or in Combination with Bortezomib in Patients with Multiple Myeloma.* National Institutes of Health: Bethesda; 2010.

Colic, M.; Pavelic, K. Molecular, Cellular and Medical Aspects of the Action of Nutraceuticals and Small Molecules Therapeutics: From Chemoprevention to New Drug Development. *Drugs Exp. Clin. Res.* **2002**, *2*(3): 23–26.

Danaei, G.; Vander-Hoorn, S.; Lopez, A. D.; Murray, C. J.; Ezzati, M. Causes of Cancer in the World: Comparative Risk Assessment of Nine Behavioural and Environmental Risk Factors. *The Lancet* **2005**, *366*: 1784–1793.

De Amicis, F.; Perri, A.; Vizza, A.; Russo, A.; Panno, M. L.; Bonofiglio, D.; Ando, S. Epigallocatechin Gallate Inhibits Growth and Epithelial-to-Mesenchymal Transition in Human Thyroid Carcinoma Cell Lines. *J. Cell Physiol.* **2013**, *228*(10): 2054–2062.

Deane, K. D. Can Rheumatoid Arthritis be Prevented? *Best Pract. Res. Clin. Rheumatol.* **2013**, *27*(4): 467–485.

Dugasani, S.; Pichika, M. R.; Nadarajah, V. D.; Balijepalli, M. K.; Tandra, S.; Korlakunta, J. N. Comparative Antioxidant and Anti-inflammatory Effects of [6]-Gingerol, [8]-Gingerol, [10]-Gingerol and [6]-Shogaol. *J. Ethnopharmacol.* **2010**, *127*:515–520.

Grimble, R. F. Nutritional Therapy for Cancer Cachexia. *Gut.* **2003**, *52*: 1391–1392.

Issa, A. Y.; Volate, S. R.; Wargovich, M. J. The Role of Phytochemicals in Inhibition of Cancer and Inflammation: New Directions and Perspectives. *J. Food Compos. Anal.* **2006**, *19*: 405–419.

Guha, P.; Dey, A.; Sarkar, B.; Dhyani, M. V.; Chattopadhyay S.; Bandyopadhyay. S. K. Improved Antiulcer and Anticancer Properties of a Trans-Resveratrol Analog in Mice. *J. Pharmacol. Exp. Ther.* **2009**, *328*: 829–838.

Hoffman, T. Ginger: An Ancient Remedy and Modern Miracle Drug. *Hawaii Med. J.* **2007**, *66*: 326–327.

Kakarala, M.; Brenner, D.; Korkaya, H.; Cheng, C.; Tazi, K. Ginestier C, et al. Targeting Breast Stem Cells with the Cancer Preventive Compounds Curcumin and Piperine. *Breast Cancer Res. Treat.* **2010**, *122*:777–785.

Kamat, A. M.; Tharakan, S. T.; Sung, B.; Aggarwal, B. B. Curcumin Potentiates the Antitumor Effects of *Bacillus* Calmette–Guerin Against Bladder Cancer Through the Downregulation of NF-kappaB and Upregulation of TRAIL Receptors. *Cancer Res.* **2009**, *69*: 8958–8966.

Kang, H. J.; Lee, S. H.; Price, J. E.; Kim, L. S. Curcumin Suppresses the Paclitaxel-Induced Nuclear Factor-kappa beta in Breast Cancer Cells and Potentiates the Growth Inhibitory Effect of Paclitaxel in a Breast Cancer Nude Mice Model. *Breast J.* **2009**, *15*: 223–229.

Kundu, J. K.; Na, H. K.; Surh, Y. J. Ginger-Derived Phenolic Substances with Cancer Preventive and Therapeutic Potential. *Forum Nutr.* **2009**; *61*: 182–192.

Kunnumakkara, A. B.; Diagaradjane, P.; Anand, P.; Harikumar, K. B.; Deorukhkar, A.; Gelovani, J.; Guha, S.; Krishnan, S.; Aggarwal, B. B. Curcumin Sensitizes Human Colorectal Cancer to Capecitabine by Modulation of Cyclin D1, COX-2, MMP-9, VEGF and CXCR4 Expression in an Orthotopic Mouse Model. *Int. J. Cancer* **2009**, *125*: 2187–2197.

Lancet, T. The Good News About Cancer in Developing Countries. *The Lancet*, **2011**, *379*: 712 https://doi.org/10.1016/S0140-6736(11)61681-4.

Lee, S. H.; Krisanapun, C.; Baek, S. J. NSAID-Activated Gene-1 as a Molecular Target for Capsaicin-Induced Apoptosis Through a Novel Molecular Mechanism Involving GSK3beta, C/EBPbeta and ATF3. *Carcinogenesis* **2010**, *31*: 719–728.

Lee, H. S.; Seo, E. Y. Kang NE, Kim WK. [6]-Gingerol Inhibits Metastasis of MDA-MB-231 Human Breast Cancer Cells. *J. Nutr. Biochem.* **2008**; *19*: 313–319.

Lin, S. S.; Lai, K. C.; Hsu SC, Yang, J. S.; Kuo, C. L.; Lin, J. P. Curcumin Inhibits the Migration and Invasion of Human A549 Lung Cancer Cells Through the Inhibition of Matrix Metalloproteinase-2 and -9 and Vascular Endothelial Growth Factor (VEGF). *Cancer Lett.* **2009**, *285*: 127–133.

Liu, P. L.; Tsai, J. R.; Charles, A. L.; Hwang, J. J.; Chou, S. H.; Ping, Y. H.; Lin, F. Y.; Chen, Y. L.; Hung, C. Y.; Chen, W. C.; Chen, Y. H.; Chong, I. W. Resveratrol Inhibits Human Lung Adenocarcinoma Cell Metastasis by Suppressing Heme Oxygenase 1-Mediated Nuclear Factor-kappaB Pathway and Subsequently Downregulating Expression of Matrix Metalloproteinases. *Mol. Nutri. Food Res.* **2010**, *54*(2): S196–S204.

Manikandan, P.; Vinothini, G.; Vidya-Priyadarsini, R.; Prathiba, D.; Nagini, S. Eugenol Inhibits Cell Proliferation via NF-kappaB Suppression in a Rat Model of Gastric Carcinogenesis Induced by MNNG. *Investig. New Drugs* **2009**, *29*(1):_110–117. s

Nakata, R.; Takahashi, S.; Inoue, H. Recent Advances in the Study on Resveratrol. *Biol. Pharm. Bull.* **2012**, *35*: 273-279.

Nigam, N,; Bhui, K.; Prasad, S.; George, J.; Shukla, Y. [6]-Gingerol Induces Reactive Oxygen Species Regulated Mitochondrial Cell Death Pathway in Human Epidermoid Carcinoma A431 Cells. *Chem.-Biol. Interact.* **2009**, *181*:77–84.

Orzechowski, A.; Ostaszewski, P.; Jank, M.; Berwid, S. J. Bioactive Substances of Plant Origin in Food—Impact on Genomics. *Reprod. Nutri. Dev.* **2002**, *42*: 461–177.

Park, I. J.; Lee, Y. K.; Hwang, J. T.; Kwon, D. Y.; Ha, J.; Park, O. J. Green Tea Catechin Controls Apoptosis in Colon Cancer Cells by Attenuation of H_2O_2-Stimulated COX-2 Expression via the AMPK Signaling Pathway at Low-Dose H_2O_2. *Ann. NY Acad. Sci.* **2009**, *1171*:538–544.

Pan, S. Y.; Zhou, S. F.; Gao, S. H.; Yu, Z. L.; Zhang, S. F.; Tang, M. K.; Sun, J. N.; Ma, D. L.; Han, Y. F.; Fong, W. F.; Ko, K. M. New Perspectives on How to Discover Drugs from Herbal Medicines: CAM's Outstanding Contribution to Modern Therapeutics. *Evid.-Based Complement. Altern. Med.* **2013**, *627375*: 1–25.

Pradeep, C. R.; Kuttan, G. Effect of piperine on the inhibition of lung metastasis induced B16F-10 melanoma cells in mice. *Clin. Exp. Metastasis* **2002**, *19*: 703–708.

Raju, J.; Bird, R. P. Diosgenin, A Naturally Occurring Steroid [Corrected] Saponin Suppresses 3-Hydroxy-3-methylglutaryl CoA Reductase Expression and Induces Apoptosis in HCT-116 Human Colon Carcinoma Cells. *Cancer Lett.* **2007**, *255*: 194–204.

Ranzato, E.; Biffo, S.; Burlando, B. Selective Ascorbate Toxicity in Malignant Mesothelioma: a Redox Trojan Mechanism. *Am. J. Respir. Cell. Mol. Biol.* **2011**, *44*: 108–117.

Ranzato, E.; Martinotti, S.; Magnelli, V.; Murer, B.; Biffo, S.; Mutti, L.; Burlando, B. Epigallocatechin-3-gallate Induces Mesothelioma Cell Death via H_2O_2-Dependent T-type Ca^{2+} Channel Opening. *J. Cell. Mol. Med.* **2012**, *16*: 2667–2678.

Ranzato, E.; Martinotti, S.; Myriam-Calabrese, C.; Calabrese, G. Role of Nutraceuticals in Cancer Therapy. *J. Food Res.* **2014**, *3*: 18–25.

Rhode, J.; Fogoros, S.; Zick, S.; Wahl, H.; Griffith, K.; Huang, J.; Liu, J. R. Ginger Inhibits Cell Growth and Modulates Angiogenic Factors in Ovarian Cancer Cells. *BMC Complement. Altern. Med.* **2007**, *7*(44): 1–9.

Roomi, M. W.; Monterrey, J. C.; Kalinovsky, T.; Rath, M.; Niedzwiecki, A. Comparative Effects of EGCG, Green Tea and a Nutrient Mixture on the Patterns of MMP-2 and MMP-9 Expression in Cancer Cell Lines. *Oncol. Rep.* **2010**; *24*:747–757.

Sandur, S. K.; Deorukhkar, A.; Pandey, M. K.; Pabon, A. M.; Shentu, S.; Guha, S.; Aggarwal, B. B.; Krishnan, S. Curcumin Modulates the Radiosensitivity of Colorectal Cancer Cells by Suppressing Constitutive and Inducible NF-kappaB Activity. *Int. J. Rad. Oncol. Biol. Phys.* **2009**, *75*: 534–542.

Schneider, G. Prediction of Drug-Like Properties. In *Madam Curie Bioscience Database, 2000–2013*. Landes Bioscience: Austin, TX; 2013.

Shishodia, S.; Aggarwal, B. B.; Raju, J.; Bird, R. P.; Srinivasan, S.; Koduru, S. Diosgenin Inhibits Osteoclastogenesis, Invasion, and Proliferation Through The Downregulation of Akt, I kappa B Kinase Activation and NF-kappa B-Regulated Gene Expression. *Oncogene* **2006**, *25*: 1463–1473.

Shukla, S. K.; Gebregiworgis, T.; Purohit, V.; Chaika, N. V.; Gunda, V.; Radhakrishnan, P.; Singh, P. K. Metabolic Reprogramming Induced by Ketone Bodies Diminishes Pancreatic Cancer Cachexia. *Cancer Metabol.* **2014**, *2*(18): 1–19 https://doi.org/10.1186/2049-3002-2-18.

Singh, C. K.; George, J.; Ahmad, N. Resveratrol-Based Combinatorial Strategies for Cancer Management. *Ann. NY Acad. Sci.* **2013**, *1290*: 113–121.

Slamenova, D.; Horvathova, E.; Wsolova, L.; Sramkova, M.; Navarova, J.; Manikandan, P. Investigation of Anti-Oxidative, Cytotoxic, DNA-Damaging and DNA-Protective Effects of Plant Volatiles Eugenol and Borneol in Human-Derived HepG2, Caco-2 and VH10 Cell Lines. *Mut. Res.* **2009**, *677*:46–52.

Srinivasan, S.; Koduru, S.; Kumar, R.; Venguswamy, G.; Kyprianou, N.; Damodaran, C. Diosgenin Targets Akt-Mediated Prosurvival Signaling in Human Breast Cancer Cells. *Int. J. Cancer* **2009**; *125*: 961–967.

Surh, Y. J. Cancer Chemoprevention with Dietary Phytochemicals. *Nat. Rev. Cancer* **2003**, *3*: 768–780.

Wargovich, M. J.; Morris, J.; Brown, V.; Ellis, J.; Weber, R. Nutraceutical Use in Late-Stage Cancer. *Cancer Metastasis Rev.* **2010**, *29*: 503–510.

Taylor, W. G.; Elder, J. L.; Chang, P. R.; Richards, K. W. Microdetermination of Diosgenin from Fenugreek (*Trigonella foenum-graecum*) Seeds. *J. Agric. Food Chem.* **2000**, *48*: 5206– 5210.

Weisburger, J. H. Antimutagens, Anticarcinogens, and Effective Worldwide Cancer Prevention. *J. Environ. Pathol. Toxicol. Oncol.* **1999**, *18*: 85–93.

Yang, K. M.; Pyo, J. O.; Kim, G. Y. ; Yu, R.; Han, I. S.; Ju, S. A.; Kim, W. H.; Kim, B. S. Capsaicin Induces Apoptosis by Generating Reactive Oxygen Species and Disrupting Mitochondrial Transmembrane Potential in Human Colon Cancer Cell Lines. *Cell. Mol. Biol. Lett.* **2009**, *14*: 497–510.

Zhang, X. L. X.; Chen, L.; Cheng, O. U.; Liu, B. Plant Natural Compounds: Targeting Pathways of Autophagy as Anti-Cancer Therapeutic Agents. *Cell Prolif.* **2012**, *45*: 466–476.

Zhang, R.; Humphreys, I.; Sahu, R. P.; Shi, Y.; Srivastava, S. K. In Vitro and in Vivo Induction of Apoptosis by Capsaicin in Pancreatic Cancer Cells is Mediated Through ROS Generation and Mitochondrial Death Pathway. *Apoptosis.* **2008**, *13*:1465–1478.

PART III
Nutraceuticals and Mental Illness

CHAPTER 8

Exploration of Nanonutraceuticals in Neurodegenerative Diseases

SWATI PUND[1], AMITA JOSHI[2], and VANDANA PATRAVALE[3*]

[1]Nanomedicine Laboratory, Department of Biosciences and Bioengineering, Indian Institute of Technology, Mumbai, India

[2]Department of Pharmaceutics, B.V. Patel PERD Centre, Ahmedabad, India

[3]Department of Pharmaceutical Science and Technology, Institute of Chemical Technology, Mumbai, India

[*]Corresponding author. E-mail: vb.patravale@ictmumbai.edu.in

ABSTRACT

Longer life span has led to rise of debilitating neurodegenerative disorders, primarily Alzheimer's disease, Parkinson's disease, Huntington's disease; thereby posing a remarkable socioeconomical burden. These diseases involve complex multifactorial neurodegenerative events leading to cognitive and behavioral dysfunction. Chronic neuroinflammation, free radical generation, and the oxidative stress are the foremost propeller for neurodegeneration.

Food-based nutraceuticals have recently gained global importance owing to their multifaceted effects and better tolerance. These food-based approaches target multiple pathways in a slow but more physiological manner without causing severe adverse effects and therefore, are favorable for long-term consumption. The nutraceuticals exert their neuroprotective action via diverse mechanisms, namely, free radical scavenging, chelation of transition metals, improvement in antioxidant reserve, and anti-inflammatory effects.

Nutraceuticals also regulate several signaling pathways that play a vital role in the cell survival and stress response, thereby improving overall neuronal health. However, metabolism, poor solubility and bioavailability and inadequate blood–brain permeability, limit the efficiency of nutraceuticals in exerting neuroprotective action. Nanointerventions are therefore employed to overcome these limitations and improve therapeutic response.

This chapter will provide an insight into the application of nutraceuticals in maintaining neurophysiology along with various scientific nanotechnological interventions to improve neuroprotective efficacy of nutraceuticals.

8.1 NUTRACEUTICALS AND NEURODEGENERATION

The term "nutraceutical" was coined to describe a combination of nutrition and pharmaceutical and is synonyms to "functional food." Broadly speaking, foods providing therapeutic benefit are known as nutraceuticals. "*Let food be your medicine, and medicine be your food,*" Hippocrates (370–460 BC) has laid the rationale of usage of nutraceuticals in prevention and treatment of diseases (Hakim and Chishti, 1988). Historically, food has been explored for its therapeutic potential by all the great ancient civilizations and is well documented in ancient medicine systems, namely, Ayurveda, Chinese medicine system, to name a few. Long and healthy life is the prime goal of research in health science. Mediterranean and Asian diets, rich in herb-based recipes, are currently considered as the healthiest traditional food effective against age-related neurodegenerative diseases. Bioflavanoid rich food with strong antioxidant capacity plays an important role in reducing the incidences of progressive pathological conditions associated with neuronal functional degeneration and death (Preethi Pallavi and Sampath Kumar, 2018; Stefani and Rigacci, 2014). Ensuring a long and healthy life has always been the main goal of medical research and, certainly, the immense advances in modern medicine have highly contributed to combating several severe chronic pathologies. As a result of significant increase in life expectancy, greater incidence of age-associated cognitive decline and other neurological disorders with high socioeconomic impact are paralleled, namely, Alzheimer's disease (AD), Parkinson's disease (PD), Huntington's disease (HD), and amyotrophiclateral sclerosis (ALS) (Figure 8.1).

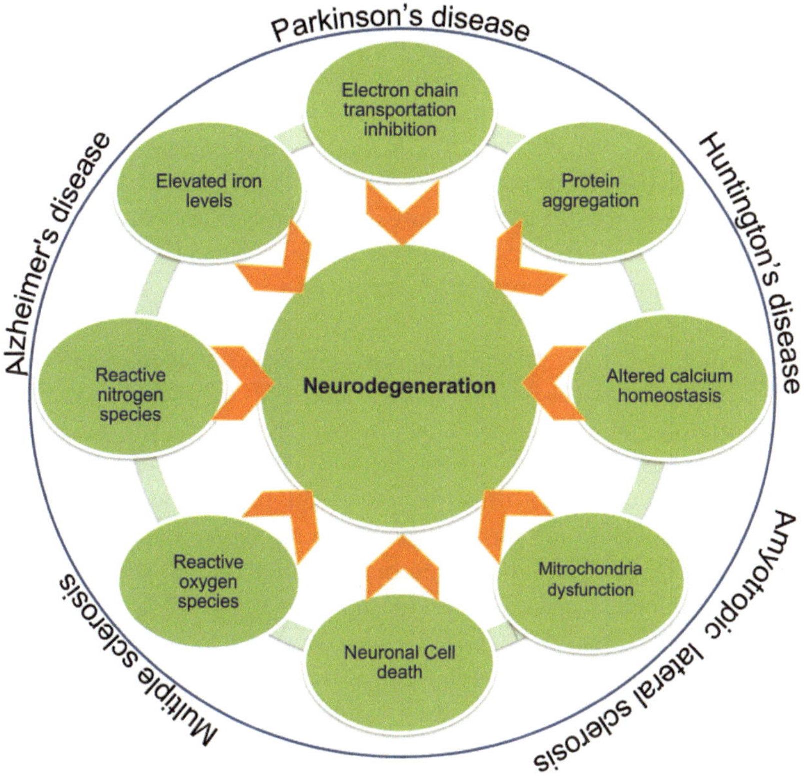

FIGURE 8.1 Neurodegenerative diseases and mechanisms of neuronal cell degeneration.

8.2 TRIGGER AND BULLETS IN NEURODEGENERATIVE PATHOGENESIS—CELLULAR AND MOLECULAR ASPECTS

The hallmarks of neurodegenerative diseases are the elevated levels of malondialdehyde as well as protein carbonyl levels, which are the outcomes of reduced mitochondrial membrane potential, and disruption of intracellular calcium signaling due to increased oxidative stress (Chowdhury et al., 2016; Domijan et al., 2014; Turillazzi et al., 2016). Oxidative stress generates reactive oxygen species (ROS) via, two main cellular organelles, namely, mitochondria and plasma membrane. The by-product of cellular metabolism generated in mitochondria is ROS. In mitochondria, unpaired electrons that escape the electron transport chain react with molecular oxygen to generate superoxide radicals. While, in plasma membrane reduced nicotinamide adenine dinucleotide phosphate (NADPH) oxidase activity results in ROS generation (Sadhukhan et al., 2018). A minimum amount of ROS is essential for normal human physiology; however, accumulation of ROS is toxic. Basal ROS homeostasis in body is maintained via various delicate balances between ROS scavengers like dietary antioxidants, glutathione, NADPH, and so forth, and ROS inducers like hypoxia, stress, metabolic defects, and so forth. Failure of this delicate redox balance leads to "oxidative stress" resulting in damaging effects seen as aging, diabetes, cognitive impairment, and neurodegenerative disorders like, AD, PD, ALS, HD (Mao et al., 2018).

High oxygen consumption and the phospholipid rich membranes make brain highly susceptible to free radical attack. The basal cellular ROS level in brain is maintained by its antioxidant system that includes, anti-oxidative enzymes, namely, superoxide dismutase, catalase, glutathione peroxidase, glutathione disulfide reductase and glutathione-*S*-transferase, as well as nonenzymatic sources, such as the antioxidant molecule glutathione, ascorbic acid, and α-tocopherol (Mao et al., 2018). Enzymes like glutamine synthetase and creatine kinase are thus, critical for the normal functioning of neuronal and glial cells and consequently, indispensable for normal brain physiology (Karthick et al., 2015; Swomley et al., 2014). The elevated levels of ROS in brain, result in activation of glutamate receptors resulting in induction of excitotoxic response, eventually leading to cell death marked as neurodegenerative disorders (Saha et al., 2014).

Histopathological abnormalities observed in most of the neurodegenerative diseases are intracellular or extracellular deposits of fibrillar proteinaceous material. These deposits originate from the polymerization of some soluble proteins and peptides due to multimerization, mis-folding and aggregation and cause cellular toxicity and proteostatic collapse (Rigacci and Stefani, 2015; Sweeney et al., 2017) (Table 8.1).

Protein toxicity is a unifying feature of both sporadic and familial cases of neurodegeneration. Neuronal degeneration causes immense cellular stress and results in the formation of stress granules. Electrostatic interactions between proteins facilitate liquid–liquid phase separation causing formation of stress granules (Chung et al., 2018). Two leading neuropathological changes of AD are the neurofibrillary tangles made of paired helical filaments containing the microtubule-associated protein, tubulin-associated unit (Tau) accumulating in neuronal perikarya and extracellular diffuse amyloid deposits containing amyloid peptide-forming neuritic senile plaques (Bloom, 2014; Giacobini and Gold, 2017).

TABLE 8.1 Major Neurological Clinical Syndromes and Corresponding Amyloid-like Fibril Component

Disease	Intrinsically Disordered Protein	Implication	References
AD	• Tau oligomers • Amyloid-β	• Synaptic Tau interacts with 26S proteasome • Mitochondrial protein import dysfunction • Binds with mitochondrial proteins to induce ROS generation • Altered protein phosphorylation states	Bonet-Costa et al., 2016; Oliveira et al., 2017; Rigacci and Stefani, 2015
PD	• α-Synuclein A53T and A30P • Mutant and WT α-synuclein	• Disruption of chaperone-mediated autophagy by blocking lysosomal translocation of substrates • Dysfunction of mitochondrial protein import • Blockage of voltage-dependent anion channel	Cuervo et al., 2004; Rostovtseva et al., 2015
HD	• PolyQ-expanded huntingtin fragment • TATA-binding protein	• Autophagy and axonal transport dysfunction • Defects in mitochondrial protein import, trafficking, mitochondrial permeability transition pore opening and calcium regulation • Formation of stress granules • Nucleocytoplasmic transport dysfunction	Chung et al., 2018; Woerner et al., 2016
ALS	• Poly-PR repeat protein • Mutant superoxide dismutase-1 • Valosin-containing protein • Poly-GR repeat protein • Mutant profilin 1 • Mutant fused in sarcoma	• Alterations in stress granules dynamics • Nucleocytoplasmic and axonal transport dysfunction • Altered stress granules assembly and dynamics	Baldwin et al., 2016; Chung et al., 2018; Taylor et al., 2016.

Inhibition of tau phosphorylation, proteolysis and aggregation, promotion of intra- and extracellular tau clearance, and stabilization of microtubules are the pathways to be explored for development of new therapeutics. Amyloid β (Aβ) peptides are 39–43 amino acid residue peptides derived from the sequential proteolytic enzymatic action of β-secretase and γ-secretase on transmembrane amyloid precursor protein. Aβ deposition and diffused plaque formation causes local microglial activation, cytokine release, reactive astrocytosis, and a multiprotein inflammatory response.

The number of therapeutic strategies based on blocking the effects of Aβ peptide is based on decreasing Aβ production by inhibition of β-secretase, inhibition of γ secretase, activation of α-secretase, inhibition of Aβ oligomerization or fibrillization and increasing Aβ degradation and clearance (Barage and Sonawane, 2015). Amyloid formation of a small intrinsically disordered protein; α-Synuclein is associated with PD (Bhasne and Mukhopadhyay, 2018; Stefanis, 2012). Counteracting intracellular and extracellular effects of α-Synuclein at the transcriptional and post-transcriptional level is an important therapeutic target in neurodegeneration. Inhibition of its aggregation or phosphorylation states, immunization or attenuation of its secretion and propagation offer additional therapeutic possibilities (Vekrellis and Stefanis, 2012). HD is a result of mutation in the IT-15 gene that expands abnormally the number of CAG nucleotide repeats causing expansions of glutamines (polyQ) (Arrasate and Finkbeiner, 2012). A triplet nucleotide repeat, cytosine-adenonsine-guanine, is found near the 5′ end of the gene's coding region and is translated into a polyQ stretch (Finkbeiner, 2011). N-terminal huntingtin peptides (Htt) with poly(Q) tracts in the pathological range (51–122 glutamines), form high molecular weight protein aggregates or misfolds with a fibrillar or ribbon-like structure suggesting amyloid fibrillogenesis in HD (Scherzinger et al., 1999). TATA-binding protein is also a potential candidate for interaction with huntington because of long polyglutamine stretch, ubiquitous expression, and nuclear localisation (Huang et al., 1998; van Roon-Mom et al., 2002).

8.3 CROSSING THE BBB FOR DELIVERY OF DRUGS TO CNS

Nature has bestowed multiple layers of barriers to protect brain from pathogens, toxins, and circulating blood cells. Major barriers include: blood cerebrospinal fluid barrier (BCSFB), the blood–brain barrier (BBB), the ependymal barrier, the blood retinal barrier, and the blood spinal cord barrier (Sun et al., 2017). BBB restrains the movement of molecules to brain parenchyma, while BCSFB restricts free diffusion of molecules from circulating blood to the cerebrospinal fluid (CSF) and ependymal barrier controls diffusion of molecules/ions from the CSF to the brain. Consequently, these three barriers collectively restrict the movement of drugs/molecules from blood and CSF to the brain.

Amongst all the barriers, the most prominent barricade is of BBB. BBB forms the periphery amidst neural tissues and circulating blood; thus controlling the molecular movement as well as maintaining brain homeostasis (Xin et al., 2018). BBB is a highly selective barrier with distinct endothelial cells connected via tight junctions and adherent junctions (Saraiva et al., 2016). The electrical resistance at tight junctions is reported to be very high (more than 1000 Ω/cm^2) physiologically (Butt et al., 1990). BBB is supported by neurovascular unit, constituted by astrocytes, neuronal endings, microglia, pericytes, and circulating immune cells. Barrier function of BBB is further armoured by efflux mechanisms of the brain endothelial cells, such as ATP-binding cassette (ABC) efflux pumps, that includes P-glycoprotein, multidrug resistance proteins 4, and so forth, which expels the xenobiotics from reaching to brain (Santander-Ortega et al., 2017; Xin et al., 2018). As a result of these barriers, tight junctions along with efflux pumps, more than 99 % of the molecules including drugs are not able to cross the BBB and reach brain. Nevertheless, hydrophobic, neutral molecules with a molecular weight of about 400–600 Da are considered to be potential candidate to cross BBB via passive diffusion, although such candidates can still be effluxed via endothelial efflux pumps (Patel et al., 2018; Santander-Ortega et al., 2017). Broadly, BBB transportation occurs via below mentioned mechanisms (Gothwal et al., 2018).

- Transcellular lipophilic route
- Paracellular aqueous route
- Efflux pump
- Receptor-mediated transcytosis
- Adsorptive transcytosis
- Lipid-based transport of drugs.

Lipophilic molecules cross BBB via transcelluar route while the proteins, peptides, and polysaccharides cross BBB via receptor-mediated transcytosis and through paracellular route. Remarkably receptor-mediated transcytosis, predominantly bi-directional, is also involved in the transportation of nanoparticles as well as biopharmaceuticals (De Boer and Gaillard, 2007). On the other hand, efflux transporters, namely, ABC transporter family, multidrug resistance proteins and brain multidrug resistance proteins, regulate the homeostasis of excitatory transmitters, like glutamate and entry of drugs to the brain. This in turn becomes a major limitation for the drug delivery to the brain for treating central nervous system (CNS) aliments (Löscher and Potschka, 2005). CNS albumin levels are controlled by adsorptive transcytosis depending upon the disease condition (Joó, 1996). Thus, a strong cohesive endothelial system protects the brain allowing a selective access to certain requisite molecules only.

8.4 NUTRACEUTICALS AS NEURO-PROTECTIVES—A MECHANISTIC INSIGHT

Complex and multifactorial mechanisms underlying the neurodegeneration suggest effective use of multifaceted multipathway intervention of nutraceuticals. To re-establish the brain physiology in neurodegenerative disorders, the basal ROS homeostasis needs to be re-established. Most nutraceuticals have a well-established antioxidant property and hence seems to be a promising resolution for the treatment of neurodegenerative disorders (Saha et al., 2017). Moreover, majority of the drugs used for the treatment of neurodegenerative disorders are administered chronically; hence the side effects of these drugs are a major concern. Nutraceuticals have been found to be effective in improving CNS disease conditions like, AD, PD, HD, dementia, schizophrenia, and so forth (Mao et al., 2018). Use of naturally occurring molecules, thus has become an attractive option for the treatment of neurodegenerative disorders (Figure 8.2, Choudhary et al., 2013; Dey et al., 2017; Hang et al., 2016).

Apart from oxidative stress, another vital suspect for neurodegenerative disorders is neuroinflammation. It has been implicated that microglial activation results in release of pro-inflammatory factors leading to neuroinflammation (Baune, 2015; Skaper et al., 2014). Hence, the nutraceuticals such as soy isoflavones, genistein, and so forth (Chinta et al., 2013; Jantaratnotai et al., 2013) that reduce the release of pro-inflammatory cytokines can be of therapeutic relevance. Oxidative stress-mediated neurodegeneration and neuroinflammation are the major pathways through which nutraceuticals can exert their neuroprotective action. However, nutraceuticals represent a large class with varied compounds, having varied chemical structure and a variety of mechanism of action/targets, which are still being elucidated. One such pathway, which is under investigation, is the involvement of sirtuins, (SIRT1-T7 proteins). Neuroprotective nutraceuticals, polyphenols, and resveratrol have been found to activate SIRT1 leading to inhibition of amyloid-ß protein and suppression of apoptosis (Kelsey et al., 2010). SIRT1 is a nuclear protein that deacetylase histones and protect cells from apoptosis, SIRT2 is a cytoplasmic protein and accumulates in neurons and rest of the SIRT proteins localize in mitochondria (Dang, 2014).

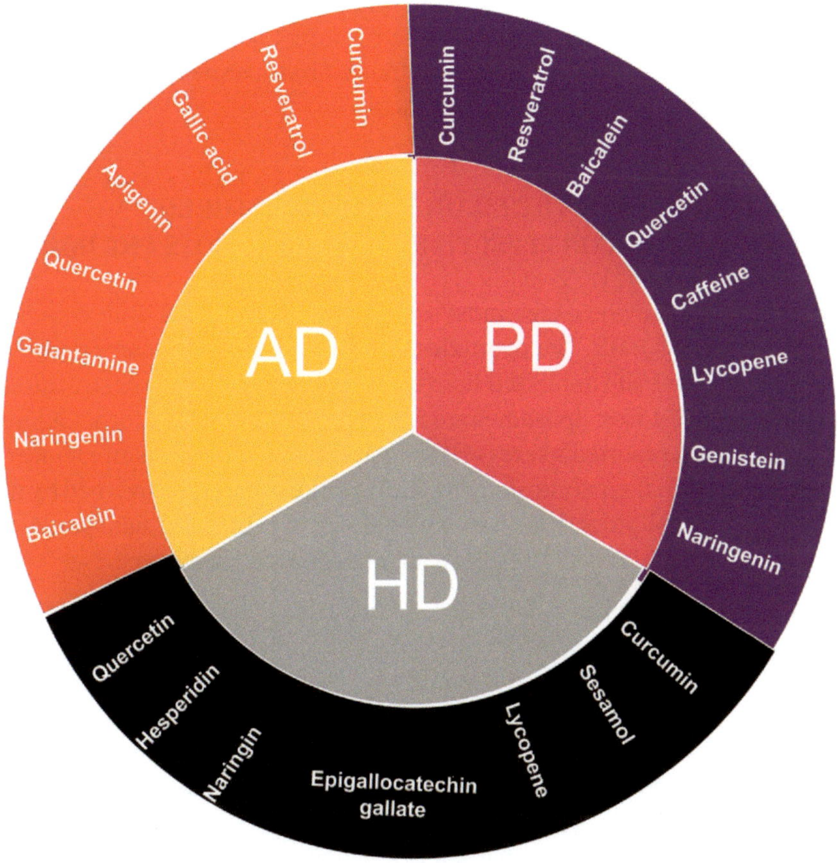

FIGURE 8.2 Neuroprotective phytochemical nutraceuticals.

Autophagy is a key cellular mechanism to eliminate undesirable/damaged cellular components via lysosomes and exert neuroprotective effect (Giordano et al., 2014; Pan et al., 2009). Normal functioning of CNS requires a regular removal of damaged mitochondria, misfolded proteins, and other cellular components of CNS. Thus, a basal autophagic activity is essential for the normal brain physiology. Consequently, deletion of the key autophagy genes (Atg5, Atg7) leads to the accumulation of damaged organelles and misfolded proteins in neuron; thus leading to neurodegenerative disorders (Komatsu et al., 2006; Mariño et al., 2011). Various nutraceuticals such as resveratrol and carnosine have been found to be stimulators of autophagy and are being investigated for their neuroprotective activity (Lin et al., 2014; Baek et al., 2014). A variety of nutraceuticals are being investigated for their neuroprotective activity at various stages of neurodegenerative disorders. However, metabolism, insufficient bioavailability, and poor blood brain permeability limits the neuroprotective potential of nutraceuticals.

8.5 GETTING INTO THE BRAIN-NANOSTRUCTURED INTERVENTIONS FOR EFFECTIVE DELIVERY ACROSS THE BBB

Nanoparticles are one of the most sort after and successful drug delivery strategy to cross BBB. Nanoparticles are typically colloidal carriers in the size range of 1–1000 nm. Solid lipid nanoparticles (SLNs), nanostructured lipid carriers (NLCs), liposomes are few of the nanovesicular formulations that have been reported to cross BBB in animals. The success of nanocarriers depends on their ability to cross BBB, which is largely governed by its physicochemical properties (Saraiva et al., 2016). Broadly, nanoparticles can cross BBB by any of the following transport mechanisms:

1. Nanoparticulate systems can stimulate the enhanced localized BBB permeability by inducing local toxic effect or open up the tight junctions of BBB endothelial cells. Thus allowing drug permeation in free form or drug conjugated with nanoparticulate system (Choi et al., 2010; Gao et al., 2014).
2. Nanoparticulate system cross BBB endothelial cells via transcytosis (Wiley et al., 2013).
3. Endocytosis of nanoparticleis another transportation approach to cross BBB (Kong et al., 2012).

4. Surfactants used in nanoparticulate formulation can cause membrane fluidization (Sekerdag, 2017).
5. P-glycoprotein efflux pump inhibition (Sekerdag, 2017).
6. A combination of above mechanisms.

Transcytosis and/or endocytosis have been reported to be the most common mode of transportation for nanoparticle (Saraiva et al., 2016). Size, shape, and charge of nanoparticles are few of the physic-chemical parameters that influence the ability and mechanism of crossing BBB. An inverse correlation between size and BBB permeation has been reported (Etame et al., 2011; Hanada et al., 2014; Sonavane et al., 2008). Studies indicate that nanoparticle with moderate zeta potential, between −1 and −15 mV (Bramini et al., 2014; Choi et al., 2010; Wiley et al., 2013), or high negative zeta potential, between −15 and −45 mV (Decuzzi et al., 2010; Kreuter et al., 2010), have higher probability to cross BBB. Although, nanoparticle with high positive zeta potential (15–45 mV) have also been reported to cross BBB in few studies (Gao et al., 2014; Jallouli et al., 2007). Spherical nanoparticles are the most common form of the nanoparticles; although cubic, rod-like, cochlear-shaped nanoformulations have also been explored. Short nanorods have been reported to be preferentially retained by liver and undergo faster clearance while long nanorods are taken up by spleen and exhibit lower clearance rate (Huang et al., 2011).

8.5.1 LIPOSOMES

Liposomes are the classical nanotransport vehicles used in brain delivery. Liposomes are basically vesicular structures made up of phospholipid and cholesterol; wherein hydrophobic groups of amphiphilic molecules assemble together to create a hydrophobic space and a hydrophilic core (Dua et al., 2012). Liposomes can thus, encapsulate both hydrophilic and hydrophobic drugs. Liposomes are simple to prepare, exhibit improved bioavailability, biodegradable, and biocompatible and have low toxicity. Phospholipid bilayers of liposomes enable them to have relatively easy penetrability in BBB. However, they are taken up by reticulo-endothelial system (RES), thus restricting their duration of circulation. Nevertheless, surface modification with polyethylene glycol (PEG) has been employed to prevent their uptake by RES and thus enhancing their duration of circulation (Xin et al., 2018). Surface modifications/functionalization of liposomes with antibodies or specific ligands, namely, magnetic targets, pH-sensitive targets, temperature-sensitive targets have been carried out to improve the targeting ability of liposomes (Vogg et al., 2016; Xin et al., 2018).

8.5.2 LIPIDIC NANOPARTICLES

Colloidal nanoparticles with lipid matrix and having size in the range of 10–1000 nm are known as lipidic nanoparticles. Based on their lipid composition, these lipidic particles can be further classified as SLNs, NLCs, and lipid drug conjugates (Patel et al., 2018).

SLNs are the submicron-sized lipidic particles, composed of physiological lipids, wherein a part of liquid lipids has been substituted with solid lipids and stabilized using a surfactant (Müller et al., 2002). Structurally SLNs can have either drug-enriched shell or drug-enriched core or may be homogeneous matrix. SLNs have excellent physical stability, protection of the entrapped drug, drug release modulation, site-specific targeting ability (Wissing et al., 2004). However, the drug loading in SLNs is limited and is governed by the solubility of drug in the lipid matrix. Also, drug expulsion from lipid matrix due to polymorphic changes has been observed in SLNs (Schwarz et al., 1994; Uner and Yener, 2007).

The type of lipids, surfactants, particle size, and inner structure determine the drug release pattern from SLNs (Cavalli et al., 2000; Fundarò et al., 2000). Size of SLNs generally ranges from 120 to 200 nm, as a result they are not cleared by RES and by pass the liver and spleen filtration (Chen et al., 2004; Saupe et al., 2006). Both hydrophilic as well as hydrophobic drug can be incorporated in SLNs and are considered to be safe as the carrier lipids are biodegradable (Patel et al., 2018). Notably, SLNs have been reported to have higher brain-blood capillary retention ensuing high drug concentration gradient across BBB (Kaur et al., 2008).

To improve upon the drug payload and increase drug retention in lipid matrix, new generation SLNs, NLCs were developed. A part of solid lipid is replaced by liquid lipid in NLCs; thus creating a less ordered lipid matrix structure resulting in higher drug payloads and reduced drug leaching from the lipid matrix on storage (Tapeinos et al., 2017). Solid-to-liquid-lipid ratio in NLCs can vary from 70:30 to 99.9:0.1 (Weber et al., 2014). NLCs can be further categorized (based on lipid and oil composition) as (Khosa et al., 2018):

- The imperfect type
- The amorphous type
- Multiple oil-in-solid fat-in-water (O/F/W) type

While, based on surface functionality, NLCs can be classified as (Patil and Deshpande, 2018):

- Plain NLCs,
- Multifunctional NLCs,
- PEGylated NLCs,
- Ligand-attached NLCs.

NLCs have a higher drug payload, longer storage stability, superior biocompatibility, and improved bioavailability (Patil et al., 2018).

8.5.3 POLYMERIC MICELLES

Micellar systems are the self-assembled nanoscale (diameter $\leq$ 100 nm) assemblies of amphiphilic polymers in an aqueous phase. Attractive forces between hydrophobic and neutral part of copolymer aid in micellization above critical micelle concentration. These micelles have a hydrophobic core, which encapsulate hydrophobic drug, and a hydrophilic outer shell (Gothwal et al., 2018). Most widely used biodegradable and biocompatible hydrophobic cores for the preparation of micellar systems are as follows (Gaucher et al., 2005; Gothwal et al., 2018):

- Polyester: poly(lactic acid) (PLA), poly(ε-caprolactone) (PCL), poly(lactic-*co*-glycolic acid) (PLGA), poly(*N*-butylcyanoacrylate) (PBCA)
- Poly(L-amino acid): Poly(aspartic acid) (PAsp), poly(glutamic acid) (PGlu)
- Phospholipids: Phosphatidylethanolamine (PE), 1-palmitoyl-2-oleoyl-phosphatidylcholine (POPC), 1,2-distearoyl-sn-glycerol-3-phosphatidylcholine (DSPC).

While, commonly used hydrophilic blocks are polyethylene oxide (PEO) and PEG. PEG is a USFDA approved biocompatible, biodegradable, nontoxic polymer with stealth capability and large number of available hydroxyl groups. PEG also helps to reduce intermicellar aggregation protein adsorption and cell adhesion (Wei et al., 2009; Yao et al., 2012).

8.6 NANONUTRACEUTICALS DELIVERY IN NEURODEGENERATIVE DISEASES

The enormous diversity in mechanism of action of herbal drugs and the nutraceuticals isolated from natural source are the new generation therapeutics for the management of age-related incapacitating neurodegeneration. Being safer than synthetic chemicals, phytochemicals are extensively explored in management of chronic diseases. Explosive growth of interest in curcumin and resveratrol as observed from extensive literature reported, we have reviewed the nanoformulations of these phytoconstituents for enhanced CNS delivery and are illustrated here.

8.6.1 NANOCURCUMIN IN NEURODEGENERATION

Curcumin and analogous curcuminoids (demethoxycurcumin and bisdemethoxycurcumin) found in southeast Asian curry spice turmeric are the most prominent dietary polyphenols that have been explored for therapeutic benefits in neurodegenerative illnesses (Darvesh et al., 2012; Maiti and Dunbar, 2018; Serafini et al., 2017). Neurodegenerative diseases of aging involve the accumulation of oxidative damage, inflammation, and protein aggregates. Curcumin exhibits neuroprotective antioxidant and anti-inflammatory activities by acting on large number of diverse signaling pathways and molecular targets like enzymes, transcription factors, and growth factors and also by chelation of iron (Hamaguchi et al., 2010; Hu et al., 2015; Min et al., 2018; Monroy et al., 2013).

Mitochondrial dysfunction and oxidative stress play an important role in the progression of neurodegenerative diseases. Mitochondria are a major source of intracellular ROS and are particularly vulnerable to oxidative stress. Oxidative damage to mitochondria has been shown to impair mitochondrial function and lead to cell death via apoptosis and necrosis (Szeto, 2006).

These diseases are likely caused by low grade inflammation driven by oxidative stress as indicated by the increase of pro-inflammatory cytokines such as IL-6, IL-1, and TNF-α, genes encoded by activation of nuclear factor kappa-B (NF-κB) (He et al., 2015).

AD is characterized by the abundance of extracellular abnormal aggregation and deposition of neurotoxic oligomer amyloid β peptide, and intracellular accumulation of neurofibrillary tangles of hyperphosphorylated τ protein and dementia (Huang et al., 2016; Selkoe, 2001). These phenomena are triggered by oxidative stress, involving increased free radicals or a decrease in antioxidant defence, increased levels of ROS and reactive nitrogen species that further react with lipids, proteins, nucleic acids, and other molecules and may also alter their structures and functions (Dumont and Beal, 2011; Liu et al., 2017; Manoharan et al., 2016). Neuron cells are susceptible to oxidative damage due to their high polyunsaturated fatty acid content in membranes, high oxygen consumption, and weak antioxidant defence (Cole et al., 2007). Curcumin is a diferuloyl methane molecule [1,7-bis (4-hydroxy-3-methoxyphenyl)-1,6-heptadiene-3,5-dione)] containing two ferulic acid residues joined by a methylene bridge (Priyadarsini, 2013).

Selective loss of the dopaminergic neurons in the substantia nigra pars compacta and the degeneration of projecting nerve fibers in the striatum in PD is also triggered by oxidative stress (Hwang, 2013). Metabolism of dopamine generates oxidative stress. In addition, during neuroinflammation, activated microglia produce nitric oxide and superoxide responses, that are aggravated by the molecules released by damaged dopaminergic neurons such as α-synuclein, neuromelanin, and matrix metalloproteinase-3. Parkinson's disease causing gene products including DJ-1, PINK1, parkin, alpha-synuclein, and LRRK2 also influence mitochondrial function leading to generation of oxidative stress (Blesa et al., 2015; Dias et al., 2013). Oxidative stress is also a primary event in HD neuropathology (Browne et al., 1999; Kumar and Ratan, 2016).

The presence of conjugated double bonds, multiple hydroxyl groups, and more than one phenolic group are the structural features responsible for the free radical quenching properties of curcumin. Curcumin co-exist with several tautomeric forms, of which two predominant forms are 1,3-diketo form and 1,3-dienol form. The keto form is predominant in acidic to neutral conditions, while the enol form is predominant in alkaline solutions, and is a potent free-radical scavenger (Lee et al., 2013). Curcumin even in low doses directly binds small amyloid β species and blocks the aggregation and fibril formation in vitro and effectively disaggregates amyloid β (Yang et al., 2005).

Lipoprotein carrier-mediated, intracerebroventricular infusion of amyloid β peptides in rats induced oxidative damage, synaptophysin loss, a microglial response and several amyloid β deposits. Curcumin suppressed oxidative damage, synaptophysin loss, reduced microgliosis in cortical layers, prevented spatial memory deficits, and reduced amyloid β deposits (Frautschy et al., 2001; Lim et al., 2001). Curcumin protects against oxidative stress-induced injury in PD rats via the Wnt/β-catenin signaling pathway (Wang et al., 2017). Rajeshwari and Sabesan (2008) showed that systemic administration of curcumin and its metabolite tetrahydrocurcumin significantly reversed the 1-methyl-4-phenyl-1,2,3,6-tetrahydropyridine (MPTP)-induced depletion of dopamine and 3,4-dihydroxy phenyl acetic acid and inhibited the MAO-B activity.

Cytokine suppressive anti-inflammatory curcumin has promising therapeutic options to reduce neuroinflammation and improve the clinical outcome in neurodegenerative diseases. The most promising action of curcumin is modulation of the activity of several transcription factors like STAT, NF-κB, AP-1, and their pro-inflammatory molecular signaling pathways (Ullah et al., 2017). However, normal curcumin preparations demonstrate low bioavailability in vivo thus requiring higher doses (Hu et al., 2015). Poor aqueous solubility at neutral or acid pH, instability at basic pH and rapid intestinal and first pass glucuronidation are the major concerns for the slow clinical development of novel curcumin formulations (Hu et al., 2015; Ji and Shen, 2014). Curcumin is metabolized by phase I and II enzymes resulting into conjugation with glucuronic acid and sulfate in enterocytes and hepatocytes. Low bioavailability is also due to active efflux of these conjugates out of the enterocyte through the multidrug resistance related proteins (Peng and Quian, 2014). To improve the bioavailability and brain targeting potential for the treatment of neurodegenerative diseases, various nanoscale preparations of curcumin have been introduced to achieve therapeutically relevant concentrations in target tissues (Table 8.2).

8.6.2 NANORESVERATROL IN NEURODEGENERATION

Resveratrol (3,5,4′-trihydroxy-trans-stilbene) is a nonflavanoid, natural phenolic stilbenoid, largely found in pistachios, peanuts, and skin of red grapes, mulberries, and blueberries (Amri et al., 2012; Pund et al., 2016). Resveratrol was isolated from the roots of plants *Polygonum cuspidatum* and *Veratrum grandiflorum*, used in traditional Chinese and Japanese medicine and is intensively investigated in recent years for its health-beneficial properties and for potential applications in the fields of pharmaceutics,

TABLE 8.2 A Glimpse at Various Nanoformulations of Curcumin Explored for Improved Therapy of Neurodegenerative Diseases

Formulation	Method of Preparation	Composition	Outcome Measures	Reference
Liposomes	Thin film hydration and sonication	Novel curcumin-lipid derivative; DPS-PEG2000-CURC. Conjugation with a BBB transport mediator (antitransferin antibody)	• Higher affinity for the amyloid deposits was noted on post-mortem brains samples of AD patients • Developed liposomes delayed Aβ1–42 peptide aggregation. • Significantly increased intake by the BBB cellular model	Mourtas et al., 2014
		DSPC, cholesterol, PEG-lipid, PEG-MAL, PEG-N3. Conjugation with thiolated-Mab with ApoE	Specific curcumin-lipid derivative as a component of multifunctional liposomes improved brain targeting capability and inhibited Aβ-peptide aggregation in vitro as well as in vivo in double transgenic double transgenic mice (APP/PS1) and WT mice.	Papadia et al., 2017a, 2017b
		DPPC, DPPG, cholesterol and pegylated curcumin lipid derivative	Nanoliposomes with curcumin, or the curcumin derivative inhibited the formation of fibrillar and/or oligomeric Aβ in vitro.	Taylor et al., 2011.
		DPPC, cholesterol, and DPS-curcumin	• Curcumin-conjugated nanoliposomes were nontoxic in vitro, downregulated the secretion of amyloid peptide and prevented Aβ-induced toxicity. • Aβ deposits in post-mortem brain tissue of AD patients and APPxPS1 mice got strongly labeled. • Injection of curcumin conjugated nanoliposomes in the hippocampus and in the neocortex of mice specifically stained the Aβ deposits in vivo	Lazar et al., 2013.
		DSPC, cholesterol, DSPE-PEG2000, DSPE-PEG2000-Biotin, anti-Ab monoclonal antibody	Aβ-MAb-liposomes were found to bind with $A\beta_{1-42}$ peptides in vitro and amyloid deposits in postmortem AD brain samples	Canovi et al., 2011

TABLE 8.2 *(Continued)*

Formulation	Method of Preparation	Composition	Outcome Measures	Reference
		Phospholipid conjugate of curcumin, 1,2-dipalmitoyl-3-(2- (1,7-bis (4-hydroxy-3-methoxyphenyl)-3,5-dioxohept-6-enylthio)ethyl phospho)-*sn*-glycerol	• Curcumin decorated nanosized liposomes demonstrated high integrity during incubation in the presence of plasma proteins • Surface plasmon resonance study, measuring the binding of flowing liposomes to immobilized $A\beta_{1-42}$, indicated extremely high affinity for $A\beta_{1-42}$ fibrils.	Mourtas et al., 2011.
	Thin film hydration in combination with sol–gel reaction and self-assembly process	Cerasome-forming lipid; *N*-[*N*-(3-triethoxysilyl) propylsuccinamoyl]-di-hexadecylamine, Tween 80 For microbubbles—DSPC, DSPE-PEG2000	• Enhanced penetration of curcumin-loaded cerasomes was observed across the BBB by transcytosis through vascular endothelial cells • Additional use of ultrasound-targeted microbubble destruction, allowed extravasation of nanoparticles into the brain from the systemic circulation, resulting in an increased uptake of curcumin in the treated brain side • MPTP-induced PD mice showed improved behavior disorder and dopamine depletion after treatment with cerasomes coupled with ultrasound-targeted microbubble destruction	Zhang et al., 2018
Polymeric micelles	Anionic polymerization	PBCA, tween 80	• Higher bioavailability of curcumin in plasma • The mean residence time of curcumin loaded nanoparticles was 14-fold that of the control solution • In brain, AUC0-∞ for curcumin loaded nanoparticles was 2.53-fold that for the control solution	Sun et al., 2010.
		PBCA Poloxamer 188. Conjugation with Apolipoprotein E3	• Significantly increased photostability of curcumin • Increased cellular uptake of ApoE3 conjugated nanoparticles in SH-SY5Y human neuroblastoma cells and decreased β amyloid induced oxidative stress in in vitro cell culture model	Mulik et al., 2010

TABLE 8.2 *(Continued)*

Formulation	Method of Preparation	Composition	Outcome Measures	Reference
	Solvent evaporation and lyophilization	PLGA-block-poly(ethylene glycol) (PLGA-PEG). Conjugation with B6 peptide	• In vitro assays indicated increased cellular uptake and good blood compatibility • Morris water maze test revealed tremendous improvement in spatial learning and memory capability of APP/PS1 mice	Fan et al., 2018
		PLGA, PVA, conjugated with Tet-1 peptide	• Nontoxic nanoparticles destroyed amyloid aggregates and exhibited antioxidant properties • Tagging the PLGA curcumin nanoparticle with Tet-1 neuropeptide increased in vitro neuronal targeting efficiency	Mathew et al., 2012a
		PLGA, PVA, conjugated with NN2 amyloid-binding aptamer	NN2 amyloid binding aptamer conjugated curcumin—PLGA nanoparticles interacted with and degraded the amyloid proteins	Mathew et al., 2012b
	Emulsification–diffusion-dialysis and cross-linking	Polyacrylamide (PAAM)-cardiolipin (CL)-PLGA, DSPE-PEG(2000)-CA, tween 80. Surface grafting of 83-14 monoclonal antibody.	• Nanocarriers increased permeability across the BBB • Enhanced the viability of SK-N-MC cells insulted with Aβ deposits • An increase in the concentration of 83-14 MAb enhanced the permeability coefficient of curcumin and rosmarinic acid	Kuo and Tsai, 2018.
	Emulsion–solvent evaporation, sonication	PLGA, Aβ generation inhibitor S1 (PQVGHL peptide), conjugation with brain targeting peptide CRT (cyclic CRTIGPSVC)	• Y-maze and new object recognition test demonstrated significantly improved spatial memory and recognition in transgenic AD mice • Nanoparticles remarkably decreased the level of Aβ, ROS, TNF-α, and IL-6, and enhanced the activities of super oxide dismutase and synapse numbers in mouse brains	Huang et al., 2017.

TABLE 8.2 *(Continued)*

Formulation	Method of Preparation	Composition	Outcome Measures	Reference
	Emulsion–diffusion–evaporation	PLGA 50:50 and 65:35	• Curcumin-loaded nanoparticles prolonged and enhanced the antioxidant, anti-inflammatory activity of curcumin while suppressing the Akt and Tau phosphorylation and gene upregulation under oxidative stress conditions • Ratio of lactic/glycolic acid in PLGA directly affected the physicochemical properties like entrapment efficiency and the in vitro drug release of the nanoparticles, thereby modulated the in vivo properties like counteracting the increase of REST and GLX expressions	Djiokeng Paka et al., 2016.
	Oil-in-water simple emulsion/solvent evaporation	PLGA, PVA, conjugation with Gly-L-PheD-Thr-Gly-L-Phe-L-Leu-LSer(O-β-D-Glucose)-$CONH_2$	• In vitro experiments showed reduced oxidative stress, inflammation and plaque load • No apparent toxicity of nanoparticles in primary hippocampal cell cultures	Barbara et al., 2017.
	Emulsion diffusion evaporation	PLGA, dimethylammonium bromide	• PLGA nanoparticles loaded with curcumin were nontoxic to human neuroblastoma SK-N-SH cells, protected cells against H_2O_2 and prevented the elevation of ROS and the consumption of glutathione • Formulation also prevented the induction of the redox-sensitive transcription factor Nrf2 in the presence of H_2O_2	Doggui et al., 2012.
	Nanoprecipitation	PLGA, PLGA-PEG2000, PLGA-PEG-Mal. Functionalization with glutathione	• Glutathione functionalization increased the uptake of formulations by the neuronal cells • Route of internalization was modified to escape the uptake by macropinocytosis and avoided lysosomal degradation	Paka and Ramassamy, 2017.

TABLE 8.2 *(Continued)*

Formulation	Method of Preparation	Composition	Outcome Measures	Reference
		PEG-PLA co-block polymer, PVP. Freeze drying with β-cyclodextrin	• Oral administration to Tg2576 AD model mice, nanocurcumin produced significantly better cue memory in the contextual fear conditioning test and tendencies toward better working memory in the radial arm maze test • Curcumin bioavailability was improved with greater plasma concentration and six-fold higher AUC and MRT in brain	Cheng et al., 2013
SLN	SLN-O/W warm microemulsion method Liosomes—thin film hydration	SLNs—stearic acid, phospholipon 90G, sodium taurocholate Liposomes—POPC, PE. sphingomyelin, cholesterol, diphosphatidyl glycerol	In vitro experiments showed high binding affinity of liposomes as well as SLNs for aggregated forms of $A\beta_{1-42}$ (fibrils and oligomers)	Gobbi et al., 2010.
	Marketed nanocurcumin	Longvida®, Verdure Sciences	• Curcumin and nanocurcumin demonstrated as an alternative to Aβ-specific antibody for labeling and imaging of Aβ plaques *ex vivo* and in vivo • Significant reduction in GFAP-IR and Iba-1-IR in PFC, in the striatum as well as CA1, CA3, DG, subicular complex of hippocampus, and the entorhinal cortex in the 5xFAD mice • Observed decrease of pyknotic, or tangle like, neurons in PFC, CA1, and CA3 areas of hippocampus	Maiti et al., 2016; 2018.
Cocrystal micelles	High shear mixing in planetary ball mill	Curcumin and antioxidant hydrophilic conformer A	• Successful development of curcumin cocrystal micelles using green scalable planetary ball milling • Superior antioxidant performance with 1.7-fold increase in the bioavailability with delayed clearance. • Significantly high brain distribution even beyond 6 h of dosing	Desai and Patravale, 2018

nutraceuticals, and functional foods (Neves et al., 2016a; Sessa et al., 2011). Several beneficial effects of resveratrol like neuroprotective, cancer-preventive, anti-inflammatory, antiobesity, and positive cardiovascular effects were traditionally thought to exert by antioxidative properties (Jardim et al., 2018; Neves et al., 2012; Pund et al., 2014). The neuroprotective effects of resveratrol in neurological diseases are related to the protection of neurons against oxidative damage and toxicity, and prevention of apoptotic neuronal death. Several intracellular biochemical pathways activated are vital for antioxidant defence, regulation of the cell cycle, mitochondrial energy production, vascular tone, oncogene suppression, and prevention of many other events leading to progression of age-related neurodegenerative diseases (Markus and Morris, 2008). Antioxidant and the free-radical scavenging action of resveratrol is due its ability to transfer hydrogen atoms or electrons to the free radicals, because of the highly reactive hydroxyl group at position 4 and the two phenol groups with conjugated double bond that easily delocalize of electrons (Stivala et al., 2001).

Brain cells, both neuronal as well as glial cells, are susceptible to mitochondrial dysfunction due to their high demand for adenosine triphosphate. Mitochondria are the highly dynamic cell organelles involved in the synthesis of adenosine triphosphate. Brain cells utilize high levels of oxygen, leading to a proportionally high mitochondrial production of reactive species (Jardim et al., 2018). The maintenance of the mitochondrial function is important for maintaining the bioenergetic requirements of cells. Impaired mitochondrial function during aging causes cellular energetic imbalance, over production of ROS and reactive nitrogen species, impaired synthesis of biomolecules like lipids, proteins and nucleic acids, destruction of genetic material like DNA and RNA and also changes cellular osmotic pressure leading to cell death (Islam, 2017).

Resveratrol interacts with a significant number of molecular targets with several different mechanisms of therapeutic actions. However, its antioxidant activity is the main responsible factor for positive effect on the human health. As oxidative stress is the major cause of neuronal pathologies like AD, HD, cerebral ischemia, PD, and ALS, treatment with resveratrol has shown beneficial effects in these neurodegenerative disorders (Tellone et al., 2015). Resveratrol triggers mitochondrial biogenesis, improving the mitochondria-related bioenergetics status in mammalian cells (Wang et al., 2014). Caloric restriction prevents diseases of aging, including AD, perhaps by activation of sirtuins. The sirtuins (e.g., mammalian SIRT1) are NAD-dependent deacetylases that regulate energy balance (NAD+/NADH) and thus gene transcription, metabolism and cell survival (Ho et al., 2010). Resveratrol is a potent activator of SIRT1 (silent mating type information regulation 2, *Saccharomyces cerevisiae*, homolog 1), and thus may mimic caloric restriction that extends longevity and increases neuronal survival and prevent diseases of aging (Anekonda, 2006; Sawda et al., 2017, Ungvari et al., 2011). Resveratrol indirectly inhibits nuclear factor κB (NF-κB) signaling by activating SIRT-1, and protect against microglia-dependent β-amyloid toxicity in a model of AD (Chen et al., 2005). It also protects neuronal cells from toxicity of β-amyloid through activation of SIRT1/Akt1 pathway (Zhang et al., 2014), through down-regulation of Rho-associated kinase 1 (ROCK1) (Feng et al., 2013) and by promoting the expression auto-poly-ADP-ribosylation of poly(ADP-ribose) polymerase 1, as well as tyrosyl transfer-RNA synthetase (Deng and Mi, 2016). Peroxisome proliferator-activated receptor gamma co-activator-1alpha (PGC-1alpha), an important substrate of SIRT1 is a principal regulator of energy metabolism, whose function is significantly impaired in HD (Ferretta et al., 2014; Pasinetti et al., 2011). Administration of resveratrol increased PGC-1α mRNA levels, reduced vacuolation in the brown adipose tissue and decreased elevated blood glucose levels in N171-82Q transgenic mice thus protecting peripheral tissues (Ho et al., 2010). Resveratrol protects dopaminergic neurons expressing polyQ-huntingtin from dopamine toxicity by preventing dopamine-induced ROS formation in dopaminergic neurons and restoring the levels of ATG4 and of LC3-II in dopamine-treated cells; thus facilitating the autophagy degradation of polyQ-huntingtin (Vidoni et al., 2018).

Gaballah et al. (2016) showed potential use of resveratrol as a neuroprotective therapeutic agent in in a rat model of rotenone-induced PD. Resveratrol improved endoplasmic reticulum stress by downregulating C/EBP homologous protein and glucose-regulated protein 78 genes expression and hampered caspase-3 activity in the brain. It also activated glutathione peroxidase and Nrf2 signaling pathway and suppressed xanthine oxidase activity and protein carbonyls formation; thus restoring redox balance. Resveratrol induces heme oxygenase-1-dependent autophagy and prevents dopaminergic cell death by regulating autophagic flux (Lin et al., 2014).

Despite potential pharmacological activities, the clinical translation of resveratrol appear strongly limited due to poor bioavailability (Amri et al., 2012; Augustin et al., 2013; Pasinetti et al., 2015; Pund et al.,

2016; Trotta et al., 2018). The poor in vivo efficacy is attributed to its poor physicochemical characteristics, namely, aqueous solubility and stability. Aqueous solubility of resveratrol is ~3 mg/100 mL; however exhibits more solubility in DMSO ~160 mg/L, in ethanol 500 mg/L, and in coconut oil 1.8 mg/L (Amri et al., 2012; Delmas et al., 2011; Hung et al., 2006). Practically water insoluble substance resveratrol, has higher value of log *P* (3.1) responsible for its lipid solubility and permeability across the biological membranes. This classifies resveratrol as BCS class II drug (low solubility, high permeability) (Amri et al., 2012; Seljak et al., 2014). Resveratrol is highly photosensitive, being rapidly converted to cis-resveratrol on exposure to light (Trela and Waterhouse, 1996)

Resveratrol is extensively metabolized into glucuronide and sulfate conjugates, which might retain some biological activity (Cottart et al., 2010, Smoliga and Blanchard, 2014). These sulfate and glucuronide conjugates improve aqueous solubility of resveratrol, allow rapid excretion, and thus reduces the absorption and interaction with essential macromolecules (Sergides et al., 2016; Walle et al., 2004). A number of nanoscale formulation strategies have been experimented to improve the bioavailability (Jeong et al., 2016; Neves et al., 2013; 2016a; Ramlingam and Ko, 2016; Vanaja et al., 2013) and brain targeting potential of resveratrol for effective management of neurodegenerative diseases (Table 8.3).

8.7 CONCLUDING REMARKS

Neurodegenerative amyloid diseases and cognitive decline are becoming more prevalent as a result of increase in life expectancy. The use of nutraceuticals for maintaining a healthy life is becoming more popular in scientific research as well as in society. However, limited bioavailability and extensive postingestion biotransformations are the major drawbacks for successful clinical translation. These limitations are further compounded by the physiological BBB. Poor BBB permeability and efflux of drugs from brain to blood are the two major reasons for failure of drug delivery to brain. These shortcomings of nutraceuticals, however, are potentially addressed using highly flexible nanovehicles in the treatment of neurodegenerative disorders. Owing to their small size, balanced hydrophilic–lipophilic character, and/or use of BBB transport mediators, nanovehicles are able to cross BBB. Additionally, nanoencapsulated drug molecule is protected from enzymatic as well as chemical degradation and solubility is

TABLE 8.3 Various Nanostructured Interventions of Resveratrol for Improved Efficacy in Neurodegenerative Diseases

Formulation	Method of Preparation	Formulation Composition	Clinical Activity or Biological Response	Reference
SLN	Melt homogenization and microemulsion dilution	Stearic acid, lecithin and taurocholate	• SLN showed 4.5 times higher levels of resveratrol in brain, improved cognitive decline observed in permanent bilateral common carotid artery occlusion model of vascular dementia, reduced mitochondrial ROS generation, lipid peroxidation, and protein carbonyls • Significantly improved redox ratio and Mn superoxide dismutase activity and reduced hypoxia-inducible factor 1a levels • Increased levels of Nrf-2 and heme oxygenase 1	Yadav et al., 2018
	Emulsion solvent evaporation	Compritol 888 ATO, Tween 80 and polyvinyl alcohol	• SLNs showed 6 fold higher uptake of resveratrol in brain • Low accumulation in liver and kidneys	Jose et al., 2014
	Melt homogenization and sonication	Cetyl palmitate and tween 80. Functionalization with anti-transferrin receptor monoclonal antibody (OX26 mAb)	• Increased transcytosis ability and cellular internalization of functionalized SLNs in human brain-like endothelial cells	Loureiro et al., 2017
	high shear homogenization followed by sonication	Cetyl palmitate and tween 80. Functionalization with apolipoprotein E	• SLNs functionalized with apolipoprotein E showed 1.8-fold higher permeability through hCMEC/D3 monolayers	Neves e al., 2016b
Chitosan-coated lipid particles	Melt oil/water emulsification by phase inversion	Phosphatidyl choline, chitosan	• Significant increased levels of resveratrol were fund in cerebrospinal fluid after nasal administration • No distribution in the systemic circulation	Trotta et al., 2018
Polymeric micelles	Antisolvent precipitation	PCL and PEG	• Biocompatible nanoparticles with controlled release of resveratrol reversed intracellular oxidative stress of Aβ treated PC12 cells • Attenuated ROS production and caspase-3 activation	Lu et al., 2009

TABLE 8.3 *(Continued)*

Formulation	Method of Preparation	Formulation Composition	Clinical Activity or Biological Response	Reference
	Single emulsion solvent evaporation	Tween 80, PLA, PVA	• Resveratrol loaded nanoparticles displayed significant neuroprotection against MPTP-induced behavioural and neurochemical changes in parkinsonian model C57BL/6 mice	da Rocha Lindner et al., 2015
Nano-suspension	Antisolvent precipitation	Nanosuspension dispersed in *in situ* gel of deacetylated gellan gum	• 2.88-times increased bioavailability in the brain by intranasal in situ gel formulation. • 458.2% drug targeting efficiency and 78.78% direct transport demonstrated direct delivery via the nose-brain pathway	Hao et al., 2016
Liposomes	Thin film hydration	Lecithin, cholesterol	• Oral treatment for 14 days with liposomal resveratrol in parkinsonian rats showed significant decrease in the abnormal rotational behaviour, the loss and apoptosis of nigral cells, and the levels of total ROS • The total antioxidant capability of nigral tissues was significantly improved	Wang et al., 2011
Nanoemulsion	Spontaneous emulsification followed by high pressure homogenization	Vitamin E, Sefsol, Tween 80, Transcutol P	• Nanoemulsion provided relatively maximum in vitro as well as significantly high ex vivo trans-nasal mucosal flux • The higher resveratrol concentration in the brain proved the targeting efficiency of nanoemulsion through intranasal route	Pangeni et al., 2014
	Spontaneous emulsification	Labrafac lipophile, labrafac PG, tween 80 cremophor RH 40, haluronic acid	• The mucoadhesive nanoemulsion of curcumin and resveratrol was safe on nasal mucosa and managed to increase the amounts of both about 7- and 9-folds in the brain, respectively • Successful carrier for enhancing the solubility, stability and brain targeting ability of polyphenols	Nasr, 2016

enhanced. Modulation of drug release and circulation time can be tailored for targeting tissue/site specific delivery of drug. Various studies have proven the stability, biocompatibility and efficacy of nanoscaled nutraceuticals in various neurodegenerative disorders. However, neurotoxicity of these formulations is still in its nascent stage of research. Extensive studies are required to establish the safety of nutraceutical loaded nanovehicles. Efficacy by nonparenteral route of administration, clinical efficiency, and potential toxicity of the developed nanoscale formulations of nutraceuticals need to be explored on bigger scale before recommending for treatment of chronic diseases. This will revolutionize the nutraceutical nanodelivery for chronic neurodegenerative disorders.

ACKNOWLEDGMENTS

The authors SP and AJ gratefully acknowledge funding from the Department of Science and Technology (DST), Government of India and Science and Engineering Research Board (SERB, DST, Government of India), respectively, during the completion of the chapter.

KEYWORDS

- **neuroprotective**
- **nanoformulations**
- **Alzheimer's disease**
- **Parkinson's disease**
- **blood-brain barrier**
- **curcumin**
- **resveratrol**

REFERENCES

Amri, A.; Sfar, J.C.S.; Charrueau, C. Administration of resveratrol: what formulation solutions to bioavailability limitations? *J. Control. Release*. **2012**, 158: 182–193.

Anekonda, T.S. Resveratrol - a boon for treating Alzheimer's disease? *Brain Res. Rev.* **2006**, 52: 316–326.

Arrasate, M.; Finkbeiner, S. Protein aggregates in Hutington's disease. *Exp. Neurol.* **2012**, 238: 1–11.

Augustin, M.A.; Sanguansri, L.; Lockett, T. Nano- and micro-encapsulated systems for enhancing the delivery of resveratrol. *Ann. N. Y. Acad. Sci.* **2013**, 1290: 107–112.

Baek, S.H.; Noh, A.R.; Kim, K.A.; Akram, M.; Shin, Y.J.; Kim, E.S.; Yu, S.W.; Majid, A.; Bae, O.N. Modulation of mitochondrial function and autophagy mediates carnosine neuroprotection against ischemic brain damage. *Stroke.* **2014**, 45: 2438–2443.

Baldwin, K.R.; Godena, V.K.; Hewitt, V.L.; Whitworth, A.J. Axonal transport defects are a common phenotype in Drosophila models of ALS. *Hum Mol Genet.* **2016**, 25: 2378–2392.

Barage, S.H.; Sonawane, K.D. Amyloid cascade hypothesis: pathogenesis and therapeutic strategies in Alzheimer's disease. *Neuropeptides.* **2015**, 52: 1–18.

Barbara, R.; Belletti, D.; Pederzoli, F.; Masoni, M.; Keller, J.; Ballestrazzi, A.; Vandelli, M.A.; Tosi, G.; Grabrucker, A.M. Novel curcumin loaded nanoparticles engineered for blood–brain barrier crossing and able to disrupt Abeta aggregates. *Int. J. Pharm.* **2017**, 526: 413–424.

Baune, B.T. Inflammation and neurodegeneration: is there still hope for therapeutic intervention? *Curr. Opin. Psychiatry.* **2015**, 28: 148–154.

Bhasne, K.; Mukhopadhyay, S. Formation of heterotypic amyloids: α-synuclein in co-aggregation. *Proteomics.* **2018**, 14: e1800059.

Blesa, J.; Trigo-Damas, I.; Quiroga-Varela, A.; Jackson-Lewis, V.R. Oxidative stress and Parkinson's disease. *Front. Neuroanat.* **2015**, 9: 91.

Bloom, G.S. Amyloid-β and tau: the trigger and bullet in Alzheimer disease pathogenesis. *JAMA Neurol.* **2014**, 71: 505–508.

Bonet-Costa, V.; Pomatto, L.C.; Davies, K.J. The proteasome and oxidative stress in Alzheimer's disease. *Antioxid. Redox Signal.* **2016,** 25:886–901.

Bramini, M.; Ye, D.; Hallerbach, A.; Nic Raghnaill, M.; Salvati, A.; Aberg, C.; Dawson, K.A. Imaging approach to mechanistic study of nanoparticle interactions with the blood–brain barrier. *ACS Nano.* **2014**, 8: 4304–4312.

Browne, S.E.; Ferrante, R.J.; Beal, M.F. Oxidative stress in Huntington's disease. *Brain Pathol.* **1999**, 9: 147–163.

Butt, A. M.; Jones, H. C.; Abbott, N. J. Electrical resistance across the blood–brain barrier in anaesthetized rats: a developmental study. *J. Physio.* **1990**, 429: 47–62.

Canovi, M.; Markoutsa, E.; Lazar, A.N.; Pampalakis, G.; Clemente, C.; Re, F.; Sesana, S.; Masserini, M.; Salmona, M.; Duyckaerts, C.; Flores, O.; Gobbi, M.; Antimisiaris, S.G. The binding affinity of anti-Aβ1-42 MAb-decorated nanoliposomes to Aβ1–42 peptides in vitro and to amyloid deposits in post-mortem tissue. *Biomaterials.* **2011**, 32: 5489–5497.

Cavalli, R.; Caputo, O.; Gasco, M.R. Preparation and characterization of solid lipid nanospheres containing paclitaxel. *Eur. J. Pharm. Sci.* **2000**, 10: 305–309.

Chen, Y.; Dalwadi, G.; Benson, H.A. Drug delivery across the blood–brain barrier. *Curr. Drug Deliv..* **2004**, 1: 361–376.

Chen, J.; Zhou, Y.; Mueller-Steiner, S.; Chen, L.F.; Kwon, H.; Yi, S.; Mucke, L.; Gan, L. SIRT1 protects against microglia-dependent amyloid-beta toxicity through inhibiting NF-kappaB signaling. *J. Biol. Chem.* **2005**, 280: 40364–40374.

Cheng, K.K.; Yeung, C.F.; Ho, S.W.; Chow, S.F.; Chow, A.H.; Baum, L. Highly stabilized curcumin nanoparticles tested in an in vitro blood–brain barrier model and in Alzheimer's disease Tg2576 mice. *AAPS J.* **2013**, 15: 324–336.

Chinta, S.J.; Ganesan, A.; Reis-Rodrigues, P.; Lithgow, G.J.; Andersen, J.K. Anti-inflammatory role of isoflavone diadzein in lipopolysaccharide-stimulated microglia: implications for Parkinson's disease. *Neurotox. Res.* **2013**, 23: 145–153.

Choi, C.H.J.; Alabi, C.A.; Webster, P.; Davis, M.E. Mechanism of active targeting in solid tumors with transferrin-containing gold nanoparticles. *Proc. Natl. Acad. Sci. U.S.A.* **2010**, 107: 1235–1240.

Choudhary, S.; Kumar, P.; Malik, J. Plants and phytochemicals for Huntington's disease. *Pharmacogn. Rev.* **2013**, 7: 81–91.

Chowdhury, S.; Sinha, K.; Banerjee, S.; Sil, P.C. Taurine protects cisplatin induced cardiotoxicity by modulating inflammatory and endoplasmic reticulum stress responses. *Biofactos*. **2016**, 42: 647–664.

Chung, C.G.; Lee, H.; Lee, S.B. Mechanisms of protein toxicity in neurodegenerative diseases. *Cell Mol. Life Sci.* **2018**, 75: 3159–3180.

Cottart, C. H.; Nivet-Antoine, V.; Laguillier-Morizot, C.; Beaudeux, J. L. Resveratrol bioavailability and toxicity in humans. *Mol. Nutr. Food Res.* **2010**, 54: 7–16.

Cole, G.M.; Teter, B.; Frautschy, S.A. Neuroprotective effects of curcumin. *Adv. Exp. Med. Biol.* **2007**, 595:197–212.

Cuervo, A.M.; Stefanis, L.; Fredenburg, R.; Lansbury, P.T.; Sulzer, D. Impaired degradation of mutant alpha-synuclein by chaperone-mediated autophagy. *Science*, **2004**, 305: 1292–1295.

da Rocha Lindner, G.; Bonfanti Santos, D.; Colle, D.; Gasnhar Moreira, E.L.; Daniel Prediger, R.; Farina, M.; Khalil N.M.; Mara Mainardes, R. Improved neuroprotective effects of resveratrol-loaded polysorbate 80-coated poly(lactide) nanoparticles in MPTP-induced Parkinsonism. *Nanomedicine* (*Lond*). **2015**, 10: 1127–1138.

Dang, W., The controversial world of sirtuins. *Drug. Discov. Today.* **2014**, e9–e17.

Darvesh, A.S.; Carroll, R.T.; Bishayee, A.; Novotny, N.A.; Geldenhuys, W.J.; Van der Schyf, C.J. Curcumin and neurodegenerative diseases: a perspective. *Expert Opin. Investig. Drugs*. **2012**, 21: 1123–1140.

De Boer, A.G.; Gaillard, P.J. Drug targeting to the brain. *Annu. Rev. Pharmacol. Toxicol.* **2007**, 47: 323–355.

Decuzzi, P.; Godin, B.; Tanaka, T.; Lee, S.Y.; Chiappini, C.; Liu, X.; Ferrari, M. Size and shape effects in the biodistribution of intravascularly injected particles. *J. Control. Release*. **2010**, 141: 320–327.

Delmas, D.; Aires, V.; Limagne, E.; Dutartre, P.; Mazué, F.; Ghiringhelli, F.; Latruffe, N. Transport, stability, and biological activity of resveratrol. *Ann. N.Y. Acad. Sci.* **2011**, 1215: 48–59.

Deng, H.; Mi, M.T. Resveratrol attenuates Aβ25–35 caused neurotoxicity by inducing autophagy through the TyrRS-PARP1-SIRT1 signaling pathway. *Neurochem. Res.* **2016**, 41: 2367–2379.

Desai, P.P.; Patravale, V.B. Curcumin cocrystal micelles-multifunctional nanocomposites for management of neurodegenerative ailments. *J. Pharm. Sci.* **2018**, 107: 1143–1156.

Dey, A.; Bhattacharya, R.; Mukherjee, A.; Pandey, D.K. Natural products against Alzheimer's disease: pharmaco-therapeutics and biotechnological interventions. *Biotechnol. Adv.* **2017**, 35: 178–216.

Dias, V., Junn, E., Mouradian, M.M. The role of oxidative stress in Parkinson's disease. *J. Parkinson's Dis.* **2013**, 3: 461–491.

Djiokeng Paka, G.; Doggui, S.; Zaghmi, A.; Safar, R.; Dao, L.; Reisch, A.; Klymchenko, A.; Roullin, V.G.; Joubert, O.; Ramassamy, C. Neuronal uptake and neuroprotective properties of curcumin-loaded nanoparticles on SK-N-SH cell line: role of poly(lactide-*co*-glycolide) polymeric matrix composition. *Mol. Pharm.* **2016**, 13: 391–403.

Doggui, S.; Sahni, J.K.; Arseneault, M.; Dao, L.; Ramassamy, C. Neuronal uptake and neuroprotective effect of curcumin-loaded PLGA nanoparticles on the human SK-N-SH cell line. *J. Alzheimers Dis.* **2012**, 30: 377–392.

Domijan, A. M.; Kovac, S.; Abramov, A.Y. Lipid peroxidation is essential for phospholipase C activity and the inositol-trisphosphate-related Ca^{2+} signal. *J. Cell Sci.* **2014**, 127: 21–26.

Dua, J.; Rana, A.; Bhandari, A. Liposome: methods of preparation and applications. *Int. J. Pharm. Sci. Res.* **2012**, 3: 14–20.

Dumont, M.; Beal, M.F. Neuroprotective strategies involving ROS in Alzheimer disease. *Free Radic. Biol. Med.* **2011**, 51: 1014–1026.

Etame, A.B.; Smith, C.A.; Chan, W.C.; Rutka, J.T. Design and potential application of PEGylated gold nanoparticles with size-dependent permeation through brain microvasculature. *Nanomedicine.* **2011**, 7: 992–1000.

Fan, S.; Zheng, Y.; Liu, X.; Fang, W.; Chen, X.; Liao, W.; Jing, X.; Lei, M.; Tao, E.; Ma, Q.; Zhang, X.; Guo, R.; Liu, J. Curcumin-loaded PLGA-PEG nanoparticles conjugated with B6 peptide for potential use in Alzheimer's disease. *Drug Deliv.* **2018**, 25: 1091–1102.

Feng, X.; Liang, N.; Zhu, D.; Gao, Q.; Peng, L.; Dong, H.; Yue, Q.; Liu, H.; Bao, L.; Zhang, J.; Hao, J.; Gao, Y.; Yu, X.; Sun, J. Resveratrol inhibits β-amyloid-induced neuronal apoptosis through regulation of SIRT1-ROCK1 signaling pathway. *PLoS One.* **2013**, 8: e59888.

Ferretta, A.; Gaballo, A.; Tanzarella, P.; Piccoli, C.; Capitanio, N.; Nico, B.; Annese, T.; Di Paola, M.; Dell'aquila, C.; De Mari, M.; Ferranini, E.; Bonifati, V.; Pacelli, C.; Cocco, T. Effect of resveratrol on mitochondrial function: implications in Parkin-associated familiar Parkinson's disease. *Biochim. Biophys. Acta.* **2014**, 1842: 902–915.

Finkbeiner, S. Huntington's disease. *Cold Spring Harb. Perspect. Biol.* **2011**, 3: a007476.

Frautschy, S.A.; Hu, W.; Kim, P.; Miller, S.A.; Chu, T.; Harris-White, M.E.; Cole, G.M. Phenolic anti-inflammatory antioxidant reversal of Abeta-induced cognitive deficits and neuropathology. *Neurobiol. Aging.* **2001**, 22: 993–1005.

Fundaro', A.; Cavalli, R.; Bargoni, A.; Vighetto, D.; Zara, G. P.; Gasco, M. R. Non-stealth and stealth solid lipid nanoparticles (SLN) carrying doxorubicin: pharmacokinetics and tissue distribution after I.V. administration to rats. *Pharmacol. Res.* **2000**, 42: 337–343.

Gaballah, H.H.; Zakaria, S.S.; Elbatsh, M.M.; Tahoon, N.M. Modulatory effects of resveratrol on endoplasmic reticulum stress-associated apoptosis and oxido-inflammatory markers in a rat model of rotenone-induced Parkinson's disease. *Chem. Biol. Interact.* **2016**, 251: 10–16.

Gao, X.; Qian, J.; Zheng, S.; Changyi, Y.; Zhang, J.; Ju, S.; Zhu, J.; Li, C. Overcoming the blood–brain barrier for delivering drugs into the brain by using adenosine receptor nanoagonist. *ACS Nano.* **2014**, 8: 3678–3689.

Gaucher, G.; Dufresne, M.H.; Sant, V.P.; Kang, N.; Maysinger, D.; Leroux, J.C. Block copolymer micelles: preparation, characterization and application in drug delivery. *J. Control. Release.* **2005**, 109: 169–188.

Giacobini, E.; Gold, G. Alzheimer disease therapy—moving from amyloid-β to tau. *Nat. Rev. Neurol.* **2013**, 9: 677–686.

Giordano, S.; Darly-Usmar, V.; Zhang, J. Autophagy as an essential cellular antioxidant pathway in neurodegenerative disease. *Redox Biol.* **2014**, 2: 82–90.

Gobbi, M.; Re, F.; Canovi, M.; Beeg, M.; Gregori, M.; Sesana, S.; Sonnino, S.; Brogioli, D.; Musicanti, C.; Gasco, P.; Salmona, M.; Masserini, M.E. Lipid-based nanoparticles with high binding affinity for amyloid-beta1–42 peptide. *Biomaterials.* **2010**, 31: 6519–6529.

Gothwal, A.; Khan, I.; Kesharwani, P.; Chourasia, M.K.; Gupta, U. Micelle-based drug delivery for brain tumors. In: Nanotechnology-based targeted drug delivery systems for brain tumors; Kesharwani, P., Gupta, U., Eds.; Academic Press: USA, **2018**; pp. 307–326.

Hakim, G.M.; Chishti, N.D. The Traditional Healer's Handbook. Healing Arts Press: Vermont, **1988;** p. 11.

Hamaguchi, T.; Ono, K.; Yamada, M. Curcumin and Alzheimer's disease. *CNS Neurosci. Ther.* **2010**, 16: 285–297.

Hanada, S.; Fujioka, K.; Inoue, Y.; Kanaya, F.; Manome, Y.; Yamamoto, K. Cell based in vitro blood–brain barrier model can rapidly evaluate nanoparticles' brain permeability in association with particle size and surface modification. *Int. J. Mol. Sci.* **2014**, 15: 1812–1825.

Hang, L.; Basil, A.H.; Lim, K. Nutraceuticals in Parkinson's disease. *Neuromolecular. Med.* **2016**, 18: 306–321.

Hao, J.; Zhao, J.; Zhang, S.; Tong, T.; Zhuang, Q.; Jin, K.; Chen, W.; Tang, H. Fabrication of an ionic-sensitive in situ gel loaded with resveratrol nanosuspensions intended for direct nose-to-brain delivery. *Colloids Surf. B Biointerfaces.* **2016**, 147: 376–386.

He, Y.; Yue, Y.; Zheng, X.; Zhang, K.; Chen, S.; Du, Z. Curcumin, inflammation, and chronic diseases: how are they linked? *Molecules.* **2015**, 20: 9183–9213.

Ho, D.J.; Calingasan, N.Y.; Wille, E.; Dumont, M.; Beal, M.F. Resveratrol protects against peripheral deficits in a mouse model of Huntington's disease. *Exp Neurol.* **2010**, 225: 74–84.

Hu, S.; Maiti, P.; Ma, Q.; Zuo, X.; Jones, M.R.; Cole, G.M.; Frautschy, S.A. Clinical development of curcumin in neurodegenerative disease. *Expert Rev. Neurother.* **2015**, 15: 629–637.

Huang, C.C.; Faber, P.W.; Persichetti, F.; Mittal, V.; Vonsattel, J.P.; MacDonald, M.E.; Gusella, J.F. Amyloid formation by mutant huntingtin: threshold, progressivity and recruitment of normal polyglutamine proteins. *Somat. Cell Mol. Genet.* **1998**, 24: 217–233.

Huang, X.; Li, L.; Liu, T.; Hao, N.; Liu, H.; Chen, D.; Tang, F. The shape effect of mesoporous silica nanoparticles on biodistribution, clearance, and biocompatibility in vivo. *ACS Nano.* **2011**, 5: 5390–5399.

Huang, W.J.; Zhang, X.; Chen, W.W. Role of oxidative stress in Alzheimer's disease. *Biomed. Rep.* **2016**, 4: 519–522.

Huang, N.; Lu, S.; Liu, X.G.; Zhu, J.; Wang, Y.J.; Liu, R.T. PLGA nanoparticles modified with a BBB-penetrating peptide co-delivering Aβ generation inhibitor and curcumin attenuate memory deficits and neuropathology in Alzheimer's disease mice. *Oncotarget.* **2017**, 8: 81001–81013.

Hung, C.F.; Chen, J.K.; Liao, M.H.; Lo, H.M.; Fang, J.Y. Development and evaluation of emulsion-liposome blends for resveratrol delivery. *J. Nanosci. Nanotechnol.* **2006**, 6: 2950–2958.

Hwang, O. Role of oxidative stress in Parkinson's disease. *Exp Neurobiol.* **2013**, 22: 11–17.

Islam, M.T. Oxidative stress and mitochondrial dysfunction-linked neurodegenerative disorders. *Neurol. Res.* **2017**, 39: 73–82.

Jallouli, Y.; Paillard, A.; Chang, J.; Sevin, E.; Betbeder, D. Influence of surface charge and inner composition of porous nanoparticles to cross blood–brain barrier in vitro, *Int. J. Pharm.* **2007**, 344: 103–109.

Jantaratnotai, N.; Utaisincharoen, P.; Sanvarinda, P.; Thampithak, A.; Sanvarinda, Y. Phytoestrogens mediated anti-inflammatory effect through suppression of IRF-1 and pSTAT1 expressions in lipopolysaccharide-activated microglia. *Int. Immunopharmacol.* **2013**, 17: 483–488.

Jardim, F.R.; de Rossi, F.T.; Nascimento, M.X.; da Silva Barros, R.G.; Borges, P.A.; Prescilio, I.C.; de Oliveira, M.R. Resveratrol and brain mitochondria: a review. *Mol. Neurobiol.* **2018**, 55: 2085–2101.

Ji, H.F.; Shen, L. Can improving bioavailability improve the bioactivity of curcumin? *Trends Pharmacol. Sci.* **2014**, 35: 265–266.

Jeong, H.; Samdani, K.J.; Yoo, D.H.; Lee, D.W.; Kim, N.H.; Yoo, I.S.; Lee, J.H. Resveratrol cross-linked chitosan loaded with phospholipid for controlled release and antioxidant activity. *Int. J. Biol. Macromol.* **2016**, 93: 757–766.

Joó, F. Endothelial cells of the brain and other organ systems: some similarities and differences. *Prog. Neurobiol.* **1996**, 48: 255–273.

Jose, S.; Anju, S.S.; Cinu, T.A.; Aleykutty, N.A.; Thomas, S.; Souto, E.B. In vivo pharmacokinetics and biodistribution of resveratrol-loaded solid lipid nanoparticles for brain delivery. *Int. J. Pharm.* **2014**, 474: 6–13.

Karthick, N.; Alwin, D.; Poornima, K.; Chitra, V.; Saravanan, A.; Balakrishnan, D.; Venkataraman, P. Neurobehavioral alterations and brain creatine kinase system changes in chronic renal failure induced male wistar rats: impact of erythropoietin supplementation. *J. Bioequiv. Bioavail.* **2015**, 7: 74–81.

Kaur, I.P.; Bhandari, R.; Bhandari, S.; Kakkar, V. Potential of solid lipid nanoparticles in brain targeting. *J. Control. Release*. **2008**, 127: 97–109.

Kelsey, N.A.; Wilkins, H.M.; Linseman, D.A. Nutraceutical antioxidants as novel neuroprotective agents. *Molecules*. **2010**, 15: 7792–7814.

Khosa, A.; Reddi, S.; Saha, R.N. Nanostructured lipid carriers for site-specific drug delivery. *Biomed. Pharmacother.* **2018**, 103: 598–613.

Komatsu, M.; Waguri, S.; Chiba, T.; Murata, S.; Iwata, J.; Tanida, I.; Ueno, T.; Koike, M.; Uchiyama, Y.; Kominami, E.; Tanaka, K. Loss of autophagy in the central nervous system causes neurodegeneration in mice. *Nature*. **2006**, 441: 880–884.

Kong, S.D.; Lee, J.; Ramachandran, S.; Eliceiri, B.P.; Shubayev, V.I.; Lal, R.; Jin, S. Magnetic targeting of nanoparticles across the intact blood–brain barrier. *J. Control. Release*. **2012**, 164: 49–57.

Kreuter, J.; Hekmatara, T.; Dreis, S.; Vogel, T.; Gelperina, S.; Langer, K. Covalent attachment of apolipoprotein A-I and apolipoprotein B-100 to albumin nanoparticles enables drug transport into the brain. *J. Control. Release*. **2007**, 118: 54–58.

Kumar, A.; Ratan, R.R. Oxidative stress and Huntington's disease: the good, the bad, and the ugly. *J. Huntingtons Dis.* **2016**, 5; 217–237.

Kuo, Y.C.; Tsai, H.C. Rosmarinic acid- and curcumin-loaded polyacrylamide-cardiolipin-poly(lactide-*co*-glycolide) nanoparticles with conjugated 83-14 monoclonal antibody to protect β-amyloid-insulted neurons. *Mater. Sci. Eng. C Mater. Biol. Appl.* **2018**, 91: 445–457.

Lazar, A.N.; Mourtas, S.; Youssef, I.; Parizot, C.; Dauphin, A.; Delatour, B.; Antimisiaris, S.G.; Duyckaerts, C. Curcumin-conjugated nanoliposomes with high affinity for Aβ deposits: possible applications to Alzheimer disease. *Nanomedicine*. **2013**, 9: 712–721.

Lee, W.H.; Loo, C.Y.; Bebawy, M.; Luk, F.; Mason, R.S.; Rohanizadeh, R. Curcumin and its derivatives: Their application in neuropharmacology and neuroscience in the 21st century. *Curr. Neuropharmacol*. **2013**, 11: 338–378.

Liu, Z.; Zhou, T.; Ziegler, A.C.; Dimitrion, P.; Zuo, L. Oxidative stress in neurodegenerative diseases: from molecular mechanisms to clinical applications. *Oxid. Med. Cell Longev.* **2017**, Article ID 2525967.

Lim, G.P.; Chu, T.; Yang, F.; Beech, W.; Frautschy, S.A.; Cole, G.M. The curry spice curcumin reduces oxidative damage and amyloid pathology in an Alzheimer transgenic mouse. *J. Neurosci.* **2001**, 21: 8370–8377.

Lin, T.K.; Chen, S.D.; Chuang, Y.C.; Lin, H.Y.; Huang, C.R.; Chuang, J.H.; Wang, P.W.; Huang, S.T.; Tiao, M.M.; Chen, J.B.; Liou, C.W. Resveratrol partially prevents rotenone-induced neurotoxicity in dopaminergic SH-SY5Y cells through induction of heme oxygenase-1 dependent autophagy. *Int. J. Mol. Sci.* **2014**, 15:1625–1646.

Löscher, W.; Potschka, H. Drug resistance in brain diseases and the role of efflux transporters. *Nat. Rev. Neurosci.* **2005**, 6: 591–602.

Loureiro, J.A.; Andrade, S.; Duarte, A.; Neves, A.R.; Queiroz, J.F.; Nunes, C.; Sevin, E.; Fenart, L.; Gosselet, F.; Coelho, M.A.; Pereira, M.C. Resveratrol and grape extract-loaded solid lipid nanoparticles for the treatment of Alzheimer's disease. *Molecules*. **2017**, 22: 277–293.

Lu, X.; Ji, C.; Xu, H.; Li, X.; Ding, H.; Ye, M.; Zhu, Z.; Ding, D.; Jiang, X.; Ding, X.; Guo, X. Resveratrol-loaded polymeric micelles protect cells from Aβ-induced oxidative stress. *Int. J. Pharm*. **2009**, 375: 89–96.

Maiti, P.; Hall, T.C.; Paladugu, L.; Kolli, N.; Learman, C.; Rossignol, J.; Dunbar, G.L. A comparative study of dietary curcumin, nanocurcumin, and other classical amyloid-binding dyes for labeling and imaging of amyloid plaques in brain tissue of 5x-familial Alzheimer's disease mice. *Histochem. Cell Biol.* **2016**, 146: 609–625.

Maiti, P.; Paladugu, L.; Dunbar, G.L. Solid lipid curcumin particles provide greater anti-amyloid, anti-inflammatory and neuroprotective effects than curcumin in the 5xFAD mouse model of Alzheimer's disease. *BMC Neurosci.* **2018**, 19: 7.

Maiti, P.; Dunbar, G.L. Use of curcumin, a natural polyphenol for targeting molecular pathways in treating age-related neurodegenerative diseases. *Int. J. Mol. Sci.* **2018**, 19: pii: E1637.

Manoharan, S.; Guillemin, G.J.; Abiramasundari, R.S.; Essa, M.M.; Akbar, M.; Akbar, M.D. The role of reactive oxygen species in the pathogenesis of Alzheimer's disease, Parkinson's disease, and Huntington's disease: a mini review. *Oxid. Med. Cell Longev.* **2016**, Article ID 8590578.

Mao, X.Y.; Jin, M.Z.; Chen, J.F.; Zhou, H.H.; Jin, W.L. Live or let die: Neuroprotective and anti-cancer effects of nutraceutical antioxidants. *Pharmacol. Therap.* **2018**, 183: 137–151.

Mariño, G.; Madeo, F.; Kroemer, G. Autophagy for tissue homeostasis and neuroprotection. *Curr. Opin. Cell Biol.* **2011**, 23: 198–206.

Markus, M.A.; Morris, B.J. Resveratrol in prevention and treatment of common clinical conditions of aging. *Clin. Interv. Aging.* **2008**, 3, 331–339.

Mathew, A.; Fukuda, T.; Nagaoka, Y.; Hasumura, T.; Morimoto, H.; Yoshida, Y.; Maekawa, T.; Venugopal, K.; Kumar, D.S. Curcumin loaded-PLGA nanoparticles conjugated with Tet-1 peptide for potential use in Alzheimer's disease. *PLoS One.* **2012**a, 7: e32616.

Mathew, A.; Aravind, A.; Brahatheeswaran, D.; Fukuda, T.; Nagaoka, Y.; Hasumura, T.; Iwai, S.; Morimoto, H.; Yoshida, Y.; Maekawa, T.; Venugopal, K.; Kumar, D.S. Amyloid-binding aptamer conjugated curcumin–PLGA nanoparticle for potential use in Alzheimer's disease. *BioNanoSci.* **2012**b, 2: 83–93.

Min, C.; Zhi-Yun, D.; Xi, Z.; Dong-Li, L.; Ren-Ping, Z.; Kun, Z. Use of curcumin in diagnosis, prevention, and treatment of Alzheimer's disease. *Neural Regen. Res.* **2018**, 13: 742–752.

Monroy, A.; Lithgow, G.J.; Alavez, S. Curcumin and neurodegenerative diseases. *Biofactors.* **2013**, 39: 122–132.

Mourtas, S.; Canovi, M.; Zona, C.; Aurilia, D.; Niarakis, A.; La Ferla, B.; Salmona, M.; Nicotra, F.; Gobbi, M.; Antimisiaris, S.G. Curcumin-decorated nanoliposomes with very high affinity for amyloid-β1–42 peptide. *Biomaterials.* **2011**, 32: 1635–1645.

Mourtas, S.; Lazar, A.N.; Markoutsa, E.; Duyckaerts, C.; Antimisiaris, S.G. Multifunctional nanoliposomes with curcumin-lipid derivative and brain targeting functionality with potential applications for Alzheimer disease. *Eur. J. Med. Chem.* **2014**, 280: 175–183.

Mulik, R.S.; Mönkkönen, J.; Juvonen, R.O.; Mahadik, K.R.; Paradkar, A.R. ApoE3 mediated poly(butyl) cyanoacrylate nanoparticles containing curcumin: study of enhanced activity of curcumin against β amyloid induced cytotoxicity using in vitro cell culture model. *Mol. Pharm.* **2010**, 7: 815–825.

Müller, R. H.; Radtke, M.; Wissing, S. A. Solid lipid nanoparticles (SLN) and nanostructured lipid carriers (NLC) in cosmetic and dermatological preparations. *Adv. Drug Deliv. Rev.* **2002**, 54: S131-S155.

Nasr, M. Development of an optimized hyaluronic acid-based lipidic nanoemulsion co-encapsulating two polyphenols for nose to brain delivery. *Drug Deliv.* **2016**, 23: 1444–1452.

Neves, A.R.; Lucio, M.; Lima, J.L.; Reis, S. Resveratrol in medicinal chemistry: A critical review of its pharmacokinetics, drug-delivery, and membrane interactions. *Curr. Med. Chem.* **2012**, 19: 1663–1681.

Neves, A.R.; Lucio, M.; Martins, S.; Lima, J.L.; Reis, S. Novel resveratrol nanodelivery systems based on lipid nanoparticles to enhance its oral bioavailability. *Int. J. Nanomedicine.* **2013**, 8: 177–187.

Neves, A.R.; Martins, S.; Segundo, M.A.; Reis, S. Nanoscale delivery of resveratrol towards enhancement of supplements and nutraceuticals. *Nutrients.* **2016**a, 8: 131.

Neves, A.R.; Queiroz, J.F.; Reis, S. Brain-targeted delivery of resveratrol using solid lipid nanoparticles functionalized with apolipoprotein E. *J. Nanobiotechnology.* **2016**b, 14: 1–27.

Oliveira, J.; Costa, M.; de Almeida, M.S.C.; da Cruz, E.; Silva, O.A.B.; Henriques, A.G. Protein phosphorylation is a key mechanism in Alzheimer's disease. *J. Alzheimers Dis.* **2017,** 58:953–978.

Pan, T.; Rawal, Y.; Wu, Y.; Xie, W.; Jankovic, J.; Le, W. Rapamycin protects against rotenone-induced apoptosis through autophagy induction. *Neuroscience.* **2009**, 164: 541–551.

Paka, G.D.; Ramassamy, C. Optimization of curcumin-loaded PEG-PLGA nanoparticles by GSH functionalization: investigation of the internalization pathway in neuronal cells. *Mol. Pharm.* **2017**, 14: 93–106.

Pangeni, R.; Sharma, S.; Mustafa, G.; Ali, J.; Baboota, S. Vitamin E loaded resveratrol nanoemulsion for brain targeting for the treatment of Parkinson's disease by reducing oxidative stress. *Nanotechnology.* **2014**, 25: 485102.

Papadia, K.; Markoutsa, E.; Mourtas, S.; Giannou, A.D.; La Ferla, B.; Nicotra, F.; Salmona, M.; Klepetsanis, P.; Stathopoulos, G.T.; Antimisiaris, S.G. Multifunctional LUV liposomes decorated for BBB and amyloid targeting. A. In vitro proof-of-concept. *Eur. J. Pharm. Sci.* **2017**a, 101: 140–148.

Papadia, K.; Giannou, A.D.; Markoutsa. E.; Bigot, C.; Vanhoute, G.; Mourtas, S.; Van der Linded, A.; Stathopoulos, G.T.; Antimisiaris, S.G. Multifunctional LUV liposomes decorated for BBB and amyloid targeting-B. In vivo brain targeting potential in wild-type and APP/PS1 mice. *Eur. J. Pharm. Sci.* **2017**b, 102: 180–187.

Pasinetti, G.M.; Wang, J.; Marambaud, P.; Ferruzzi, M.; Gregor, P.; Knable, L.A.; Ho, L. Neuroprotective and metabolic effects of resveratrol: therapeutic implications for Huntington's disease and other neurodegenerative disorders. *Exp. Neurol.* **2011**, 232: 1–6.

Pasinetti, G.M.; Wang, J.; Ho, L.; Zhao, W.; Dubner, L. Roles of resveratrol and other grape-derived polyphenols in Alzheimer's disease prevention and treatment. *Biochim. Biophys. Acta.* **2015**, 1852: 1202–1208.

Patel, S.G.; Patel, M.D.; Patel, A.J.; Chougule, M.B.; Choudhury, H. Solid lipid nanoparticles for targeted brain drug delivery. In Nanotechnology-based Targeted Drug Delivery Systems for Brain Tumors; Kesharwani, P., Gupta, U., Eds.; Academic Press: USA, 2018; pp. 191–244.

Patil, T.S.; Deshpande, A.S. Nanostructured lipid carriers-based drug delivery for treating various lung diseases: a state-of-the-art review. *Int. J. Pharm.* **2018**, 547: 209–225.

Patil, T.S.; Deshpande, A.S.; Deshpande, S.; Shende, P. Targeting pulmonary tuberculosis using nanocarrier-based dry powder inhalation: current status and futuristic need. *J. Drug Target.* **2018**, 1–16.

Peng, J.R.; Qian, Z.Y. Drug delivery systems for overcoming the bioavailability of curcumin: not only the nanoparticle matters. *Nanomedicine (Lond).* **2014**, 9: 747–750.

Preethi Pallavi, M.C.; Sampath Kumar, H.M. Nutraceuticals in prophylaxis and therapy of neurodegenerative diseases. Chapter 8. In Discovery and Development of Neuroprotective Agents from Natural Products; Bramhachari, G. Eds.; Elsevier: The Netherlands. **2018;** pp. 359–376.

Priyadarsini, K.I. Chemical and structural features influencing the biological activity of curcumin. *Curr. Pharm. Des.* **2013**, 19: 2093–2100.

Pund, S.; Thakur, R.; More, U.; Joshi, A. Lipid based nanoemulsifying resveratrol for improved physicochemical characteristics, in vitro cytotoxicity and in vivo antiangiogenic efficacy. *Colloids Surf. B Biointerfaces.* **2014**, 120: 110–117.

Pund, S.; Joshi, A.; Patravale, V. Improving bioavailability of nutraceuticals by nanoemulsification. In Nutraceuticals. Nanotechnology in the Agri-Food Industry; Grumezescu, A., Eds.; Elsevier Inc., the Netherlands, **2016**; Vol. 4, pp 481–534.

Rajeshwari, A.; Sabesan, M. Inhibition of monoamine oxidase-B by the polyphenolic compound, curcumin and its metabolite tetrahydrocurcumin, in a model of Parkinson's disease induced by MPTP neurodegeneration in mice. *Inflammopharmacology*. **2008**, 16: 96–99.

Ramalingam, P.; Ko, Y.T. Improved oral delivery of resveratrol from N-trimethyl chitosan-g-palmitic acid surface-modified solid lipid nanoparticles. *Colloids Surf. B Biointerfaces*. **2016**, 139: 52–61.

Rigacci, S.; Stefani, M. Nutraceuticals and amyloid neurodegenerative diseases: a focus on natural phenols. *Expert Rev. Neurother.* **2015**, 15: 41–52.

Rostovtseva, T.K.; Gurnev, P.A.; Protchenko, O.; Hoogerheide, D.P.; Yap, T.L.; Philpott, C.C.; Lee, J.C.; Bezrukov, S.M. α-Synuclein shows high affinity interaction with voltage-dependent anion channel, suggesting mechanisms of mitochondrial regulation and toxicity in Parkinson disease. *J. Biol. Chem.* **2015**, 290:18467–18477.

Sadhukhan, P.; Saha, S.; Dutta, S.; Mahalanobish, S.; Sil, P.C. Nutraceuticals: An emerging therapeutic approach against the pathogenesis of Alzheimer's disease. *Pharm. Res.* **2018**, 129: 100–114.

Saha, S.; Sadhukhan, P.; Sil, P. C. Genistein: a phytoestrogen with multifaceted therapeutic properties. *Mini Rev. Med. Chem.* **2014**, 14: 920–940.

Saha, S.; Sadhukhan, P.; Sil, P.C. Beneficial upshots of naturally occurring antioxidant compounds against neurological disorders. In Neuroprotective Natural Products: Clinical Aspects and Mode of Action; Bramhchari G., Eds.; Wiley-VCH Verlag GmbH and Co.: Germany, **2017** pp 23–56.

Santander-Ortega, M.J.; Plaza-Oliver, M.; Rodríguez-Robledo, V.; Castro-Vázquez, L.; Villaseca-González, N.; González-Fuentes, J.; Marcos, P.; Arroyo-Jiménez, M.M.; Lozano, M.V. Colloids for drug delivery to the brain. *J. Drug Deliv. Sci. Technol*. **2017**, 42: 193–206.

Saraiva, C.; Praca, C.; Ferreira, R.; Santos, T.; Ferreira, L.; Bernardino, L. Nanoparticle-mediated brain drug delivery: Overcoming blood–brain barrier to treat neurodegenerative diseases. *J. Control. Release*. **2016**, 235: 34–47.

Saupe, A.; Gordon, K. C.; Rades, T. Structural investigations on nanoemulsions, solid lipid nanoparticles and nanostructured lipid carriers by cryo-field emission scanning electron microscopy and Raman spectroscopy. *Int. J. Pharm.* **2006**, 314: 56–62.

Sawda, C.; Moussa, C.; Turner, R.S. Resveratrol for Alzheimer's disease. *Ann. N.Y. Acad. Sci.* **2017**.

Schwarz, C.; Mehnert, W.; Lucks, J. S.; Müller, R. H. Solid lipid nanoparticles (SLN) for controlled drug delivery. I. Production, characterization and sterilization. *J. Control. Release*. **1994**, 30: 83–96.

Scherzinger, E.; Sittler, A.; Schweiger, K.; Heiser, V.; Lurz, R.; Hasenbank, R.; Bates, G.P.; Lehrach, H.; Wanker, E.E. Self-assembly of polyglutamine-containing huntingtin fragments into amyloid-like fibrils: implications for Huntington's disease pathology. *Proc. Natl. Acad. Sci. USA.* **1999**, 96: 4604–4609.

Sekerdag, E. Challenges of the nose-to-brain route. In Nanotechnology Methods for Neurological Diseases and Brain Tumors; Özdemir, Y.G., Pehlivan, S.B., Sekerdag, E., Eds; Elsevier Inc.: the Netherlands, **2017,** pp 103–113.

Selkoe, D.J. Alzheimer's disease: genes, proteins, and therapy. *Physiol. Rev.* **2001**, 81: 741–766.

Seljak, K.B.; Berginc, K.; Trontelj, J.; Zvonar, A.; Kristl, A.; Gašperlin, M.A self-microemulsifying drug delivery system to overcome intestinal resveratrol toxicity and presystemic metabolism. *J. Pharm. Sci.* **2014**, 103: 3491–3500.

Serafini, M.M.; Catanzaro, M.; Rosini, M.; Racchi, M.; Lanni, C. Curcumin in Alzheimer's disease: Can we think to new strategies and perspectives for this molecule? *Pharmacol. Res.* **2017**, 124: 146–155.

Sergides, C.; Chirilă, M.; Silvestro, L.; Pitta, D.; Pittas, A. Bioavailability and safety study of resveratrol 500 mg tablets in healthy male and female volunteers. *Exp. Ther. Med.* **2016**, 11: 164–170.

Sessa, M.; Tsao, R.; Liu, R.H.; Ferrari, G.; Donsì, F. Evaluation of the stability and antioxidant activity of nanoencapsulated resveratrol during in vitro digestion. *J. Agric. Food Chem.* **2011**, 59: 12352–12360.

Skaper, S.D.; Facci, L.; Giusti, P. Neuroinflammation, microglia and mast cells in the pathophysiology of neurocognitive disorders: a review. *CNS Neurol. Disord. Drug Targets.* **2014**, 13: 1654–1666.

Smoliga, J.M.; Blanchard, O. Enhancing the delivery of resveratrol in humans: if low bioavailability is the problem, what is the solution? *Molecules.* **2014**, 19: 17154–17172.

Sonavane, G.; Tomoda, K.; Makino, K. Biodistribution of colloidal gold nanoparticles after intravenous administration: effect of particle size. *Colloids Surf B Biointerfaces.* **2008**, 66: 274–280.

Stefanis, L. α-Synuclein in Parkinson's disease. *Cold Spring Harb. Perspect. Med.* **2012**, 2: a009399.

Stefani, M.; Rigacci, S. Beneficial properties of natural phenols: highlight on protection against pathological conditions associated with amyloid aggregation. *Biofactors.* **2014**, 40: 482–493.

Stivala, L.A.; Savio, M.; Carafoli, F.; Perucca, P.; Bianchi, L.; Maga, G.; Forti, L.; Pagnoni, U.M.; Albini, A.; Prosperi, E.; Vannini, V. Specific structural determinants are responsible for the antioxidant activity and the cell cycle effects of resveratrol. *J. Biol. Chem.* **2001**, 276: 22586–22594.

Sun, M.; Gao, Y.; Guo, C.Y.; Cao, F.; Song, Z.; Xi, Y.; Yu, A.; Li, A.; Zhai, G. Enhancement of transport of curcumin to brain in mice by poly(*N*-butylcyanoacrylate) nanoparticle. *J. Nanopart. Res.* **2010**, 12: 3111–3122.

Sun, C.; Ding, Y.; Zhou, L.; Shi, D.; Sun, L.; Webster, T.J.; Shen, Y. Noninvasive nanoparticle strategies for brain tumor targeting. *Nanomedicine.* **2017**, 13: 2605–2621.

Sweeney, P.; Park, H.; Baumann, M.; Dunlop, J.; Frydman, J.; Kopito, R.; McCampbell, A.; Leblanc, G.; Venkateswaran, A.; Nurmi, A.; Hodgson, R. Protein misfolding in neurodegenerative diseases: implications and strategies. *Transl. Neurodegener.* **2017**, 13: 6.

Swomley, A.M.; Förster, S.; Keeney, J.T.; Triplett, J.; Zhang, Z.; Sultana, R.; Butterfield, D.A. A beta oxidative stress in Alzheimer disease: evidence based on proteomics studies. *Biochim. Biophys. Acta Mol. Basis Dis.* **2014**, 1842: 1248–1257.

Szeto, H.H. Mitochondria-targeted peptide antioxidants: novel neuroprotective agents. *AAPS J.* **2006**, 8: E521-E531.

Tapeinos, C.; Battaglini, M.; Ciofani, G. Advances in the design of solid lipid nanoparticles and nanostructured lipid carriers for targeting brain diseases. *J. Control. Release.* **2017**, 264: 306–332.

Taylor, M.; Moore, S.; Mourtas, S.; Niarakis, A.; Re, F.; Zona, C.; La Ferla, B.; Nicotra, F.; Masserini, M.; Antimisiaris, S.G.; Gregori, M.; Allsop, D. Effect of curcumin-associated and lipid ligand-functionalized nanoliposomes on aggregation of the Alzheimer's Aβ peptide. *Nanomedicine.* **2011**, 7: 541–550.

Taylor, J.P.; Brown, R.H.; Cleveland, D.W. Decoding ALS: from genes to mechanism. *Nature*. **2016**, 539: 197–206.

Tellone, E.; Galtieri, A.; Russo, A.; Giardina, B.; Ficarra, S. Resveratrol: a focus on several neurodegenerative diseases. *Oxid. Med. Cell Longev*. **2015**, Article ID 392169.

Trela, B.C.; Waterhouse, A.L. Resveratrol: isomeric molar absorptivities and stability. *J. Agric. Food Chem.* **1996**, 44: 1253–1257.

Trotta, V.; Pavan, B.; Ferraro, L.; Beggiato, S.; Traini, D.; Des Reis, L.G.; Scalia, S.; Dalpiaz, A. Brain targeting of resveratrol by nasal administration of chitosan-coated lipid microparticles. *Eur. J. Pharm. Biopharm*. **2018**,127: 250–259.

Turillazzi, E.; Neri, M.; Cerretani, D.; Cantatore, S.; Frati, P.; Moltoni, L.; Busardò, F.P.; Pomara, C.; Riezzo, I.; Fineschi, V. Lipid peroxidation and apoptotic response in rat brain areas induced by long-term administration of nandrolone: the mutual crosstalk between ROS and NF- kB. *J. Cell Mol. Med.* **2016**, 20: 601–612.

Ullah, F.; Liang, A.; Rangel, A.; Gyengesi, E.; Niedermayer, G.; Münch, G. High bioavailability curcumin: an anti-inflammatory and neurosupportive bioactive nutrient for neurodegenerative diseases characterized by chronic neuroinflammation. Arch. Toxicol. **2017**, 91: 1623–1634.

Uner, M.; Yener, G. Importance of solid lipid nanoparticles (SLN) in various administration routes and future perspectives. *Int. J. Nanomedicine*. **2007**, 2: 289–300.

Ungvari, Z.; Sonntag, W.E.; de Cabo, R.; Baur, J.A.; Csiszar, A. Mitochondrial protection by resveratrol. *Exerc. Sport Sci. Rev*. **2011**, 39: 128–132.

van Roon-Mom, W.M.; Reid, S.J.; Jones, A.L.; MacDonald, M.E.; Faull, R.L.; Snell, R.G. Insoluble TATA-binding protein accumulation in Huntington's disease cortex. *Brain Res. Mol. Brain Res*. **2002**, 109: 1–10.

Vanaja, K.; Wahl, M.A.; Bukarica, L.; Heinle, H. Liposomes as carriers of the lipid soluble antioxidant resveratrol: evaluation of amelioration of oxidative stress by additional antioxidant vitamin. *Life Sci.* **2013**, 93: 917–923.

Vekrellis, K.; Stefanis, L. Targeting intracellular and extracellular alpha-synuclein as a therapeutic strategy in Parkinson's disease and other synucleinopathies. *Expert Opin. Ther. Targets*. **2012**, 16: 421–432.

Vidoni, C.; Secomandi, E.; Castiglioni, A.; Melone, M.A.B.; Isidoro, C. Resveratrol protects neuronal-like cells expressing mutant huntingtin from dopamine toxicity by rescuing ATG4-mediated autophagosome formation. *Neurochem. Int.* **2018**, 117: 174–187.

Vogg, A.T. J.; Doczi, R.; Szigeti, K.; Horvath, I.; Mathe, D.; Roller, M.; Varga, Z. Targeted liposomal drug delivery system for radionucleosides—a theranostic approach. *Eur. J. Nucl. Med. Mol. Imaging*. **2016**, 43: S468.

Walle, T.; Hsieh, F.; DeLegge, M.H.; Oatis, J.E. Jr.; Walle, U.K. High absorption but very low bioavailability of oral resveratrol in humans. *Drug Metab. Dispos*. **2004**, 32: 1377–1382.

Wang, Y.; Xu, H.; Fu, Q.; Ma, R.; Xiang, J. Protective effect of resveratrol derived from *Polygonum cuspidatum* and its liposomal form on nigral cells in Parkinsonian rats. *J. Neurol. Sci.* **2011**, 304: 29–34.

Wang, B.; Sun, J.; Ma, Y.; Wu, G.; Tian, Y.; Shi, Y.; Le, G. Resveratrol preserves mitochondrial function, stimulates mitochondrial biogenesis, and attenuates oxidative stress in regulatory T cells of mice fed a high-fat diet. *J. Food Sci.* **2014**, 79: H1823-H1831.

Wang, Y.L.; Ju, B.; Zhang, Y.Z.; Yin, H.L.; Liu, Y.J.; Wang, S.S.; Zeng, Z.L.; Yang, X.P.; Wang, H.T.; Li, J.F. Protective effect of curcumin against oxidative stress-induced injury in rats with Parkinson's disease through the Wnt/β-Catenin signaling pathway. *Cell Physiol. Biochem.* **2017**, 43: 2226–2241.

Weber, S.; Zimmer, A.; Pardeike, J. Solid lipid nanoparticles (SLN) and nanostructured lipid carriers (NLC) for pulmonary application: a review of the state of the art. *Eur. J. Pharm. Biopharm.* **2014**, 86: 7–22.

Wei, Z.; Hao, J.; Yuan, S.; Li, Y.; Juan, W.; Sha, X.; Fang, X. Paclitaxel loaded pluronic P123/F127 mixed polymeric micelles: Formulation. Optimization and in vitro characterization. *Int. J. Pharm.* **2009,** 376: 176–185.

Wiley, D.T.; Webster, P.; Gale, A.; Davis, M.E. Transcytosis and brain uptake of transferrin-containing nanoparticles by tuning avidity to transferrin receptor. *Proc. Natl. Acad. Sci. USA*. **2013**, 110: 8662–8667.

Wissing, S. A.; Kayser, O.; Muller, R. H., Solid lipid nanoparticles for parenteral drug delivery. *Adv. Drug Deliv. Rev.* **2004,** 56: 1257–1272.

Woerner, A. C.; Frottin, F.; Hornburg, D.; Feng, L. R.; Meissner, F.; Patra, M.; Tatzelt, J.; Mann, M.; Winklhofer, K. F.; Hartl, F. U.; Hipp, M. S.; Cytoplasmic protein aggregates interfere with nucleocytoplasmic transport of protein and RNA. *Science*. **2016**, 351: 173– 176.

Xin, H.; Jiang, Y.; Lv, W.; Xu, J. Liposome based drug delivery for brain tumor theranostics, In Nanotechnology-Based Targeted Drug Delivery Systems for Brain Tumors; Kesharwani, P., Gupta, U. Eds.; Academic Press, USA, **2018;** pp. 245–266.

Yadav, A.; Sunkaria, A.; Singhal, N.; Sandhir, R. Resveratrol loaded solid lipid nanoparticles attenuate mitochondrial oxidative stress in vascular dementia by activating Nrf2/HO-1 pathway. *Neurochem. Int.* **2018**, 112: 239–254.

Yang, F.; Lim, G.P.; Begum, A.N.; Ubeda, O.J.; Simmons, M.R.; Ambegaokar, S.S.; Chen, P.P.; Kayed, R.; Glabe, C.G.; Frautschy, S.A.; Cole, G.M. Curcumin inhibits formation of amyloid beta oligomers and fibrils, binds plaques, and reduces amyloid in vivo. *J. Biol. Chem.* **2005**, 280: 5892–5901.

Yao, J.; Zhou, J.; Dahmani, F. Z.; Yang, H.; Zhang, T.; Zhan, Q. Enhanced oral bioavailability of paclitaxel in pluronic/LHR mixed polymeric micelles: preparation, in vitro and in vivo evaluation. *Eur. J. Pharm. Sci.* **2012**, 47: 179–189.

Zhang, J.; Feng, X.; Wu, J.; Xu, H.; Li, G.; Zhu, D.; Yue, Q.; Liu, H.; Zhang, Y.; Sun, D.; Wang, H.; Sun, J. Neuroprotective effects of resveratrol on damages of mouse cortical neurons induced by β-amyloid through activation of SIRT1/Akt1 pathway. *Biofactors.* **2014**, 40: 258–267.

Zhang, N.; Yan, F.; Liang, X.; Wu, M.; Shen, Y.; Chen, M.; Xu, Y.; Zou, G.; Jiang, P.; Tang, C.; Zheng, H.; Dai, Z. Localized delivery of curcumin into brain with polysorbate 80-modified cerasomes by ultrasound-targeted microbubble destruction for improved Parkinson's disease therapy. *Theranostics*. **2018,** 8: 2264–2277.

CHAPTER 9

Preclinical Perspectives of Mechanisms Underlying the Neuroprotective Effects of Selected Phyto-Constituents in Alzheimer's Disease

MARCO CASCELLA[1*], MARIA ROSARIA MUZIO[2], DOMENICO CARBONE[3], ARTURO CUOMO[1], and SABRINA BIMONTE[1]

[1]*Istituto Nazionale Tumori—IRCCS—Fondazione "G. Pascale," Via Semmola, Naples, Italy*

[2]*Division of Infantile Neuropsychiatry, UOMI-Maternal and Infant Health, Naples, Italy*

[3]*Department of Emergency Medicine, Umberto I Hospital, Nocera Inferiore, Salerno, Italy*

[*]*Corresponding author. E-mail: m.cascella@istitutotumori.na.it*

ABSTRACT

Alzheimer's disease (AD), the most common form of dementia, accounts for approximately two-thirds of all cases. At the pathological level, the disease is characterized by extracellular amyloidal beta (Aβ) protein deposits and intracellular neurofibrillary tangles (NT). The brain deposits of Aβ initiate a series of pathological reactions leading to microglia-mediated chronic inflammation. Neuroinflammation and oxidative stress processes are responsible for the impairment of the neurovascular working leading to axonal demyelination, local hypoxia-ischemia, white matter damages, and finally progressive loss of neurons and synapsis.

Despite the availability of huge data on its pathogenesis, a limited number of therapies for AD have emerged until now. Natural compounds could represent an important strategy for preventing or delaying the functional decline in AD's patients due to their antioxidants and anti-inflammatory properties. It is worth noting that many in vitro and in vivo studies using natural compounds combined with antioxidants for AD treatment have been conducted so far. Among the wide range of nutraceuticals studied for this scope, the anti-inflammatory, antioxidant, antiamyloid, and anti-tau properties of *Nigella sativa* and its constituent thymoquinone, (−)-epigallocatechin-3-gallate, and curcumin may offer interesting neuroprotective benefits and translational perspectives into clinical practice.

9.1 INTRODUCTION

The term dementia is referred to as a heterogeneous group of diseases featuring a progressive decline in mental abilities, including memory, thinking, and judgment, as well as personality changes. Alzheimer's disease (AD) is the most common form of dementia, accounting for approximately two-thirds of all cases (Rizzi et al., 2014). The main features of AD pathophysiology are the progressive loss of neurons and synapsis, and the presence of two main microscopic neuropathological hallmarks: extracellular amyloidal beta (Aβ) protein deposits in the brain parenchyma and the cerebral blood vessels contributing to senile plaques and cerebral amyloid angiopathy, and intracellular NFT composed largely of paired helical filaments with hyper phosphorylated tau (P-tau) proteins, a group of microtubule-associated proteins expressed predominantly in axons. The cellular and molecular mechanisms involved are multiple. According to the amyloid cascade hypothesis, the brain deposits of Aβ initiate a series of pathological reactions causing chronic inflammation (Zotova et al., 2010). At the beginning stages of the cascade the amyloid precursor protein (APP), bypassing the physiological pathway of the α-secretase, is aberrantly processed by β-secretase amyloid cleaving enzyme (BACE), and γ-secretase complex (Zanh et al., 2014) in various isoforms of Aβ peptides, the most common being 40 and 42 amino acids in length (i.e., Aβ40 and Aβ42). In particular, the BACE1 isoform is responsible for the first step in APP proteolysis and, thus, specific inhibition of BACE1 by small molecules—including natural compounds such as neferine, ginsenoside Rg1, and hispidin—is a current focus for AD research and therapy (Yan, 2017). The effect of this dysregulation is an imbalance between the production and clearance of Aβ peptide. Current pieces of evidence underline that the role of Aβ in the initiation and/or progression of AD is mediated by oxidative damage and microglia-induced neuro-inflammatory responses (Mosher and Wyss-Coray, 2014). The microglial activation (i.e., M1 phenotype) induces a complex cascade that involves the secretion of reactive oxygen species (ROS), reactive nitrogen species (NOS) and many pro-inflammatory molecules such as tumor necrosis factor-α (TNF-α), interleukin (IL) IL-6, IL-1β (Agostinho et al., 2010). In turn, the oxidative stress and the neuroinflammation processes are responsible for an increased calcium influx, altered kinase and phosphatase activities, irregular regulation of transcription factors, and disruptions in ion channel and receptor gene expression. All these processes culminate in progressive impairment of the neurovascular working that leads to axonal demyelination, local hypoxia-ischemia, and white matter damages (Iadecola, 2010).

Despite a large number of investigations on AD pathogenesis that have been reported in the literature, a limited number of therapies for AD are available. Natural compounds could represent an important strategy for preventing or delaying the functional decline in AD's patients due to their antioxidants and anti-inflammatory properties. Thus, many clinical investigations have been performed to evaluate nutraceuticals as potential neuroprotective agents. For example, it has been demonstrated that a diet rich in antioxidants (e.g., flavonoids, vitamin E, and omega-3 fatty acids) could reduce the risk of AD (Andrade and Radhakrishnan, 2009). On the contrary, high dosages of B vitamin supplements (folate, B6, and B12) or docosahexaenoic acid did not slow the rate of cognitive and functional decline in patients with mild to moderate AD (Aisen et al., 2008; Quinn et al., 2010). In addition to the clinical investigations, most experimental studies have been conducted in vitro or on animal models of AD (De Domenico and Giudetti, 2017). Among the wide range of nutraceuticals studied for this purpose, the anti-inflammatory, antioxidant, anti-amyloid, and anti-tau properties of *Nigella sativa* (NS) and its constituents Thymoquinone (TQ) (Cascella et al., 2018), (−)-epigallocatechin-3-gallate (Cascella et al., 2017a), and curcumin (Goozee et al., 2016) may offer interesting neuroprotective benefits and could be considered as potential chemicals for the future research focused on their potential role on prevention and treatment of minimal cognitive impairment and AD.

9.2 *NIGELLA SATIVA* AND TQ

NS, also named black seed or black cumin, is a flowering plant that belongs to the family *Ranunculaceae.* Regarding its chemistry, it is composed of 30% fixed oils (mainly fatty acids), 0.40%–0.45% volatile oils, vitamins, amino acids, proteins, carbohydrates, alkaloids, saponins, crude fiber, and minerals. NS oil (NSO) is rich in polyunsaturated fatty acids, phytosterols, carvacrol, t-anethole, sesquiterpenelongifolene, and 4 terpinol. Moreover, a significant component of NS is TQ that accounts for 25% of volatile oil (Akram Khan and Afzal, 2016). Several studies revealed that NS and TQ have relevant anti-inflammatory and antioxidant properties (Ragheb et al., 2009). In particular, it has been demonstrated that TQ could be effective against immunological, infective, and degenerative diseases due to its redox properties and its ability to bypass the biological barriers (Badary et al, 2003). Moreover, different studies provide evidence for the anticancer activities of NS and TQ (Kundu et al., 2014) as well as on their role on chemotherapy-induced nephrotoxicity (Cascella et al., 2017b). Then, other preclinical findings suggested that TQ could be considered as an add-on treatment for neurological, and neurosurgical conditions, for example for glial brain tumors (Elmaci and Altinoz, 2016).

In clinical investigations on healthy volunteers, it has been shown that NS enhanced memory and attention (Bin Sayeed et al., 2013), and stabilized mood, decreased anxiety and modulated cognition positively when compared to patients treated with placebo (Bin Sayeed et al., 2014). Despite clinical data are very limited, a large number of scientific evidences have been collected (Figure 9.1).

9.2.1 *IN VITRO STUDIES*

In vitro studies on the neuroprotective effects of NS and TQ have been carried out on different cells lines, such as primary rat cerebellar granule neurons (CGNs), undifferentiated pheochromocytoma rat cell line (PC-12), rat primary neurons (E18), human-induced pluripotent stem cells (hi-PSC), human neuroblastoma N2a cells, neuroblastoma derived cells (SH-Y5Y), and immortalized murine microglial cell line BV-2. Results from these studies, strongly indicated that TQ is able to counteract the neurotoxicity induced by Aβ peptide. Ismail et al. (2008) reported data on the neuroprotective role of NSO and its fractions (e.g., hexane fraction, ethyl acetate fraction and

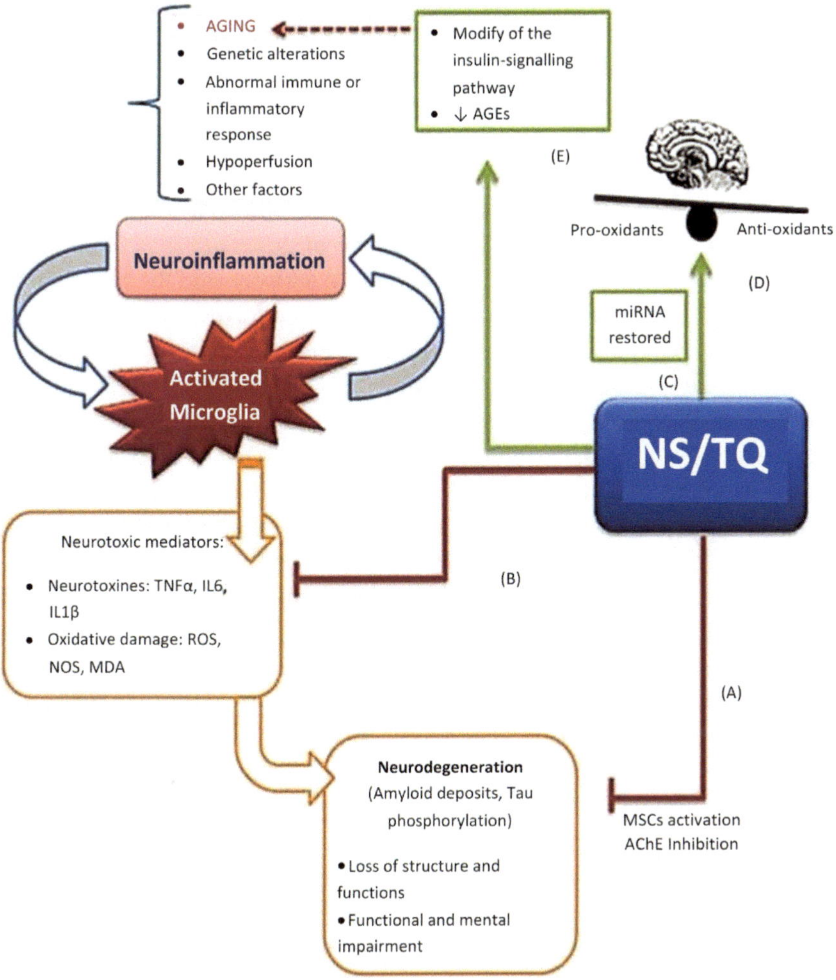

FIGURE 9.1 Neuroprotective mechanisms of NS and TQ. These compounds are able to limit the loss of structure and improve memory impairments (A) by inducing the activation of mesenchymal stem cells (MSCs), and by inducing AChE inhibition. They also interfere with the activity of hippocampal neuroinflammatory cytokines (TNFα, and IL6, IL1β) and markers of oxidative damage such as reactive oxygen species (ROS), reactive nitrogen species (NOS), MDA in brain tissues (B). On the contrary, they restore the level of AD correlate miRNA (C) and increase the activity of the anti-oxidants glutathione, glutathione peroxidase (GPx), superoxide dismutase (SOD) and catalase in hippocampus and cortex (reduced in aging), thus balancing the oxidative process (D). Again, they modify the insulin-signaling pathway and decrease the levels of advanced glycation end products (AGEs) involved in aging and AD (E).

water fraction [WF]) in primary rat CGNs. In these in vitro studies, WF and NSO had a better protective effect on Aβ-induced toxicity than NSO fractions, whereas the effect was obtained by acting on the antioxidant pathway. The neuroprotective properties of TQ in CGNs cells were also confirmed after exposition to the Aβ1-40 sequence (Aβ1-40) (Ismail et al., 2013). The authors showed that the pretreatment with TQ (0.1 and 1 μM) prevented the generation of free radicals and the apoptosis through caspase pathways, thus resulting in an improvement of cell viability. Additionally, TQ was able to preserve the extensive neurite network and the morphology of cell bodies damages. In another in vitro model, Khan et al. (2012) proved that TQ reduced the Aβ-induced neurotoxicity in PC-12 cells by inhibiting the oxidative stress and the mitochondrial dysfunction through the re-establishment of the abnormal levels of matrix metalloproteinases and ROS. Alhebshi et al. (2013) reported similar findings. In E18 neurons exposed to the neurotoxic Aβ, the authors showed that TQ had a neuroprotection role induced by the inhibition of ROS formation and mitochondrial membrane depolarization. Subsequently, the underlying neuroprotective mechanisms were depicted. Specifically, it has been shown that TQ protected E18 and hi-PSC against α-synuclein (αSN)-induced synaptic toxicity by acting on synaptic vesicle recycling (Alhebshi et al., 2014). Interesting findings were also reported by Kennedy et al. (2014) in a study on SH-SY5Y cells. The authors showed that TQ and other antioxidants [e.g., (−)-epigallocatechin-3-gallate (EGCG), dilinoleoylphosphatidylcholine] prevented the Aβ-induced oxidation by acting on the TNF-α-mediated signaling pathway, through the downregulation of nitric oxide (NO) and the upregulation of glutathione (GSH). It is important to underline that other studies were focused on TQ's antioxidant properties. Recently, Cobourne-Duval et al. (2016) demonstrated that TQ had antioxidant effects, inactivated BV-2 murine microglial cells, by acting on ROS and pro-inflammatory cytokines, which are highly expressed in several neurodegenerative disorders, including AD. It is worth noting that the authors showed that the pretreatment with TQ impaired the expression profiles of specific genes involved in the oxidation process. Specifically, TQ downregulated the expression of inducible nitric oxide synthase 2 (NOS2), chemokine (C–C) motif ligand 5 (Ccl5), prostaglandin-endoperoxide synthase 2 (Ptgs2), thioredoxin-interacting protein (Txnip) and upregulated the expression of peroxiredoxin 1 (Prdx 1) and sulfiredoxin 1 (Sxrn1). Interestingly, by using a technology based on co-encapsulated NSO and plasmid DNA (pDNA) with the application of polylactic-*co*-glycolic acid nanoparticles, Doolaanea et al. (2016) showed that the encapsulated NSO promoted the outgrowth of murine neuroblastoma cells BV-2, thus resulting in a fascinating translational perspective for AD's treatment based on gene therapy.

9.2.2 PRECLINICAL IN VIVO STUDIES

Several in vivo experiments were performed on a rat model with cognitive defects to prove the neuroprotective effects of NS and/or its constituent TQ. The first study was conducted in a rat model with neural damage induced by transient forebrain ischemia (Al-Majed et al., 2006). Data that emerged from these studies indicated that pretreatment with TQ was able to reduce the loss of hippocampal neuronal cells ($P < 0.001$), in the CA1 region. This effect was correlated to the superoxide dismutase (SOD) activities and the antioxidant activities of TQ on the levels of malondialdehyde (MDA), GSH, and catalase. Hosseinzadeh et al. (2007) showed that TQ and NSO had a similar neuroprotective effect on rat hippocampal cells, correlated with the inhibition of lipid peroxidation after cerebral ischemia.

More recently, Hobbenaghi et al. (2014) by administrating NS in rats during the carotid clamp proved that the strongest neuroprotective effect of NS, identified as the prevention of intracellular edema of hippocampal interneurons and astrocytes, was obtained with the highest dose (50 mg/kg). Later on, animal behavioral studies were conducted by using the two-vessel occlusion (2VO) murine model (Azzubaidi et al., 2012). This model is generated by a ligation and a cut of the common carotid arteries, thus representing an optimal strategy to demonstrate the so-called vascular hypothesis of AD responsible for neurodegeneration (de la Torre, 2010). In these studies, the Morris water maze (MWM) was applied to long-term memory, short-term memory, and working memory. Data emerged from these studies showed that the oral administration of NSO prior and post the 2VO surgeries were able to preserve the spatial cognitive functions. Azizi et al. (2012), in a rat model of AD, conducted an interesting study on the efficacy of thymol and carvacrol (a monoterpenic phenols present in NS) against cognitive deficits. Results suggested that these phenols were able to reduce the cognitive defects, thanks to their anti-inflammatory and antioxidant properties. Recently, a murine model of AD was used lipopolysaccharide (LPS) in combination with PNU-120596, a positive allosteric modulator (PAM) of $\alpha7$ nicotine acetyl choline receptor (nAChR), with the specific nAChR agonist PNU-282987 or with TQ (Ibrahim AbdEl Fattah et al., 2016). Data showed that TQ, or PNU-282987, combined with the PAM attenuated the neuroinflammation, and to enhance the activities of mesenchymal stem cells (MSCs). Indeed, by using molecular [anti-phospho-cyclic adenosine phosphate (cAMP) response element-binding protein, and CD44 Ab], and morphometric analyses (the evaluation of the area of deformed neurons, and Aβ plaque a reduction in a number of deformed neurons and glial cells as well as a rising number of pyramidal cells, were observed. Furthermore, the combination of TQ, or α7 nAChR agonist, with PAM had a stronger effect if compared to TQ alone.

The role of the brain insulin resistance and reduced insulin signaling in the AD's pathogenesis were also described by several studies focused on the molecular similarities in the inflammatory pathways involved in both AD and type 2 diabetes mellitus (Aliev et al., 2014). Baalba et al. (2016) showed, in streptozotocin-induced diabetic (STZ-T2DM) rats, that the daily intake of NSO altered the expression of insulin-like growth factor-1 and phosphoinositide-3 kinase and induced the expression of insulin receptor and (Balbaa et al., 2016). Based on these results, subsequently, the authors showed that NSO (2.0 mL/kg) and antidiabetic drugs (metformin and glimepiride) alone and/or in combination, suppressed the brain levels of oxidative stress markers such as xanthine oxidase and NOS, by producing lipid peroxidation (e.g., thiobarbituric acid reactive substances) (Balbaa et al., 2017). On the contrary, they detected increased activity of antioxidant agents (e.g., GTH) and enzymes such as GTH peroxidase (GPx), SOD, and glutathione S-transferase. Moreover, the NSO administration lowered the insulin receptor inhibitory effect of IOMe-AG538 by modifying the insulin-signaling pathway and by reducing the levels of advanced glycation end products (AGEs) that are proteins or lipids involved in aging and the pathogenesis of degenerative diseases, including AD. Again, the authors studied the modification of brain AD-related miRNA expression profile and demonstrated that NSO was able to restore the brain tissue levels and serum values of miRNA, suggesting a role of miRNAs in the early diagnosis of AD.

Recently, other studies on this topic have been performed. Bargi et al. (2017) reported that TQ improved the memory impairments induced by LPS in rats, yet at lower doses. These behavioral results were associated with reduced expression of the hippocampal neuroinflammatory cytokines TNF-α and IL6, a decreased level of markers of oxidative damage in brain tissues such as NOS, MDA, and increased activity of SOD and catalase in hippocampus and cortex. Finally, in a recent paper, Ismail et al. (2017a) tested TQ rich-fraction nanoemulsion (TQRFNE) compared to TQ conventional emulsion and TQ nanoemulsion, in rats fed with a diet rich of high fat cholesterol for 6 months. The data were very interesting since TQRFNE was able to ameliorate the behavioral changes, the lipid peroxidation and the soluble Aβ levels. The same authors showed that the reduction in Aβ levels could be associated with the modulation of β- and γ-secretase enzyme activity, and to the Aβ degradation and transportation in/out of the brain (Ismail et al., 2017b). Furthermore, there was an improved radical scavenging activity and increased antioxidants genes expression levels in the brain cortex and the hippocampus. These findings offer evidence that the presence of different compounds in the preparation (e.g., camphor, carvone, limonene, and thymol) may improve the effect of TQ alone. Moreover, oral bioavailability and brain delivery could be enhanced by the nanoemulsion.

9.3 (–)-EPIGALLOCATECHIN-3-GALLATE

The principal bioactive polyphenol in the solid green tea extract (GTE) is represented by the ester of epigallocatechin and gallic acid, EGCG [(2*R*,3*R*)-5,7-dihydroxy-2-(3,4,5-trihydroxyphenyl)-3,4-dihydro-2H-1-benzopyran-3-yl 3,4,5-trihydroxybenzoate] (PubChem CID: 65,064). EGCG is known as a natural compound with preventive and therapeutic effects in many clinical conditions, including cancer (Rahmani et al., 2015), metabolic and cardiovascular diseases (e.g., obesity and diabetes) (Eng et al, 2018) and neuropathic pain (Bimonte et al., 2017). The neuroprotective effects of EGCG are predominantly associated with the regulation of different molecular signaling pathways involved in the neuro-inflammation cascade. Furthermore, this compound reduced brain edema and neuronal damage after unilateral cerebral ischemia (Lee et al., 2004).

9.3.1 IN VITRO STUDIES

In vitro studies on the antineuroinflammatory effects of EGCG have been performed on different cells lines including human neurons that contains the C-99 fragment of APP (MC65), mouse-derived microglial cells (EOC

13.31), murine neuron-like cells (N2a) transfected with the human "Swedish" mutant APP (SweAPP N2a), N2a/APP695, differentiated neuronal cultures (DIV8), Chinese hamster ovary (CHO) cells, and neurons overexpressing human mutated presenilin-1 (M146 L) cells. Results from these studies showed that the antineuroinflammatory ability of EGCG is mainly related to the inhibition of microglia-induced cytotoxicity.

Regarding the role of EGCG against amyloid damages, Lin et al. (2009) showed that EGCG may attenuate the neurotoxicity induced by Aβ and P-tau, through the activation of the glycogen synthase kinase-3β (GSK-3β) and the inhibition of c-Abl/FE65 nuclear translocation. This was a significant result, since it is of note that c-Abl is a cytoplasmic nonreceptor tyrosine kinase involved in the development of the nervous system and in the regulation of cell apoptosis, whereas the β-isoform of GSK3 is a proline-directed serine–threonine kinase involved in neuronal cell development and energy metabolism and regulates the phosphorylation of tau (Wang, 2005). Interestingly, related to AD's pathogenesis, it has been shown that the c-Abl/GSK-3β signaling is involved in the neuronal loss, the neuroinflammation and the gliosis through the proteolytic processing of the transmembrane glycoprotein APP, responsible for the Aβ's origin (Sinha and Lieberburg, 1999). Other investigations have been performed to evaluate the effect of EGCG on Aβ-induced inflammatory responses in microglia. In this regard, researchers investigated the inhibitory effects of EGCG in Aβ-stimulated EOC 13.31 cell lines. Results proved that EGCG suppressed the expression of TNFα, IL-1β, IL-6, and inducible NOS (iNOS) and restored the levels of intracellular antioxidants against free radical-induced pro-inflammatory effects in microglia, the nuclear erythroid-2 related factor 2 (Nrf2) and the heme oxygenase-1 (HO-1). Additionally, EGCG suppressed Aβ-induced cytotoxicity by reducing ROS-induced nuclear factor κB (NF-κB) activation and mitogen-activated protein kinase (MAPK) signaling, including c-Jun N-terminal kinase (JNK) and p38 signaling (Cheng-Chung Wei et al., 2016). Taken together, these data suggest that EGCG is able to inhibit the neuroinflammatory response of microglia induced by Aβ and to protect against indirect neurotoxicity by acting with several mechanisms. Moreover, it seems that EGCG was able to reduce the production of amyloid since Rezai-Zadeh et al. (2005) showed, in SweAPP N2a cells, that GTE reduced the generation of Aβ and activated the nonamyloidogenic processing of APP, by promoting its physiological α-secretase cleavage. It is worth noting that one of the processes involved in the amyloid formation cascade, is the β-sheet formation. This event is frequently associated with cellular toxicity in many human protein misfolding diseases, including AD. It has been shown that EGCG interfered with this cascade, by redirection of prone polypeptides' aggregation into off-pathway protein assemblies (Ehrnhoefer et al., 2008). In another study, Bieschke et al. (2010) investigated the Aβ cleavage, showing that EGCG was able to convert the large mature Aβ fibrils into smaller forms featuring no toxicity for the mammalian cell. Subsequently, Chesser et al. (2016) showed, in DIV8 primary rat cortical neurons, that EGCG was able to improve the clearance of AD-relevant P-tau species. Similar results were obtained by Chang et al. (2015) in both M146 L and CHO cell lines. Other investigations showed that the nuclear receptor peroxisome proliferator-activated receptor-gamma (PPARγ) could be considered a potential target for AD treatment, due to its potent anti-oxidant and inhibitory effects on Aβ production, by negative regulation of the BACE1 pathway. Specifically, it has been recently proved that EGCG was able to attenuate the production of Aβ in N2a/APP695 cells such as the PPARγ agonist, pioglitazone, by suppressing the transcription and translation of BACE1. Moreover, it was proved that its effect was attenuated by a specific PPARγ inhibitor (GW9662) (Zhang et al., 2017). In other words, the potential therapeutic efficacy of EGCG in AD could be explained in terms of the up-regulation of PPARγ transcription.

9.3.2 PRECLINICAL IN VIVO STUDIES

The neuroprotective effects of EGCG have been dissected by in vivo experiments on several animal models. In an AD's mouse model generated by subcutaneous injection of D-gal (150 mg/kg for 6 weeks), the authors showed that EGCG (intragastric given, 2–6 mg/kg for 4 weeks) decreased significantly the hippocampal expression of APP and Aβ ($P < 0.01$). (He et al., 2012). Moreover, in a transgenic (Tg) mouse model of "Swedish" mutant APP, Rezai-Zadeh et al. (2005) showed that EGCG (injected intraperitoneally at a dose of 20 mg/kg) decreased the levels of Aβ and the formation of plaques, in a significant manner. Similar results were then confirmed by the same researchers, when they demonstrated that EGCG reduced the deposition of Aβ even if orally administered (EGCG in drinking water, 50 mg/kg) (Rezai-Zadeh et al., 2008). In another study conducted on APP Tg mouse models of AD, the authors investigated the role of EGCG on the deposition of Aβ in different brain areas. Data that emerged by immunohistochemistry showed that EGCG, given orally 20 mg/kg/day for 3 months, reduced Aβ deposits by 60% in the frontal cortex and by 52% in the hippocampus. A reduction of Thioflavin-S histochemistry labeling compact plaques was also detected in both regions. Moreover, the percentage of a marker of microglial activation, CD45, was lower in the cortex and the hippocampus with respect to those observed in the controls (Li et al., 2006). Smith et al. (2010) provide new evidences for the role of EGCG in AD's treatment, despite the poor bioavailability of poor delivery of EGCG to the central nervous system (Feng, 2006). The authors engineered nano lipidic EGCG particles (NanoEGCG, 30–80 nm diameter nanoparticle complexes) to improve EGCG's oral bioavailability and showed by using a rat model, which NanoEGCG bioavailability was more enhanced with respect to rats treated with free EGCG. Furthermore, the nanoparticles promoted the α-secretase activity in SweAPP N2a cells, even at reduced concentrations.

The role of EGCG in AD's treatment was also described in vivo (Tg2576) by Giunta et al. (2010), who tested fish oil (8 mg/kg/day) in combination with EGCG (62.5 mg/kg/day or 12.5 mg/kg/day). Results obtained from these studies showed a synergetic effect of these substances on the inhibition of cerebral Aβ deposits. Yet, fish oil enhanced the bioavailability of EGCG versus EGCG treatment alone ($P<0.001$). More recently, Lee et al. (2013) tested the effects of EGCG on neuroinflammation and amyloid genesis in mice with systemic inflammation. According to these data, EGCG prevented the memory impairment induced by LPS and the apoptotic neuronal cell death. Moreover, EGCG prevented LPS-induced activation of astrocytes and increased the expression of cytokines (TNF-α, IL-1β, IL-6). In another murine model of dementia, Biasibetti et al. (2013) assumed that EGCG (delivered by oral gavage at the dose of 10 mg/kg/day for 4 weeks) completely abrogated the cognitive deficit (through the evaluation with the MWM apparatus). Molecular tests indicated that EGCG was able to influence the GSH peroxidase activity, the glial-specific calcium-binding protein S100B content in the hippocampus, the acetylcholinesterase (AChE) activity, the NO metabolites, and ROS content. On the contrary, EGCG was not able to modify GSH content and the uptake of glucose. Behavioral investigations were conducted by Chang et al. (2015) in senescence-accelerated mice P8 (SAMP8). The authors showed that EGCG attenuated the cognitive deterioration by upregulation of neprilysin expression, a zinc-dependent metalloprotease that cleaves peptides (e.g., substance P, neurotensin, oxytocin, and bradykinin) and, related to AD, Aβ aggregates.

Regarding the relationship between metabolic processes and AD pathogenesis, it is of note that insulin signaling plays a significant role in the regulation of synaptic activities involved in learning and memory processes. Thus, insulin resistance—expressed as increased phosphorylation levels of insulin receptor substrate-1 (IRS-1)—and subsequent alteration in signaling pathways, may cause the cognition impairment in AD's patients (de La Monte, 2012). Interestingly, Jia et al. (2013) showed that EGCG reduced the spatial memory impairment (AD-related cognitive deficit) in APP/PS1 mice (bearing brain insulin resistance) by inducing IRS-1 signaling defects in the hippocampus, in a dose-dependent manner.

9.4 CURCUMIN

Curcumin (1,7-bis(4-hydroxy-3-methoxyphenyl)-1,6-heptadiene-2,5-dione) is an orange–yellow colored lipophilic polyphenol compound derived from turmeric (*Curcuma longa*) that is a perennial herbaceous plant belonging to the ginger family (*Zingiberaceae*) and widely cultivated in the south and the southeast Asia, especially China and India. It is acquired from the rhizomes of the herb and has been used as a food additive or dietary pigment and in traditional medicine; today, curcumin is one of the most popular herbal supplements. This pigment is chemically a diferuloylmethane with a polyphenolic molecular structure similar to other pigments (e.g., resveratrol), catechins, or compounds found in certain fruit juices (e.g., blueberries, strawberries, pomegranates, etc.). *Curcuma longa* contains several congeners of curcumin (curcuminoids) such as demethoxycurcumin and bisdemethoxycurcumin. Another chemical in turmeric tested in lab-studies is aromatic-turmerone. Although this compound possesses anti-inflammatory properties resulting from the blockade of microglia signaling (Park et al., 2012), further studies are needed to prove its efficacy on neurodegenerative conditions like AD.

In vitro and in vivo studies have demonstrated interesting antioxidant, anti-inflammatory, and anticancer activities of curcumin (Bimonte et al., 2016) and its favorable toxicity profile. About the neuroprotective effects, the molecular structure of curcumin suggested potential Aβ binding. Moreover, because of its antioxidant, anti-inflammatory (e.g., inhibition of NF-kB-induced iNOS and inflammatory cytokine production) and anti-amyloid properties, a large number of preclinical studies have been conducted to explore the mechanisms of the neuroprotective role of curcumin in AD (Tang and Taghibiglou, 2017). From the clinical point of view, daily oral curcumin may improve memory and attention in nondemented adults (Small et al., 2018). Furthermore, an 18-month study of memory effects of curcumin, given twice daily (90 mg) to subjects with mild cognitive impairment or any age-related memory decline, is ongoing (ClinicalTrials.gov Identifier: NCT01383161). These results may provide more precise details on the utility of using this nutraceutical for AD prevention.

9.4.1 IN VITRO STUDIES

Accumulating pieces of evidence underlined the role of curcumin in AD pathogenesis. Zhang et al. (2006) proved, by in vitro studies, that curcumin may induce immune clearance of amyloidosis in the AD brain. To demonstrate this hypothesis, the macrophages of AD patients were treated with curcuminoids (curcumin complex) and showed that after the treatment (curcuminoids 0.1 μM), Aβ uptake by macrophages was significantly enhanced ($P < 0.001$ to 0.081). Later on, Park et al. (2008) conducted in vitro studies to dissect the neuroprotective role of curcumin on amyloid toxicity and demonstrated, in neuron-like PC12 rat cells exposed to 10 μM Aβ (25–35) for 24 h, that pretreatment with 10 μg/mL curcumin for 1 h, significantly reversed the Aβ-induced toxicity by decreasing the oxidative stress and the DNA damage. Later on, studies on the effects of curcuminoids on Aβ dynamics showed a dose-dependent inhibition of Aβ aggregation and secretion (up to 20%) in SweAPP N2A cells, although the results were influenced by the turmeric extracts used (Shytle et al., 2009). Zhang et al. (2010) demonstrated that curcumin may affect the endocytosis of APP from the cell surface, delaying the exit of immature APP from the endoplasmic reticulum, and enhancing the stability of that dangerous compound. The ratio of mature APP to immature was decreased up to 67% in the presence of curcumin that also markedly attenuated the effects of brefeldin A, a specific Golgi-disrupting agent, on APP maturation and trafficking. These mechanisms may lead to a significant decrease in Aβ40 and Aβ42 levels. The protective role was also manifested as attenuation of intracellular calcium levels and P-tau reduction via direct inhibition of GSK-3β (Xiong et al., 2011), or by interfering on its signaling pathway (Huang et al, 2014). In other investigations, curcumin (0.1–1 μM) reduced fibril formation and extension, and destabilized Aβ fresh fibrils, in a dose-depending manner, promoted the inhibition of P-tau (Ma et al., 2009) and decreased P-tau neurotoxicity by inducing its cleavage and by improving microtubule stabilization (Miyasaka et al., 2016). Furthermore, several studies showed that curcumin may modulate the genomic instability events (Thomas et al., 2009) and mitochondrial functions by upregulating the expression of genes encoding for antioxidant proteins, such as heme oxygenase-1, SOD, catalase as well as GSH reductase, GSH peroxidase, and GSH transferase (Gibellini et al., 2015). Again, curcumin may also induce immune clearance of amyloidosis in the AD brain. Recently, researchers from the Yat-sen University in China (Shi et al., 2018) investigated the effects of curcumin on acrolein-induced AD-like pathologies in HT22 mouse hippocampal cells. The acrolein-induced neurotoxicity is mostly expressed as a reduction in A-disintegrin and metalloprotease combined with an increment of APP, BACE, and receptor for advanced glycation end products. In this study curcumin (5 μg/mL for 30 min) significantly reversed acrolein-induced damages by acting on oxidative stress. Moreover, it was observed a regulation in the brain-derived neurotrophic factor (BDNF)/tropomyosin-related kinase B (TrkB) signaling which is involved in the production of Aβ, P-tau inclusions and, finally, in cognition decline (Zhang et al., 2012). Curcumin may directly act on the microglial activation by attenuating the production of NO and inflammatory cytokines, and by modulating the expression of iNOS. Cianculli et al. (2016) proved, in LPS-activated microglia, that these significant effects of curcumin are mediated by the phosphoinositide 2 kinase (PI3k)/Akt signaling pathway and NF-κB modulation. Taken together, these results indicate that curcumin is able to interfere with Aβ and P-tau levels mostly by mitigating Aβ-induced ROS damages (Wang et al., 2009), and microglial activation (Table 9.1).

9.4.2 PRECLINICAL IN VIVO STUDIES

Research on the ability of curcumin to reduce the formation of Aβ plaques in Tg2576 mice showed that low doses of dietary curcumin (160 ppm) for 6 months decreased the formation of both insoluble (39.2%) and soluble (43%) amyloids. Moreover, it was observed a reduction in amyloid plaque burdens in different brain areas (Lim et al., 2001). It is noted that a higher dose of dietary curcumin did not reduce Aβ levels. In aged

TABLE 9.1 A Summary of In Vitro Studies on the Role of Curcumin in AD

Cell lines	Dose of Curcumin	Effects	Ref.
Macrophages of AD patients	Curcuminoids 0.1 µM	Reversion of the Aβ-induced toxicity	Zhang et al., 2006
PC12	10 µg/mL		Park et al, 2008
SweAPP N2A	5 µg/mL.	Inhibition of Aβ aggregation and secretion	Shytle et al, 2009
H4 cells	0, 1, 2.5, 7.5, 10, and 20 µm	Impairment of endocytosis of APP. Decreased levels of Aβ40 and Aβ42	Zhang et al, 2010
SH-SY5Y	0, 1.25, 5.0, or 20.0 µM	Decreased Aβ production by inhibiting GSK-3β-mediated PS1 activation. Depression of Aβ-induced upregulation of PTEN induced by Aβ	Xiong et al., 2011; Huang et al, 2014
Neuronal cultures*	0.1–1 µM	Inhibition of P-tau and decreasing of P-tau neurotoxiity	Ma et al., 2009; Miyasaka et al., 2016.
HT22	5 µg/mL	Reversion of acrolein-induced damages by acting on the oxidative stress	Shi et al., 2018
BV2 microglia	10–50 µM	Attenuation of LPs-induced inflammatory responses in microglial cells and downregulation of PI3K/Akt signaling	Cianculli et al., 2016

Legend: *prepared from embryonic day 18 Sprague Dawley rat fetuses); Abbreviations: APP, amyloid-β precursor protein; PTEN, phosphatase and tensin homolog

Sprague–Dawley rats, dietary curcumin (2000 ppm) suppressed oxidative damage and synaptophysin loss, attenuated microgliosis in cortical layers, and increased microglial labeling within and adjacent to plaques. Furthermore, at the lower dosage (160 ppm) curcumin prevented spatial memory deficits in the MWM tests (Frautschy et al., 2001). Subsequently, Yang et al. (2005) proved in aged Tg2576 mice with advanced amyloid accumulation, that curcumin, injected peripherally, crossed the blood-brain barrier and disrupted existing plaques. However, Hamaguchi et al. (2009) showed in the same murine model that curcumin (5000 ppm) did not impact amyloid deposition in brain tissues and, consequently, hypothesized that curcumin may prevent Aβ polymerization but not its deposition. However, in another study, low doses of curcumin (500 ppm), orally administered for 4 months, induced a significant reduction in central nervous system levels of IL-1β and isoprostanes (an index of oxidative stress) combined with a decrease in plaque burden and amyloids deposits (Begum et al., 2008). Probably, these discordant results depend on the different bioavailability and/or efficacy of the turmeric extract or curcuminoids used. Indeed, Shytle et al. (2012) used orally a specific extract (HSS-888) in Tg2576 mice for a period of 6 months and demonstrated a reduced Aβ deposition throughout the entorhinal cortex and hippocampus combined with an attenuation of microglial inflammation. The authors compared HSS-888 to tetrahydrocurcumin (THC), a product of bacterial or intestinal metabolism of curcumin. However, studies underlined that curcumin differs from THC for molecular targets, signaling pathways and cellular responses (Aggarwal et al., 2014). Furthermore, it is also known that curcumin is unstable at physiological pH in vitro and undergoes a rapid autoxidative transformation, whereas its efficacy in models of neuroinflammation has been debated (Begun et al., 2008). Recently, Joseph et al. (2018) proved, in RAW264.7 cells, that the products resulting from the oxidative transformation of the curcumin analog 2,6-dimethyl-curcumin contribute to the anti-inflammatory activity (e.g., inhibition of NF-κB activity) in addition to its enhanced resistance against the enzymatic reduction. These findings should be better investigated in vivo. Moreover, the major curcumin metabolites, hexahydrocurcumin, exhibited similar or more potent bioactivity than curcumin (Huang et al., 2018). About the histopathology, it was proved that curcumin restored a considerable number of distorted neurites and, in turn, enhanced plasticity (Garcia-Alloza e al., 2007). Electrically, the effects of curcumin on synaptic plasticity were also investigated by Ahmed et al. (2011) who showed that in rats exposed to Aβ peptide curcuminoids can restore susceptibility for plastic changes in CA1 hippocampal neurons, by enhancing long-term potentiation.

Although the pharmacokinetics of curcumin is often a limitation of in vivo studies on the topic, Garcia-Alloza et al. (2006), by used multiphoton microscopy in vivo, demonstrated that curcumin was able to cross the blood-brain barrier. This research was conducted on the APP-swe/PS1dE9 murine model, by the intravenous administration of curcumin.

KEYWORDS

- **Alzheimer disease**
- **dementia**
- **curcumin**
- ***Curcuma longa***
- **green tea**
- **epicatechin**
- ***Nigella sativa***

REFERENCES

Aggarwal, B. B.; Deb, L.; Prasad, S. Curcumin differs from tetrahydrocurcumin for molecular targets, signaling pathways and cellular responses. *Molecules*. **2014**, *20*, 185–205.

Agostinho P.; Cunha, R. A.; Oliveira, C. Neuroinflammation, oxidative stress and the pathogenesis of Alzheimer's disease. *Curr. Pharm. Des*. **2010**, *16*, 2766–2778.

Ahmed, T.; Gilani, A. H.; Hosseinmardi, N.; Semnanian, S.; Enam, S. A.; Fathollahi, Y. Curcuminoids rescue long-term potentiation impaired by amyloid peptide in rat hippocampal slices. *Synapse*. **2011**, *65*, 572–582.

Aisen P. S.; Schneider L. S.; Sano, M.; Diaz-Arrastia, R.; van Dyck, C. H.; Weiner M. F.; Bottiglieri, T.; Jin, S.; Stokes, K.T.; Thomas, R.G.; Thal, L.J.: Alzheimer disease cooperative study. High-dose B vitamin supplementation and cognitive decline in Alzheimer disease: a randomized controlled trial. *JAMA*. **2008**, *300*, 1774–1783.

Akram Khan, M.; Afzal, M. Chemical composition of Nigella sativa Linn: Part 2. Recent advances. *Inflammopharmacology*. **2016**, *24*, 67–79.

Al-Majed, A. A.; Al-Omar, F. A.; Nagi, M. N. Neuroprotective effects of thymoquinone against transient forebrain ischemia in the rat hippocampus. *Eur. J. Pharmacol*. **2006**, *543*, 40–47.

Alhebshi, A. H.; Gotoh, M.; Suzuki, I. Thymoquinone protects cultured rat primary neurons against amyloid beta-induced neurotoxicity. *Biochem. Biophys. Res. Commun.* **2013**, *433*, 362–367.

Alhebshi, A. H.; Odawara, A.; Gotoh, M.; Suzuki, I. Thymoquinone protects cultured hippocampal and human induced pluripotent stem cells-derived neurons against alpha-synuclein-induced synapse damage. *Neurosci. Lett.* **2014**, *570*, 126–131.

Aliev, G.; Shahida, K.; Gan, S. H.; Firoz, C.; Khan, A.; Abuzenadah, A. M.; Kamal, W.; Kamal, M. A.; Tan, Y.; Qu, X.; Reale, M. Alzheimer disease and type 2 diabetes mellitus: the link to tyrosine hydroxylase and probable nutritional strategies. *CNS Neurol. Disord. Drug Targets.* **2014**, *13*, 467–477.

Andrade C.; Radhakrishnan, R. The prevention and treatment of cognitive decline and dementia: an overview of recent research on experimental treatments. *Indian J. Psychiatry.* **2009**, *51*, 12–25.

Azizi, Z.; Ebrahimi, S.; Saadatfar, E.; Kamalinejad, M.; Majlessi, N. Cognitive-enhancing activity of thymol and carvacrol in two rat models of dementia. *Behav. Pharmacol.* **2012**, *23*, 241–249.

Azzubaidi, M. S.; Saxena, A. K.; Talib, N. A.; Ahmed, Q. U.; Dogarai, B. B.Protective effect of treatment with black cumin oil on spatial cognitive functions of rats that suffered global cerebrovascular hypoperfusion. *Acta Neurobiol. Exp.* **2012**, *72*, 154–165.

Badary, O. A.; Taha, R. A.; Gamal el-Din, A. M.; Abdel-Wahab, M. H. Thymoquinone is a potent superoxide anion scavenger. *Drug Chem. Toxicol.* **2003**, *26*, 87–98.

Balbaa, M.; Abdulmalek, S. A.; Khalil, S. Oxidative stress and expression of insulin signaling proteins in the brain of diabetic rats: role of Nigella sativa oil and antidiabetic drugs. *PLoS One*. **2017**, *12*, e0172429. 10.1371/journal.pone.0172429

Balbaa, M.; El-Zeftawy, M.; Ghareeb, D.; Taha, N.; Mandour, A. W. Nigella sativa relieves the altered insulin receptor signaling in streptozotocin-induced diabetic rats fed with a high-fat diet. *Oxid. Med. Cell. Longev*. **2016**, *2016*, 2492107. 10.1155/2016/2492107

Bargi, R.; Asgharzadeh, F.; Beheshti, F.; Hosseini, M.; Sadeghnia, H. R.; Khazaei, M. The effects of thymoquinone on hippocampal cytokine level, brain oxidative stress status and memory deficits induced by lipopolysaccharide in rats. *Cytokine*. **2017**, *96*, 173–184.

Begum, A.N.; Jones, M.R.; Lim, G. P.; Morihara, T.; Kim, P.; Heath, D. D.; Rock, C. L.; Pruitt, M. A.; Yang, F.; Hudspeth, B.; Hu, S.; Faull, K. F.; Teter, B.; Cole, G.M.; Frautschy, S. A. Curcumin structure-function, bioavailability, and efficacy in models of neuroinflammation and Alzheimer's disease. *J. Pharmacol. Exp. Ther*. **2008**, *326*: 196–208.

Biasibetti, R.; Tramontina, A. C.; Costa, A. P.; Dutra, M. F.; Quincozes-Santos, A.; Nardin, P.; Bernardi, C. L.; Wartchow, K. M.; Lunardi, P. S.; Gonçalves, C. A. Green tea (−) epigallocatechin-3-gallate reverses oxidative stress and reduces acetylcholinesterase activity in a streptozotocin-induced model of dementia. *Behav. Brain Res*. **2013**, *236*, 186–193.

Bieschke, J.; Russ, J.; Friedrich, R. P.; Ehrnhoefer, D. E., Wobst, H.; Neugebauer, K.; Wanker, E. E. EGCG remodels mature α-synuclein and amyloid-β fibrils and reduces cellular toxicity. *Proc. Natl. Acad. Sci. U S A*. **2010**, *107*, 7710–7715.

Bimonte, S.; Barbieri, A.; Leongito, M.; Piccirillo, M.; Giudice, A.; Pivonello, C.; de Angelis, C; Granata, V.; Palaia, R.; Izzo, F. Curcumin anticancer studies in pancreatic cancer. *Nutrients*. 2016, *8*, 433.

Bimonte, S.; Cascella, M.; Schiavone, V.; Mehrabi-Kermani, F.; Cuomo, A. The roles of epigallocatechin-3-gallate in the treatment of neuropathic pain: an update on preclinical in vivo studies and future perspectives. *Drug. Des. Devel. Ther.* **2017**, *11*, 2737–2742.

Bin Sayeed, M. S.; Asaduzzaman, M.; Morshed H., Hossain, M. M.; Kadir, M. F.; Rahman, M. R. The effect of Nigella sativa Linn. seed on memory, attention and cognition in healthy human volunteers. *J. Ethnopharmacol.* **2013**, *148*, 780–786.

Bin Sayeed, M. S.; Shams, T.; Fahim Hossain, S.; Rahman, M. R.; Mostofa, A.; Fahim Kadir, M., Mahmood, S.; Asaduzzaman, M. Nigella sativa L. seeds modulate mood, anxiety and cognition in healthy adolescent males. *J. Ethnopharmacol.* **2014**, *152*, 156–162.

Cascella, M.; Bimonte, S.; Barbieri, A.; Del Vecchio, V.; Muzio, M. R.; Vitale, A.; Benincasa, G.; Ferriello, A. B.; Azzariti, A.; Arra, C.; Cuomo, A. Dissecting the potential roles of Nigella sativa and Its constituent thymoquinone on the prevention and on the progression of Alzheimer's disease. *Front. Aging Neurosci.* **2018**, *10*, 16. 10.3389/fnagi.2018.00016.

Cascella M.; Bimonte S.; Muzio M. R.; Schiavone V.; Cuomo A. The efficacy of epigallocatechin-3-gallate (green tea) in the treatment of Alzheimer's disease: an overview of pre-clinical studies and translational perspectives in clinical practice. *Infect. Agents Cancer.* **2017**, *12*:36.

Cascella, M.; Palma, G.; Barbieri, A.; Bimonte, S.; Amruthraj, N. J.; Muzio, M. R.; del Vecchio, V.; Rea, D.; Falco, M.; Luciano, A.; Cuomo, A. Role of *Nigella sativa* and its constituent thymoquinone on chemotherapy-induced nephrotoxicity: evidences from experimental animal studies. *Nutrients.* **2017**, *9*(6), 625. 10.3390/nu9060625.

Chang, X.; Rong, C.; Chen, Y.; Yang, C.; Hu, Q.; Mo, Y.; Zhang, C.; Gu, X.; Zhang, L.; He, W.; Cheng, S.; Hou, X.; Su, R.; Liu, S.; Dun, W.; Wang, Q.; Fang, S. (−)-Epigallocatechin-3-gallate attenuates cognitive deterioration in Alzheimer's disease model mice by upregulating neprilysin expression. *Exp. Cell. Res.* **2015**, *334*, 136–145.

Cheng-Chung Wei, J.; Huang, H. C.; Chen, W. J.; Huang, C. N.; Peng, C. H.; Lin, C. L. Epigallocatechin gallate attenuates amyloid β-induced inflammation and neurotoxicity in EOC 13.31 microglia. *Eur. J. Pharmacol.* **2016**, *770*, 16–24.

Chesser, A. S.; Ganeshan, V.; Yang, J.; Johnson, G. V. Epigallocatechin-3-gallate enhances clearance of phosphorylated tau in primary neurons. *Nutr. Neurosci.* **2016**, *19*, 21–31.

Cianciulli, A.; Calvello, R.; Porro, C.; Trotta, T.; Salvatore, R.; Panaro, M. A. PI3k/Akt signalling pathway plays a crucial role in the anti-inflammatory effects of curcumin in LPS-activated microglia. *Int. Immunopharmacol.* **2016**, *36*, 282- 290.

Cobourne-Duval, M. K.; Taka, E.; Mendonca, P.; Bauer, D.; Soliman, K. F. The antioxidant effects of thymoquinone in activated BV-2 murine microglial cells. *Neurochem. Res.* **2016**, *41*, 3227–3238.

De Domenico S.; Giudetti, A. M. Nutraceutical intervention in ageing brain. *J.G.G.*, **2017**, *65*, 79–92.

de la Monte, S. M. Brain insulin resistance and deficiency as therapeutic targets in Alzheimer's disease. *Curr. Alzheimer Res.* **2012**, *9*, 35–66.

de la Torre, J. C. The vascular hypothesis of Alzheimer's disease: bench to bedside and beyond. *Neurodegener. Dis.* **2010**, *7*, 116–21.

Doolaanea, A. A.; Mansor, N.; Mohd Nor, N. H.; Mohamed, F. Co-encapsulation of Nigella sativa oil and plasmid DNA for enhanced gene therapy of Alzheimer's disease. *J. Microencapsul.* **2016**, *33*, 114–126.

Elmaci, I.; Altinoz, M. A. Thymoquinone: an edible redox-active quinone for the pharmacotherapy of neurodegenerative conditions and glial brain tumors. A short review. *Biomed. Pharmacother*. **2016**, *83*, 635–640.

Feng, W. Y. Metabolism of green tea catechins: an overview. *Curr. Drug. Metab*. 2006, *7*, 755–809.

Frautschy, S. A.; Hu, W.; Kim, P.; Miller, S. A.; Chu, T.; Harris-White, M. E.; Cole, G. M. Phenolic anti-inflammatory antioxidant reversal of Aβ-induced cognitive deficits and neuropathology. *Neurobiol. Aging*. **2001**, *22*, 993–1005.

Garcia-Alloza, M.; Borrelli, L. A.; Rozkalne, A.; Hyman, B. T.; Bacskai, B. J. Curcumin labels amyloid pathology in vivo, disrupts existing plaques, and partially restores distorted neurites in an Alzheimer mouse model. *J. Neurochem*. 2007, *102*, 1095–104.

Garcia-Alloza, M.; Robbins, E. M.; Zhang-Nunes, S. X.; Purcell, S. M.; Betensky, R. A.; Raju, S.; Prada, C.; Greenberg, S. M.; Bacskai, B. J.; Frosch, M. P. Characterization of amyloid deposition in the APPswe/PS1dE9 mouse model of Alzheimer disease. *Neurobiol. Dis*. 2006, *24*, 516–524.

Gibellini, L.; Bianchini, E.; De Biasi, S.; Nasi, M.; Cossarizza, A; Pinti, M. Natural compounds modulating mitochondrial functions. *Evid. Based Complement. Alternat. Med*. 2015, 1–13.

Giunta, B.; Hou, H.; Zhu, Y.; Salemi, J.; Ruscin, A.; Shytle, R. D.; Tan, J. Fish oil enhances anti-amyloidogenic properties of green tea EGCG in Tg2576 mice. *Neurosci. Lett*. 2010, *471*, 134–138.

Goozee, K. G.; Shah, T. M.; Sohrabi, H. R.; Rainey-Smith, S. R.; Brown, B.; Verdile, G.; Martins, R. N. Examining the potential clinical value of curcumin in the prevention and diagnosis of Alzheimer's disease. *Br. J. Nutr*. 2016, *115*, 449–65.

Ehrnhoefer, D. E.; Bieschke, J.; Boeddrich, A., Herbst, M.; Masino, L.; Lurz, R.; Engemann, S.; Pastore, A.; Wanker, E. E. EGCG redirects amyloidogenic polypeptides into unstructured, off-pathway oligomers. *Nat. Struct. Mol. Biol*. 2008, *15*, 558–566.

Eng, Q. Y.; Thanikachalam, P. V.; Ramamurthy, S. Molecular understanding of Epigallocatechin gallate (EGCG) in cardiovascular and metabolic diseases. J. Ethnopharmacol. 2018, *210*, 296–310.

Hamaguchi, T.; Ono, K.; Murase, A.; Yamada, M. Phenolic compounds prevent Alzheimer's pathology through different effects on the amyloid-β aggregation pathway. *Am. J. Pathol*. 2009, *175*, 2557–2565.

He, M.; Liu, M. Y.; Wang, S.; Tang, Q. S.; Yao, W. F.; Zhao, H. S.; Wei, M. J. Research on EGCG improving the degenerative changes of the brain in AD model mice induced with chemical drugs. *Zhong Yao Cai*. 2012, *35*, 1641–1644.

Hosseinzadeh, H.; Parvardeh, S.; Asl, M. N.; Sadeghnia, H. R.; Ziaee, T. Effect of thymoquinone and Nigella sativa seed oil on lipid peroxidation level during global cerebral ischemia-reperfusion injury in rat hippocampus. *Phytomedicine*. 2007, *14*, 621–627.

Huang, Y.; Cao, S.; Zhang, Q.; Zhang, H.; Fan, Y.; Qiu, F.; Kang, N. Biological and pharmacological effects of hexahydrocurcumin, a metabolite of curcumin. *Arch. Biochem. Biophys*. 2018, *646*, 31–37.

Huang, H-C.; Tang, D.; Xu, K.; Jiang, Z-F. Curcumin attenuates amyloid--induced tau hyperphosphorylation in human neuroblastoma SH-SY5Y cells involving PTEN/Akt/GSK-3 signaling pathway. *J. Recept. Signal. Transduct. Res*. 2014, *34*, 26–37.

Iadecola, C. The overlap between neurodegenerative and vascular factors in the pathogenesis of dementia. *Acta Neuropathol.* 2010, *120*, 287–296.

Ibrahim AbdEl Fattah, L.; Zickri, M. B.; Aal, L. A.; Heikal, O.; Osama, E. The effect of thymoquinone, alpha7 receptor agonist and alpha7 receptor allosteric modulator on the cerebral cortex in experimentally induced Alzheimer's disease in relation to MSCs activation. *Int. J. Stem Cells.* 2016, *9*, 230–238.

Ismail, N.; Ismail, M.; Latiff, L. A.; Mazlan, M.; Mariod, A. A. Black cumin seed (Nigella sativa linn.) oil and its fractions protect against beta amyloid peptide-induced toxicity in primary cerebellar granule neurons. *J. Food Lipids.* 2008, *15*, 519–533.

Ismail, N.; Ismail, M.; Mazlan, M.; Latiff, L. A.; Imam, M. U.; Iqbal, S.; Azmi, N. H.; Ghafar, S. A.; Chan, K. W. Thymoquinone prevents beta-amyloid neurotoxicity in primary cultured cerebellar granule neurons. *Cell. Mol. Neurobiol.* 2013, *33*, 1159–1169.

Ismail, N.; Ismail, M.; Azmi, N. H.; Bakar, M. F. A.; Yida, Z.; Stanslas, J.;Sani, D.; Basri, H.; Abdullah, M. A. Beneficial effects of TQRF and TQ nano- and conventional emulsions on memory deficit, lipid peroxidation, total antioxidant status, antioxidants genes expression and soluble Aβ levels in high fat-cholesterol diet-induced rats. *Chem. Biol. Interact.* 2017a, *275*, 61–7359.

Ismail, N.; Ismail, M.; Azmi, N. H.; Bakar, M. F. A.; Yida, Z.; Abdullah, M. A.; Basri, H. Thymoquinone-rich fraction nanoemulsion (TQRFNE) decreases Aβ40 and Aβ42 levels by modulating APP processing, up-regulating IDE and LRP1, and down-regulating BACE1 and RAGE in response to high fat/cholesterol diet-induced rats. *Biomed. Pharmacother.* 2017b, *95*, 780–788.

Jia, N.; Han, K.; Kong, J. J.; Zhang, X. M.; Sha, S.; Ren, G. R.; Cao, Y. P. (−)-Epigallocatechin-3-gallate alleviates spatial memory impairment in APP/PS1 mice by restoring IRS-1 signaling defects in the hippocampus. *Mol. Cell. Biochem.* 2013, *380*, 211–218.

Joseph, A. I.; Edwards, R. L.; Luis, P. B.; Presley, S. H.; Porter, N. A.; Schneider, C. Stability and anti-inflammatory activity of the reduction-resistant curcumin analog, 2,6-dimethyl-curcumin. *Org. Biomol. Chem.* 2018, *16*, 3273–3281.

Kennedy, K.; Tucci M. A.; Benghuzzi, H. A. Comparison of potential preventive therapeutic agents green tea, thymoquinone, and dilinoleoylphosphatidylcholine on human neuroblastoma cells. *Biomed. Sci. Instrum.* 2014, *50*, 132–139.

Khan, A.; Vaibhav, K.; Javed, H.; Khan, M. M.; Tabassum, R.; Ahmed, M. E.; Srivastava, P.; Khuwaja, G.; Islam, F.; Siddiqui, M. S.; Safhi, M. M.; Islam, F. Attenuation of Abeta-induced neurotoxicity by thymoquinone via inhibition of mitochondrial dysfunction and oxidative stress. *Mol. Cell. Biochem.* 2012, *369*, 55–65.

Kundu, J.; Chun, K.-S.; Aruoma, O. I.; Kundu J. K. Mechanistic perspectives on cancer chemoprevention/chemotherapeutic effects of thymoquinone. *Mutat. Res.* 2014, *768*, 22–34.

Lee, H.; Bae, J. H.; Lee, S. R. Protective effect of green tea polyphenol EGCG against neuronal damage and brain edema after unilateral cerebral ischemia in gerbils. *J. Neurosci. Res.* 2004, *77*, 892–900.

Lee, Y.J.; Choi, D.Y.; Yun, Y. P., Han, S. B.; Oh, K. W., Hong, J. T. Epigallocatechin-3-gallate prevents systemic inflammation-induced memory deficiency and amyloidogenesis via its anti-neuroinflammatory properties. *J. Nutr. Biochem.* 2013, *24*, 298–310.

Li, Q.; Gordon, M.; Tan, J., Morgan, D. Oral administration of green tea epigallocatechin-3-gallate (EGCG) reduces amyloid beta deposition in transgenic mouse model of Alzheimer's disease. *Exp. Neurol.* 2006, *198*, 576.

Lim, G. P.; Chu, T.; Yang, F.; Beech, W.; Frautschy, S.A.; Cole, G. M. The curry spice curcumin reduces oxidative damage and amyloid pathology in an Alzheimer transgenic mouse. *J. Neurosci.* 2001, *21*, 8370–8377.

Lin, C. L.; Chen, T. F.; Chiu, M. J.; Way, T. D.; Lin, J. K. Epigallocatechin gallate (EGCG) suppresses beta-amyloid-induced neurotoxicity through inhibiting c-Abl/FE65 nuclear translocation and GSK3 beta activation. *Neurobiol. Aging.* 2009, *30*, 81–92.

Ma, Q. L.; Yang, F.; Rosario, E. R.; Ubeda, O. J.; Beech, W.; Gant, D. J.; Chen, P. P.; Hudspeth, B.; Chen, C.; Zhao, Y.; Vinters, H. V.; Frautschy, S. A.; Cole, G. M. Beta-amyloid oligomers induce phosphorylation of tau and inactivation of insulin receptor substrate via c-Jun N-terminal kinase signaling: suppression by omega-3 fatty acids and curcumin. *J. Neurosci.* 2009, *29*, 9078–9089.

Miyasaka, T.; Xie, C; Yoshimura, S.; Shinzaki, Y.; Yoshina, S.; Kage-Nakadai, E.; Mitani, S.; Ihara, Y. Curcumin improves tau-induced neuronal dysfunction of nematodes. *Neurobiol. Aging.* 2016, 39, 69–81.

Mosher K. I.; Wyss-Coray, T. Microglial dysfunction in brain aging and Alzheimer's disease. *J. Biochemical. Pharmacol.* 2014, *88*, 594–604.

Ono, K.; Hasegawa, K.; Naiki, H.; Yamada, M. Curcumin has potent anti-amyloidogenic effects for Alzheimer's β-amyloid fibrils in vitro. *J. Neurosci. Res.* 2004, *75*, 742–750.

Park, S.-Y.; Kim, H.-S.; Cho, E.-K.; Kwon, B. Y.; Phark, S.; Hwang, K. W.; Sul, D. Curcumin protected PC12 cells against beta-amyloid-induced toxicity through the inhibition of oxidative damage and tau hyperphosphorylation. *Food Chem. Toxicol.* 2008, *46*, 2881–2887.

Park, S. Y.; Kim, Y. H.; Kim, Y.; Lee, S. J. Aromatic-turmerone's anti-inflammatory effects in microglial cells are mediated by protein kinase A and heme oxygenase-1 signaling. *Neurochem. Int.* 2012, *61*, 767–77.

Quinn, J. F.; Raman, R.; Thomas, R. G.; Yurko-Mauro, K.; Nelson, E. B.; Van Dyck, C.; Galvin, J. E.; Emond, J.; Jack, C. R.; Weiner, M.; Shinto, L.; Aisen, P. S. Docosahexaenoic acid supplementation and cognitive decline in Alzheimer disease: a randomized trial. *JAMA*, 2010, *304*, 1903–1911.

Ragheb, A.; Attia, A.; Eldin, W.; Elbarbry, F.; Gazarin, S.; Shoker A. The protective effect of thymoquinone, an anti-oxidant and anti-inflammatory agent, against renal injury: a review. *Saudi J. Kidney Dis. Transpl.* 2009, *20*, 741–752.

Rahmani, A. H.; Al shabrmi Fahad, M.; Allemailem, K. S.; Aly, S. M.; Khan, M. A. Implications of green tea and its constituents in the prevention of cancer via the modulation of cell signalling pathway. *Biomed. Res. Int.* 2015; *2015*:925640. doi:10.1155/2015/925640.

Rezai-Zadeh, K.; Arendash, G. W.; Hou, H.; Fernandez, F.; Jensen, M.; Runfeldt, M.; Shytle, R. D.; Tan, J. Green tea epigallocatechin-3-gallate (EGCG) reduces beta-amyloid mediated cognitive impairment and modulates tau pathology in Alzheimer transgenic mice. *Brain Res.* 2008, *1214*, 177–187.

Rezai-Zadeh, K.; Shytle, D.; Sun, N.; Mori, T.; Hou, H.; Jeanniton, D.; Ehrhart, J.; Townsend, K.; Zeng, J.; Morgan, D.; Hardy, J.; Town, T.; Tan, J. Green tea epigallocatechin-3-gallate (EGCG) modulates amyloid precursor protein cleavage and reduces cerebral amyloidosis in Alzheimer transgenic mice. *J. Neurosci.* 2005, *25*, 8807–14.

Rizzi, L.; Rosset, I.; Roriz-Cruz, M. Global epidemiology of dementia: Alzheimer's and vascular types. *Biomed. Res. Int.* 2014; *2014*:908915.

Shi, L. Y.; Zhang, L.; Li, H.; Liu, T. L.; Lai, J. C.; Wu, Z. B.; Qin, J. Protective effects of curcumin on acrolein-induced neurotoxicity in HT22 mouse hippocampal cells. *J Pharmacol Rep*. 2018, *70*, 1040–1046.

Shytle, R. D.; Bickford, P.C.; Rezai-zadeh, K.; Hou, L.; Zeng, J.; Tan, J.; Sanberg, P. R.; Sanberg, C. D.; Roschek, B. Jr.; Fink, R. C.; Alberte, R. S. Optimized turmeric extracts have potent anti-amyloidogenic effects. *Curr. Alzheimer Res*. 2009, *6*, 564–571.

Shytle, R. D.; Tan, J.; Bickford, P. C.; Rezai-Zadeh, K.; Hou, L.; Zeng, J.; Sanberg, P. R.; Sanberg, C. D.; Alberte, R. S.; Fink, R. C.; Roschek, B. Jr. Optimized turmeric extract reduces β-amyloid and phosphorylated Tau protein burden in Alzheimer's transgenic mice. *Curr. Alzheimer Res*. 2012, *9*, 500–506.

Sinha, S.; Lieberburg, I. Cellular mechanisms of beta-amyloid production and secretion. *Proc. Natl. Acad. Sci. U S A*. 1999, *96*, 11049–11053.

Small, G. W.; Siddarth, P.; Li, Z.; Miller, K. J.; Ercoli, L.; Emerson, N. D.; Martinez, J.; Wong, K. P., Liu, J.; Merrill, D. A.; Chen, S. T.; Henning, S. M.; Satyamurthy, N.; Huang, S. C.; Heber, D.; Barrio, J. R. Memory and brain amyloid and Tau effects of a bioavailable form of curcumin in non-demented adults: a double-blind, placebo-controlled 18-month trial. *Am. J. Geriatr. Psychiatry*. 2018, *26*, 266–277.

Smith, A.; Giunta, B.; Bickford, P. C.; Fountain, M.; Tan, J.; Shytle, R. D. Nanolipidic particles improve the bioavailability and alpha-secretase inducing ability of epigallocatechin-3-gallate (EGCG) for the treatment of Alzheimer's disease. *Int. J. Pharm*. 2010, *389*, 207–212.

Tang, M.; Taghibiglou, C. The mechanisms of action of curcumin in Alzheimer's disease. *J. Alzheimer's Dis*. 2017, *58*, 1003–1016.

Thomas, P.; Wang, Y. J.; Zhong, J. H.; Kosaraju, S., O'Callaghan, N. J.; Zhou, X. F.; Fenech, M. Grape seed polyphenols and curcumin reduce genomic instability events in a transgenic mouse model for Alzheimer's disease. *Mutat. Res*. 2009; *661*, 25–34.

Wang, J. Y. Nucleo-cytoplasmic communication in apoptotic response to genotoxic and inflammatory stress. *Cell Res*. 2005, *15*, 43–48.

Wang, Y. J.; Thomas, P.; Zhong, J. H.; Bi, F. F.; Kosaraju, S.; Pollard, A.; Fenech, M.; Zhou, X. F. Consumption of grape seed extract prevents amyloid-beta deposition and attenuates inflammation in brain of an Alzheimer's disease mouse. *Neurotox. Res*. 2009, *15*, 3–14.

Xiong, Z.; Hongmei, Z.; Lu, S.; Yu, L. Curcumin mediates presenilin-1 activity to reduce -amyloid production in a model of Alzheimer's disease. *Pharmacol. Rep*. 2011, *63*, 1101–1108.

Yan, R. Physiological functions of the β-site amyloid precursor protein cleaving enzyme 1 and 2. *Front. Mol. Neurosci*. 2017, *10*, 97.

Yang, F.; Lim, G. P.; Begum, A. N.; Ubeda, O. J.; Simmons, M. R.; Ambegaokar, S. S.; Chen, P. P.; Kayed, R.; Glabe, C. G.; Frautschy, S. A.; Cole, G. M. Curcumin inhibits formation of amyloid β oligomers and fibrils, binds plaques, and reduces amyloid in vivo. *J. Biol. Chem*. 2005, *280*, 5892–5901.

Zhang, C.; Browne, A.; Child, D.; Tanzi, R. E. Curcumin decreases amyloid-beta peptide levels by attenuating the maturation of amyloid-beta precursor protein. *J. Biol. Chem*. 2010, *285*, 28472–28480.

Zhang, F.; Kang, Z.; Li, W.; Xiao, Z.; Zhou, X. Roles of brain-derived neurotrophic factor/tropomyosin-related kinase B (BDNF/TrkB) signalling in Alzheimer's disease. *J. Clin. Neurosci*. 2012, *19*, 946–949.

Zhang, L.; Fiala, M.; Cashman, J.; Sayre, J.; Espinosa, A.; Mahanian, M.; Zaghi, J.; Badmaev, V.; Graves, M. C.; Bernard, G.; Rosenthal, M. Curcuminoids enhance amyloid-β uptake by macrophages of Alzheimer's disease patients. *J. Alzheimers Dis*. 2006, *10*, 1–7.

Zhang, Z. X.; Li, Y. B.; Zhao, R. P. Epigallocatechin Gallate Attenuates β-Amyloid Generation and Oxidative Stress Involvement of PPARγ in N2a/APP695 Cells. *Neurochem. Res*. 2017, 42, 468–480.

Zhang, X.; Li, Y.; Xu, H.; Zhang, Y. The γ-secretase complex: from structure to function. *Front. Cell. Neurosci*. 2014, *8*, 427.

Zotova, E.; Nicoll, J. A.; Kalaria, R.; Holmes, C.; Boche, D. Inflammation in Alzheimer's disease: relevance to pathogenesis and therapy. *Alzheimers Res. Ther*. 2010, *2*(1), 1.

PART IV

Specialized Applications of Nutraceuticals

CHAPTER 10

Science Behind Maca: A Traditional Crop from the Central Andes

GUSTAVO F. GONZALES[1,2], CINTHYA VASQUEZ-VELASQUEZ[1,2], and DULCE ESPERANZA ALARCÓN-YAQUETTO[1,2*]

[1]*Research and Development Laboratories (LID), Faculty of Sciences and Philosophy, Universidad Peruana Cayetano Heredia. Lima, Peru*

[2]*High Altitude Research Institute, Universidad Peruana Cayetano Heredia, Lima, Peru*

[*]*Corresponding author. E-mail: dulce.alarcon@upch.pe*

ABTRACT

Lepidium meyenii, better known as maca, is a plant that grows 4000 m above sea level in the Central Andes. It has been used in Peruvian traditional medicine for centuries, with reports of its use to increase the offspring of domestic animals dating back from the 16th century. Traditional knowledge about its handling, preparation methods, and beneficial properties have survived until today, and guided the scientific community to conduct experimental trials to test the veracity behind these claims. The main health effects of maca as an energy booster and fertility enhancer herb has been confirmed in animal models and clinical trials. Other properties unknown to native Peruvians have also been discovered. This chapter aims to cover the tradition behind the medicinal usage of this nutritional plant and the scientific research that demonstrates its properties.

10.1 INTRODUCTION

It is estimated that there are as many as 1400 plant species currently used in traditional Peruvian medicine (Lock et al., 2016). Currently, the main interest in the national and international market is the plant known as maca (*Lepidium meyenii*) (Figure 10.1). In several reports, it is possible to find the description that Peru is the land of superfoods, being characterized by a variety of Peruvian agricultural products that promote a healthy life. Among these, *Lepidium meyeni*i (maca) is included. (Gonzales, 2017).

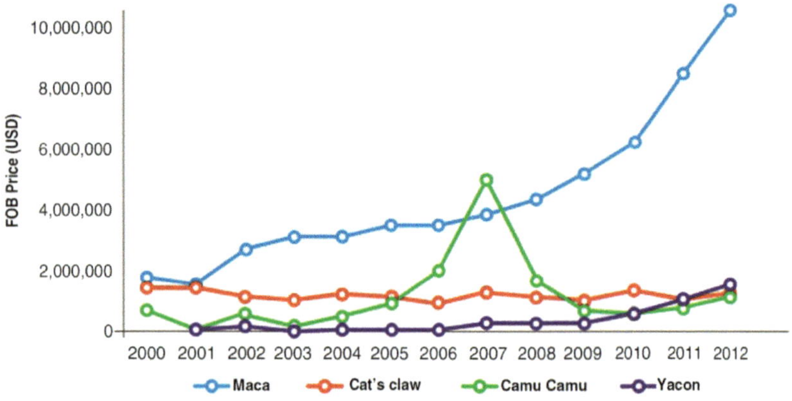

FIGURE 10.1 Export of Peruvian nutraceuticals (Free on Board, FOB, price, 2000–2012 period. (Source: Gonzales et al., 2017: IANAS)

Lepidium meyenii (maca) is a plant belonging to the Brassicaceae family cultivated exclusively between 4000 and 4500 m above the sea level (Gonzales, 2017). Maca is traditionally used because of its nutritional and allegedly medicinal properties and is now a famous food supplement, with growing international interest over the last two decades.

The main bioactive compounds are glucosinolates, macamides, macaenes, alkaloids, and a series of novel thiohydantoins. Interestingly, macamides, thio-hydantoins, and some alkaloids may originate from glucosinolates (Esparza et al., 2015; Huang et al., 2018).

The functional properties of maca have made it popular in several places and also became a species of panacea since many properties have been attributed to its consumption. Several publications describe the beneficial effects of maca. One of them says "traditionally, maca has been used for medicinal purposes and as a source for energy, stamina, and endurance. It also has been used as an aphrodisiac and tonic for postmenopausal problems" (Balick and Lee, 2002).

Similarly, many papers claim properties attributed to traditional medicine in Peru. However, most of them lack support. For instance, Balick and Lee (2002) suggest that resurgence of interest in maca is due to its impact on postmenopausal women. However, this is difficult to believe since the expectancy of life in pre-Hispanic times was below the age of menopause. Survival beyond age 65 of a large percentage of the female population is a very recent phenomenon in terms of evolutionary time (Pollycove et al., 2011). Several years ago, internet adverts stated "do not use maca if you have hypertension." Science has demonstrated that maca reduces blood pressure (Stojanovska et al., 2015). We also found statements as "maca increases testosterone and it is dangerous in man suffering from prostate disease." Scientific evidence demonstrated that maca does not modify serum testosterone levels (Gonzales et al., 2003) and red maca may reduce prostate hyperplasia (Gonzales et al., 2012) and apoptotic activity induced by Taxol and 2-methoxy estradiol in LNCaP cells (prostate cancer cell lines) was not altered with red maca treatment (Díaz et al., 2016). In the nineties of the last century, maca was popularized in Asian markets as "Andean Viagra" or "Andean Gingseng." However, according to scientific evidence, maca does not have properties like Viagra (Gonzales and Alarcón-Yaquetto, 2017).

This chapter will be focused on the nutraceutical properties of maca as traditionally described and most of the medicinal properties as the scientific evidence has demonstrated.

10.2 BOMBON PLATEAU AND THE ORIGIN OF MACA

10.2.1 TOUGH CONDITIONS OF THE ANDEAN HIGHLANDS

Lepidium meyenii is a plant growing in the highlands of Peru. This plant whose hypocotyl is the edible part was botanically described in the 19th century by Wilhelm Gerhard Walpers, a German botanist. In 1843, Wilhelm Gerhard Walpers deposited the first specimen of *Lepidium meyenii* in the Herbarium of the Universidad Nacional Mayor de San Marcos in Lima, Peru (Meissner et al., 2015). Walpers described *Lepidium meyenii* from a plant growing in the southern part of Peru in Pisacoma, Chucuito at the highlands of Puno, in the Peruvian Southern Andes (Figure 10.2).

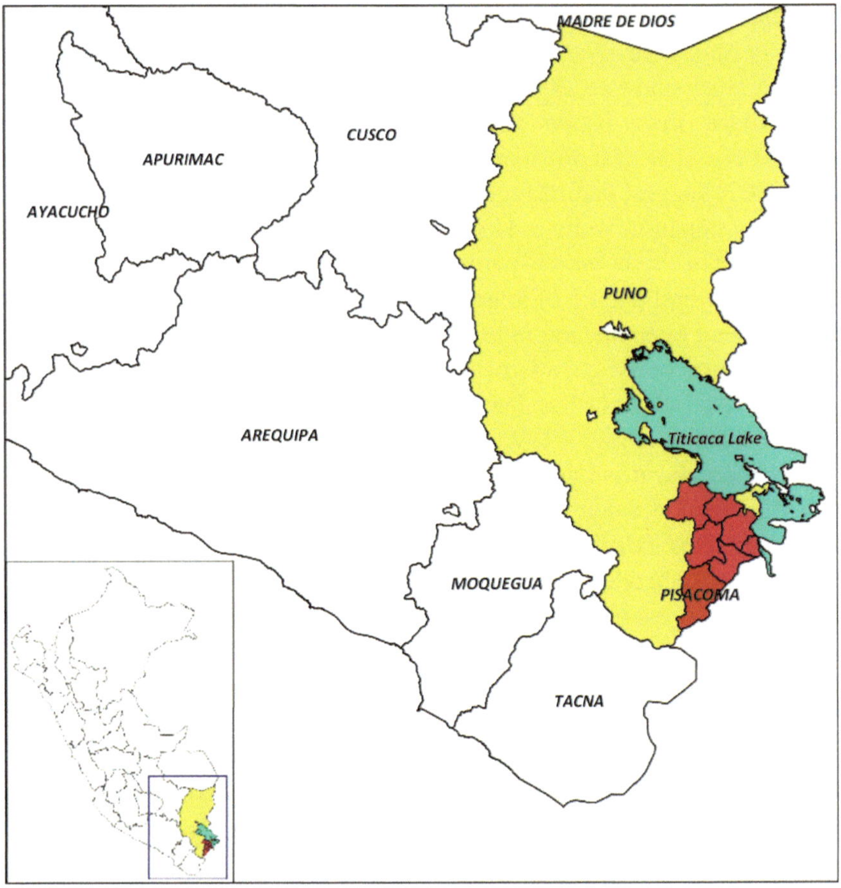

FIGURE 10.2 Pisacoma in the province of Chucuito, near Titicaca Lake in Peruvian Southern Andes. In this place, Gerard Walpers described the specimen denominated *Lepidium meyenni* Walpers.

This description, however, generates some doubts about what we know about maca today and what is traditionally known about maca. The name of "maca" was presented to the literature by the Spaniard chroniclers in the 16th century. Probably the first description of maca in the Bombon plateau was written by Pedro Cieza de Leon in 1554. In his book *Chronicles of Peru*, he described the province of Chinchaycocha (currently Carhuamayo and Junin). He describes this place as "this land is flat or level, very cold and mountains are some distance from the lake." He distinguishes a lake "which is more than ten leagues round." (Cieza de Leon, 1554). At that time, the lake was called Lake of Chinchaycocha or Lake of Bombon (currently named as Lake of Junin) (Figure 10.3). Natives are considered very warlike and they never were conquered by the Incas. Maize, abundant food in Peru, did not grow so much in Chinchaycocha and natives used roots as food. It is considered that these roots are hypocotyls of maca. However, Cieza de Leon never used the name "maca" in his chronicles.

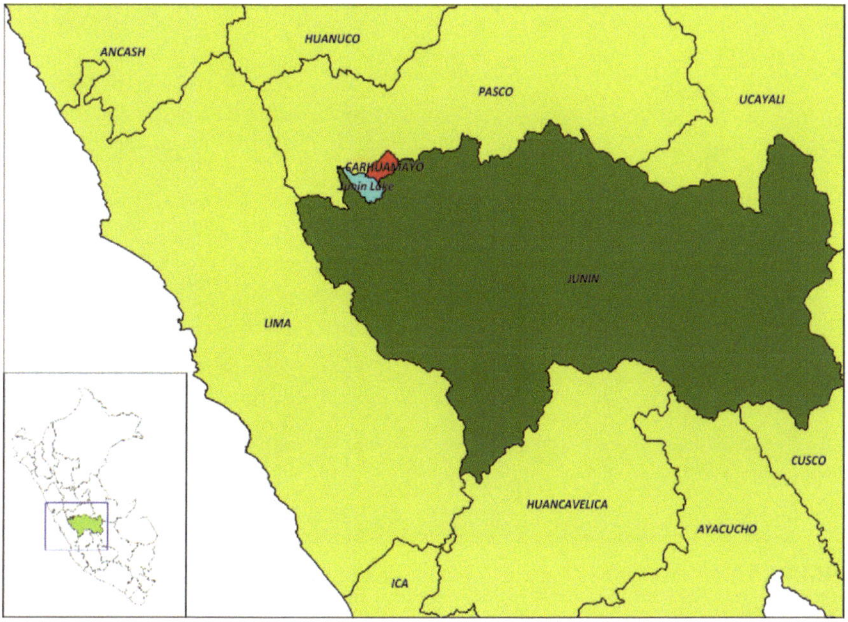

FIGURE 10.3 Junin Lake. Around the lake, maca is cultivated in Junin, Peru over 4000 m of altitude.

The name "maca" referring to the hypocotyls, we now know in Junin and Pasco, was first described in 1549 when Juan Tello Sotomayor was named by Pedro de la Gasca as Spanish Encomendero. Pedro de la Gasca asked Juan Tello Sotomayor to receive potatoes and maca as a kind of tribute by the natives of Chinchaycocha.

From the Spaniard chroniclers, we knew that maca harvest was focalized in the Peruvian Central Andes particularly in Bombon plateau, currently known as districts of Junin and border of Pasco. This place is located at 300 km from Lima, the capital of Peru, and includes the Departments of Junin and Pasco and it is far from the place in which G. Walpers described botanically *Lepidium meyenii* (Figure 10.3). Morphologically, *Lepidium meyenii* described by G. Walpers in Pisacoma, Puno represents a plant with a small hypocotyl, whereas the plant in the central highlands has succulent hypocotyl (Figures 10.4 and 10.5).

FIGURE 10.4 Hypocotyl of maca, the edible part of the plant.

Traditionally, maca is known to be used as a crop and also to treat infertility and as an energizer (Cobo, 1956; Ruiz, 1952; Vasquez de Espinoza, 1969). The chroniclers Vásquez de Espinoza and Cobo described maca in 1628 and 1653, respectively, whereas Botanist Hipólito Ruiz described maca from Carhuamayo in the 18th century. Other properties are attributed to maca but only fertility and energy-boosting properties were reported by chroniclers.

Nowadays, it is common to distinguish maca properties by the color of the hypocotyl. Several publications state that red maca, yellow maca, and black maca have different biological properties that were not described traditionally. Therefore, the knowledge about the differential properties of the different phenotypes of maca constitutes a contribution of scientific endeavor (Gonzales et al., 2010; Gonzales-Arimborgo et al., 2016).

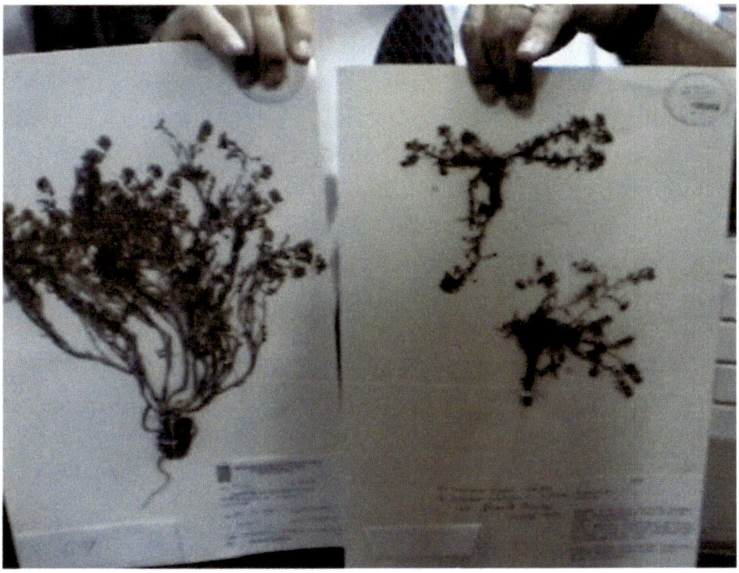

FIGURE 10.5 Two vouchers deposited in the herbarium of the Universidad Nacional Mayor de san Marcos in Lima, Peru. In the right part is the specimen deposited by Gerhard Walpers from Pisacoma, whereas the second at the left part of the figure is a specimen collected in the central Andes. The morphological difference is clear. (Source: Courtesy of Meissner et al., 2015).

The use of gas chromatography–mass spectrometry analysis in the hypocotyls of three maca phenotypes allowed us to identify 79 compounds. Among them, 62 chemical compounds have distinct profiles among three maca phenotypes (Chen et al., 2018). Recently, it has been demonstrated that compounds present in maca may have different and antagonist properties (Fano et al., 2017; Inoue et al., 2016).

10.2.2 HARVESTING ATTEMPTS AROUND THE GLOBE

Maca is original from Peru, but due to its reported properties, there have been different harvesting attempts in other parts of the world. Most have failed (Lebeda et al., 2003; Marthe et al., 2003; Melnikovova et al., 2012). However, in the last decade, several publications refer that maca is also cultivated in China using Peruvian maca seeds. One of the places described in China in which maca grows is Yunnan (Feng et al., 2009; Zhang et al., 2015).

Zhang et al. (2015) published a paper comparing eight minerals in Peruvian and Chinese maca. Since mineral content in the maca hypocotyls will depend on the soil, Peruvian maca has low or nil sodium but a high amount of potassium (Valerio and Gonzales, 2005). Although maca cultivated in China used seeds from Peru, the mineral content of the hypocotyls was different from Peruvian maca. This is probably due to the quality of the soil. Sodium content was high in hypocotyls from China. This cation may become toxic for the plant (Kronzucker et al., 2013). No data on potassium content in maca from China were presented. This is very important since food containing sodium could be inadequate for subjects suffering from hypertension. Peruvian maca after consumption may reduce arterial blood pressure (Stojanovska et al., 2015). This is probably due in part to the low sodium and high potassium contents of Peruvian maca hypocotyls (Valerio and Gonzales, 2005). A healthy diet is based on the intake of low-sodium-containing food (Méndez et al., 2014).

Other concerns were the elevated amounts of copper (Cu) and nickel (Ni) in Chinese maca. Both are excessively high over recommended values. Although these minerals could be functionally important, they become toxic when present in high amounts. This could be dangerous for health and might be related to the reported high metal pollution of Chinese farmland soil (Zhang et al., 2015).

The functional primary role for copper is catalytic, with many copper metalloenzymes acting as oxidases to achieve the reduction of molecular oxygen. The recommended dietary allowance for adults is 900 μg/day (Institute of Medicine (US) Panel on Micronutrients, 2001). High consumption of copper may produce liver damage. In maca from China, copper content amounted up to 31,000 μg/kg dw, whereas it was absent in Peruvian maca.

Elevated contents of nickel (Ni) in soils and foodstuffs can threaten human health. As the content of this metal in soil, water, and food may vary from place to place, the associated risks will also be different (Yeganeh et al., 2013). This is particularly important in China since food safety is regarded as one of its main concerns in public health (Zhang et al., 2015). Comparing among different food commodity groups, nickel concentrations are highest in nuts and legumes (128 and 55 μg/100 g, respectively). A previous study showed a potential health risk from nickel via dietary intake of foodstuffs for Chinese consumers (Feng et al., 2009). The content of nickel in Chinese maca as that in Yunnan, ranging from 4500 (Zhang et al., 2015) to 11,300 μg/kg dw (Wang et al., 2014) must be considered as the highest among foodstuffs and must be of concern, particularly in the Chinese population (Feng et al., 2009).

10.2.3 APPELLATION OF ORIGIN: MACA JUNIN

Since the first description of maca by chroniclers of the Spaniard conquest, maca has been focalized in Carhuamayo, Óndores, Junin, and Ninacaca. These places belong to the current Departments of Junin and Pasco in the Peruvian Central Andes. The description of the biological properties in the Chronicles is referred to maca from Carhuamayo. In the eighties of the 20th century, maca harvest was focalized in the Departments of Junin and Pasco, although some production was from Castrovirreyna, Huancavelica, and other places in the Central and Northern Andes. Currently, the highest maca production is observed in the Junin-Pasco Region. San Blas in Junin (currently, Óndores), is the place where evidence of first maca cultivation was found with a date of 600 years before Christ. Additionally, most of the scientific studies have been performed using maca from Carhuamayo, Junin (Junin), and Ninacaca (Pasco).

For these reasons, the National Institute for the Defense of Competition and Intellectual Property (INDECOPI), a Peruvian institution, conceded to Maca, Junin-Pasco as an appellation of origin (INDECOPI, 2001).

10.3 THE SCIENCE BEHIND TRADITION

10.3.1 THE TRADITIONAL DRYING PROCESS OF MACA ROOTS

People from the Central Andean Highlands warn against the consumption of fresh unboiled maca, although its effects on health have not been assessed yet (Gonzales, 2012). Traditional knowledge preserved until today states that maca needs to be left to dry before consumption.

Immediately after harvesting, the aerial parts of the plant are separated by hand by local farmers and hypocotyls are put to dry under sun. Generally, plastic tents are used to protect the roots from water damage from rain (Brinckmann and Smith, 2004). This drying process can last up to 3 months where maca roots are exposed to repeatedly freeze–thaw cycles (Esparza et al., 2015).

10.3.2 IMPORTANCE OF THE DRYING PROCESS IN THE BIOLOGICAL ACTIVITY OF MACA

As stated previously, the natural environment where maca grows is tough making the drying process maca is subject to, a tissue-damaging one. Research has found that the mechanical damage and repeatedly freeze–thaw cycles prompt the release of hydrolytic enzymes that result in an increase of free fatty acids, amides, and amines, mainly benzylamine (Esparza et al., 2015). These compounds are the precursors of macamides, fatty acid alkylamides that are only found in *Lepidium meyenii*, and thus are considered chemical markers of this plant (McCollom et al., 2005; Zhao et al., 2005).

These compounds are absent in fresh maca samples and, interestingly, they have been found to inhibit fatty amidic acid hydrolase (FAAH), an enzyme in charge of the hydrolysis of endocannabinoids further suggesting involvement of maca in this signaling pathway that could account for its beneficial effects.

Esparza et al. (2005) further showed that fresh hypocotyls have higher levels of benzylisothiocyanate and benzyl glucosinolate.

Several drying methods were compared to see if compounds present in maca differed depending on the drying method. It was found that oven-dry at 35 ° C rendered higher levels of macamides compared to air dry at 22 ° C. The way the hypocotyls were dried also impacted the results with whole hypocotyls showing higher macamide content when dried than when cut to pieces prior to drying (Pan et al., 2016). It is important to note that the air-drying method used in this study was not traditionally used in the Andean Highlands, the hypocotyls were left in room temperature until they dried (43 days for whole hypocotyls and 21 days for pieced hypocotyls). The traditional air-drying process involves exposure to highland conditions and usually lasts up to 3 months.

10.3.3 A 500-YEAR TRADITIONAL MEDICINE

The fertility-enhancing properties of *Lepidium meyenii* appear in some chronicles dating Spanish Conquest times (circa 1550). Chroniclers emphasized that this plant thrived in environmental conditions other plants were unable to bear. Furthermore, they noted that locals consumed this plant regularly on their diet due to its alleged medicinal properties and that population was numerous despite the apparently difficult conditions found in the highlands (Gonzales, 2012).

The Botanist Hipólito Ruiz wrote in 1777–1788, that locals consumed maca after boiling and reported the fertility-enhancing effects of the plant and how women unable to conceive used this plant as treatment. He also reported stimulant effect after high quantity consumption (Ruiz, 1952)

The reproductive rates of domestic animals brought from Europe after the conquest were lower in the Andean Highlands than in Spain and locals introduced maca to the Spaniards as a solution to this problem, with reportedly good results (León, 1964).

Nowadays, in the Central Andes, maca consumption is high. People start consuming maca since childhood, the average years consuming maca is 25.8 ± 3.2 years and the average quantity consumed is 100 g daily. Locals are aware of the health benefits of maca and affirm it is the main reason of its consumption (Gonzales et al., 2013).

10.4 MACA IN THE 21ST CENTURY

10.4.1 MACA: MYTHS VS SCIENTIFIC EVIDENCE

10.4.1.1 THE ANDEAN VIAGRA?

Maca was marketed as the Andean Viagra due to alleged effects improving libido, although the evidence to support this statement is scarce. One of the first studies that draw widespread attention towards maca reported that purified lipidic maca extracts increased the number of complete intromissions in normal rats. This was achieved after a 22-day treatment (Zheng et al., 2000). A more recent study had similar results with ethanol and aqueous extracts in a 6-day treatment in healthy rats (Zhang et al., 2016).

Several studies assessed the aphrodisiac effects in humans. Dording et al. (2008) showed a significant increase in two self-reported scales assessing sexual libido, in subjects with sexual dysfunction caused by selective serotonin reuptake inhibitors intake only with high doses of maca, while the lower dose did not show a significant increase on such parameters (Dording et al., 2008). Zenico and colleagues conducted a trial with dry maca extract on patients suffering from mild erectile dysfunction, the subjective perception of overall sexual health life increased after a 12-week treatment (Zenico et al., 2009).

Another clinical trial on both male and female healthy subjects found that red maca increased sexual desire in a 12-week treatment. The increase was noted since week two (Gonzales-Arimborgo et al., 2016).

Taking these results altogether, it is safe to say that the libido-enhancing effect of maca is dependent on the length of treatment; no immediate effect is shown after consumption, therefore calling maca as the Andean Viagra is misleading and might undermine its value as a nutraceutical.

10.4.1.2 MACA AND ANEMIA

It is believed that maca is effective for anemia treatment mainly because it contents iron. A randomized, placebo-controlled trial showed that spray-dried hidroalcoholic maca extract did not have significant effects on hemoglobin concentration over a 12-week treatment. Furthermore, hemoglobin levels significantly diminished compared to placebo in high altitude natives (Gonzales-Arimborgo et al., 2016). This is important since, at highlands, excessive erythrocytosis is a highly prevalent condition characterized by excessively high hemoglobin levels and with implications in the quality of life of the patient (León-Velarde et al., 2005). Maca might be used to treat this condition.

10.4.2 INSIGHTS INTO MACA'S MECHANISMS OF ACTION

Maca is a natural product that has gained a great interest in the last century. its medicinal properties and value, which increases continuously thanks to the many studies around the world, the scientific community is currently trying to elucidate the molecular pathways and action mechanisms that maca employs to exert its benefits.

The biological effects of maca on the body highly depend on the phenotype, the most studied being the yellow, black, and red phenotypes. Each of them generates differentiated biological effects. In the case of yellow maca, it has been mainly related to boosting physical performance. While the red and black phenotypes are those of greater biological and physiological action. The black maca is related to the improvement of the sperm capacity, since it not only increases the sperm count but also guarantees a better sperm yield for the favorable result of fertilization. Its action does not solely improve reproduction, but it can extend to other systems, acting as a regulator and homeostatic agent, Black maca has shown to reverse neurotoxic processes.

Red maca has an anti-inflammatory effect, it has been demonstrated to reverse the progress of benign prostatic hyperplasia. In regard to beneficial effects over women's health, it improves processes such as osteoporosis, scarring, and embryo implantation among others.

10.4.2.1 REPRODUCTIVE EFFECTS

Maca reproductive effects have been mostly studied in males due to the myth created around its effect of sexual performance. Testosterone is a male hormone related to the increase of libido in both sexes. It is administered as a treatment in pathologies such as erectile dysfunction, premature ejaculation among others, although not without a discussion around its suitability (Ahmadi and Daneshmand, 2013).

Effects of maca extracts with lipid characteristics on sexual behavior were evaluated in rodents with great results (Zheng et al., 2000; Zenico et al., 2009), the number of intromissions increased, reducing the time interval between ejaculations, among other parameters. In humans, the administration of gelatinized maca (1.5 or 3.0 g/day) improved sexual desire after 8 weeks (Gonzales et al., 2002), and in adult athletes, this improvement is observed after 2 weeks of administration (Stone et al., 2009). It cannot be asserted that this improvement is related to the increase in testosterone levels since several studies find no significant testosterone increment (Brooks et al., 2008; Dording et al., 2008).

On the other hand, spermatogenesis is the process by which male sex gametes are produced for fertilization. A 14-day treatment with maca in rats showed an increase in the stages of mitosis (Gonzales, et al., 2001; Gonzales et al., 2003). A 42-day treatment with black maca improved daily sperm production and increased sperm motility (Gonzales et al., 2006). This effect is not seen with red maca, but for the three phenotypes, a longer effect is the spermatic count in the epididymis (84 days) (Gasco et al., 2007).

In humans, the administration of gelatinized maca for 4 months generates an increase in seminal volume, counting, and mobility of sperm (Gonzales et al., 2001). Likewise, the levels of luteinizing hormone, follicle-stimulating hormone, prolactin, estradiol, and testosterone have shown not to be altered by the treatment with maca (Gonzales et al., 2003).

It has been argued that because there is no significant response of serum testosterone, the actions of androgenic receptors are not being stimulated, for which other mechanisms such as the activation of estrogen receptors would be more linked and respond to this male reproductive improvement (Brooks et al., 2008; Dording et al., 2008).

On the other hand, effects on the female reproductive effect have also been studied. Mice that received supplementation of maca's aqueous extract had a higher number of offspring than the control group. Similar results were reported in guinea pigs (Alvarez, 1993). Studies have shown that maca consumption does not alter parameters such as vaginal opening (Thigpen et al., 2003), embryo implantation (Oshima et al., 2003; Ruiz-Luna et al., 2005), nor ovulation (Gasco et al., 2008).

10.4.2.2 CICATRIZATION

Tissue repair process comprises a sequence that involves three main events: inflammation, proliferation, and remodeling (Reinke and Sorg, 2012). Each of these events is essential for the process to be adequate (Landén et al., 2016), in the case of the alteration of proliferation, the formation of nonhealing wounds and/or the formation of excessive fibrotic tissue, leads to the development of keloid or hypertrophic scars.

Studies with the methanolic extract of red maca and its butanolic and aqueous fractions have shown a differential effect on the formation of wounds. The methanolic extract and the aqueous fraction of red maca accelerate wound closure, while the butanolic fraction has a delayed effect on healing. The aqueous fraction of red maca had an anti-inflammatory and proliferative effect, which could be especially useful for tissue repair processes in diabetic people's wounds. While the butanolic fraction affected the level of remodeling, being useful in cases of potential fibrosis or keloids. The methanolic extract containing both fractions had an adequate effect on normal skin (Nuñez et al., 2017).

10.4.2.3 BENIGN PROSTATIC HYPERPLASIA

Benign prostatic hyperplasia (BPH) is a relevant disease in public health issues, given that globally its prevalence exceeds 50% in men over 60 years (Parsons, 2010), reaching 90% in men from Europe, America, and Asia, older than 80 years (Song et al., 2008).

Clinically, it is characterized by symptoms of the lower urinary tract (LUTS), nocturia (Izumi et al., 2014), and in severe cases sepsis, irreversible damage to the bladder, kidney failure, and even death (Parsons, 2010).

The current treatment of BPH is carried out through the administration of drugs, especially finasteride, whose use came into effect after 1990, being established as the standard treatment. This drug is an inhibitor of the enzyme 5-α reductase type 2, which is responsible for converting testosterone to dihydrotestosterone (DHT), an androgen more potent than testosterone in the activation of androgen receptors (ARs), due to its greater affinity for these (Grino et al., 1990). Although finasteride is efficient in reducing the size of the prostate and, therefore, reversing the LUTS (Gormley et al., 1992), it is not exempt from side effects that greatly affect the quality of life of the patient, as erectile dysfunction, sexual impotence, decreased libido, and ejaculatory disorders are the processes that are highly dependent on testosterone.

When BPH cannot be reversed by pharmacological treatment, surgical measures such as transurethral resection of the prostate are taken, which due to its invasive nature lead to unfavorable side effects for the patient (Gaut and Ortiz, 2015).

Natural medicine to treat the growth of the prostate gland is also used. Different cultures have used various plant-based preparations to treat various ailments, such as cancer, stomach pains, infertility, headaches, and so forth. (Khonche et al., 2017; Kim et al., 2017); however, not all medicinal treatments have scientific support. That is why various researchers seek to determine the effects of herbal treatments at the molecular level, with the aim of being able not only to provide scientific sustenance but to establish mechanisms of action, helping to clarify the etiology of different diseases and new possible pathways of treatment.

One of the most studied herbal extracts to treat BPH is Serenoa repens or saw palmetto, used in traditional medicine by Native Americans, demonstrating a favorable activity against LUTS and cystitis (Buck, 2004). Commercially it is known as Permixon, and it contains compounds such as long-chain fatty acids, phytosterols such as stigmasterol, β-sitosterol, and campesterol (Buck, 2004), which inhibit isoenzymes 5-α reductase (Van Coppenolle et al., 2000); they also inhibit the binding of DHT with RA (Nicholson and Ricke, 2011), which confers antiapoptotic and antiproliferative effects (Van Coppenolle et al., 2000).

Another plant tested in the treatment of BPH is *Pygeum africanum* or African plum tree. Among its effects is the inhibition of fibroblastic proliferation, anti-estrogenic, and anti-inflammatory effects, through the inhibition of leukotrienes. One of the active compounds identified is N-butylbenzenesulfonamide, which acts as an antagonist, preventing its translocation to the nucleus (Papaioannou et al., 2010).

Two recently studied plants are Uvaria rufa and Swertia mileensis. In the first case, it was shown that the extract with ethanol–acetic acid contains a large content of phenols and flavonoids, which confer a high antioxidant power, which would lead to anti-inflammatory effects and avoid oxidative damage; however, the compound to which the antiandrogenic effect is attributed is β-sitosterol, a component that would inhibit the expression of AR (Buncharoen et al., 2016).

In the second case, this plant is also known as Qing Ye Dang (QYD), it is used in traditional Chinese medicine and swartiamarin has been identified as the main component, which has anti-inflammatory properties, inhibits collagen deposition, and hepatoprotective. When performing studies on BPH using traditional treatment (whole plant) or only swertiamarin, they were shown to attenuate overexpression of endothelial growth factor, epidermal growth factor, fibroblast growth factor-β, ER-α, and AR; however, only QYD managed to increase ER-β expression levels (Wu et al., 2017).

Going back to *Lepidium meyenii*, it has shown favorable effects against BPH, specifically the hydroalcoholic, methanolic, and butanolic fraction derived from the methanolic extract (Gonzales et al., 2008; Fano et al., 2017), reducing the stromal area, the height of the epithelial cells and inhibiting the expression of AR in the epithelium. It has been shown that the hydroalcoholic extract modulates the inflammatory process of the prostate during the development of BPH, decreasing the levels of tumor necrosis factor alpha and considerably increasing those of interferon gamma, indicating that it favors cellular immune response through TH1 cells, unlike finasteride, which favors the humoral immune response through TH2 cells since it generates an increase in the levels of interleukin-4 (Gasco Tantachuco, 2014).

In a recently published study, it was shown that the methanolic extract of red maca and its butanolic and aqueous fractions have a favorable estrogenic effect on a model of induction of BPH in the rat since it restored ERβ expression levels. However, only the aqueous fraction, in addition to not being effective in reducing the weight of the prostate and the size of the acinar epithelium and the stromal area, caused overexpression of AR and

ERα (Fano et al., 2017). This reinforces the idea that the activation of the AR stimulates the expression of ERα, and when acting or being activated together, they mask the Erβ signaling pathways (Figure 10.6).

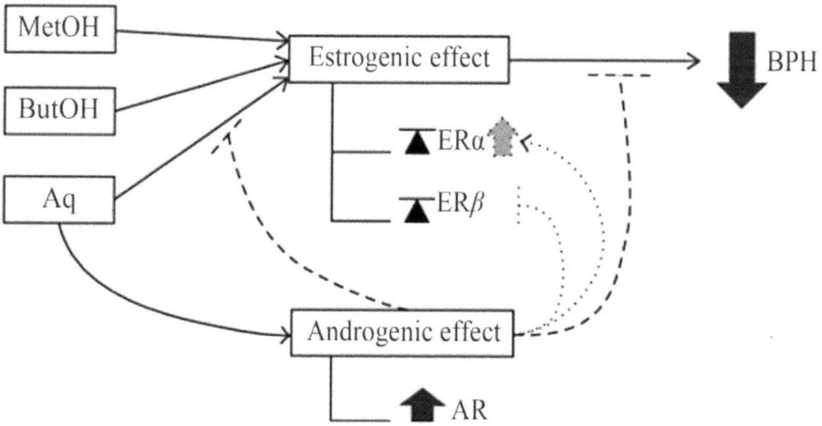

FIGURE 10.6 Red maca extracts have a positive estrogenic effect by balancing ERα and ERβ expression levels, but aqueous fraction also possesses an androgenic effect increasing AR levels, and, due to this, its positive estrogenic effect and thus its action against BPH would be blocked (dashed truncated arrows), due to an acute increment of ERα levels and by interfering with ERβ pathways (dotted truncated and normal arrow). MetOH: methanolic extract, ButOH: butanolic fraction, Aq: aqueous fraction (Source: Courtesy of Fano et al., 2017).

10.4.2.4 MACA AS A MODULATOR OF THE CANNABINOID RECEPTORS

Cannabinoid receptors 1 and 2 (CB1 and CB2) are expressed in mammalian tissues and upon activation trigger effects such as the release of neurotransmitters. Endogenous fatty acid compounds such as *N*-arachidonoylethanolamine and anandamide are able to bind to these receptors. Due to their lipid nature, these compounds cannot be stored (Ahn et al., 2008), so upon release, they are degraded by an enzyme called the FAAH.

Pino-Figueroa et al. (2011) tested if pentane extract of maca had FAAH inhibitory properties. A sigmoidal curve was observed with extract concentration-inhibitory response. The maximum inhibitory percentage achieved was 74 ± 4% with a concentration of 30 μg/mL (Pino-Figueroa et al., 2011). The authors suggest these might be the reason behind the effects of maca as a neuroprotectant.

Hajdu et al. (2014) showed that some macamides are also able to selectively bind the CB1 receptor besides their FAAH inhibition activity (Hajdu et al., 2014). Furthermore, following the pharmacological blockade of the CB1 receptor, black maca showed no effects in sperm count suggesting maca exerts its effects through these receptors (Alarcón Yaquetto, 2016).

10.5 CONCLUDING REMARKS

The Central Peruvian Highlands are home to a hypocotyl that has been used for millennia to treat infertility and to increase energy. Its relevance in past times was such that it was used as a sort of currency to pay taxes. Traditional knowledge passed on through generations led researchers around the world to assess its alleged medicinal properties. The energy booster and fertility enhancer properties were corroborated, but several more were found. Interestingly, science has shown that its properties are highly dependent on maca's phenotype easily distinguishable by the color of its outer layer. Properties such as wound healing and prostatic benign hyperplasia reversal have been described. These led to an increase in the overall production of maca and led several attempts of harvesting around the globe, although much of them have failed since maca appears to thrive only in the unique and radical conditions found on the central Andean heights. Traditionally, maca is sun dried for 3 months before its consumption. Science has proven that this process is essential for maca to be biologically active. Therefore, the science behind maca is always closely related to the traditional knowledge that not only allowed this crop to survive time despite the specific set of the environmental conditions it needs but to preserve the traditions around its preparation and consumption that enable its biological properties.

KEYWORDS

- ***Lepidium meyenii***
- **nutraceutical**
- **macamides**
- **herbal medicine**

REFERENCES

Ahmadi, H.; Daneshmand, S. Androgen deprivation therapy: Evidence-based management of side effects. *BJU Int*. **2013**, *111* (4), 543–548.

Ahn, K.; McKinney, M. K.; Cravatt, B. F. Enzymatic pathways that regulate endocannabinoid signaling in the nervous system. *Chem. Rev.* **2008**, *108* (5), 1687–1707.

Alarcón Yaquetto, D. E. Efecto Modular de La Maca (*Lepidium meyenii*) Sobre El Sistema Endocannabinoide: Estudios En Humanos y Animales, *Universidad Peruana Cayetano Heredia*, **2016**.

Alvarez CJ. Utilización de Diferentes Niveles de Maca En La Fertilidad de Cobayos, *Universidad Nacional Daniel Alcides Carrión,* **1993**.

Balick, M. J.; Lee, R. Maca: From traditional food crop to energy and libido stimulant. *Altern. Ther. Health Med.* **2002**, *8* (2), 96–98.

Brinckmann, J.; Smith, E. Maca culture of the Junín Plateau. *J. Altern. Complement. Med.* **2004**, *10* (3), 426–430.

Brooks, N. A.; Wilcox, G.; Walker, K. Z.; Ashton, J. F.; Cox, M. B.; Stojanovska, L. Beneficial effects of *Lepidium meyenii* (maca) on psychological symptoms and measures of sexual dysfunction in postmenopausal women are not related to estrogen or androgen content. *Menopause* **2008**, *15* (6), 1157–1162.

Buck, A. C. Is there a scientific basis for the therapeutic effects of Serenoa repens in benign prostatic hyperplasia? Mechanisms of action. *J. Urol.* **2004**, *172* (5 Pt 1), 1792–1799.

Buncharoen, W.; Saenphet, K.; Saenphet, S.; Thitaram, C. Uvaria rufa blume attenuates benign prostatic hyperplasia via inhibiting 5α-reductase and enhancing antioxidant status. *J. Ethnopharmacol.* **2016**, *194*, 483–494.

Chen, Q.; Li, M.; Wang, C.; Li, Z.; Xu, J.; Zheng, Q.; Liu, P.; Zhou, H. Combining targeted metabolites analysis and transcriptomics to reveal chemical composition difference and underlying transcriptional regulation in maca (*Lepidium meyenii* Walp.) ecotypes. *Genes (Basel).* **2018**, *9* (7), 335.

Cobo, B. History of the New World, *Biblioteca. Sevilla,* **1956**.

Díaz, P.; Cardenas, H.; Orihuela, P. A. Red Maca (*Lepidium meyenii*) Did not affect cell viability despite increased androgen receptor and prostate-specific antigen gene expression in the human prostate cancer cell line LNCaP. *Andrologia* **2016**, *48* (8), 922–926.

Dording, C. M.; Fisher, L.; Papakostas, G.; Farabaugh, A.; Sonawalla, S.; Fava, M.; Mischoulon, D. A double-blind, randomized, pilot dose-finding study of maca root (L. Meyenii) for the management of SSRI-induced sexual dysfunction. *CNS Neurosci. Ther.* **2008**, *14* (3), 182–191.

Esparza, E.; Hadzich, A.; Kofer, W.; Mithöfer, A.; Cosio, E. G. Bioactive maca (*Lepidium meyenii*) alkamides are a result of traditional andean postharvest drying practices. *Phytochemistry* **2015**, *116*, 138–148.

Fano, D.; Vásquez-Velásquez, C.; Gonzales-Castañeda, C.; Guajardo-Correa, E.; Orihuela, P. A.; Gonzales, G. F. N-Butanol and aqueous fractions of red maca methanolic extract exerts opposite effects on androgen and oestrogens receptors (alpha and beta) in rats with testosterone-induced benign prostatic hyperplasia. *Evidence-Based Complement. Altern. Med.* **2017**, *2017*, 1–10.

Feng, Y.; He, Z.; Xu, L.-F.; Zhang, Z.-H.; Shi, L.; Chen, X.-M. Nutritive element analysis and evaluation of maca (*Lepidium meyenii*) cultivated in Yunnan. *For. Res.* **2009**, *22* (5), 696–700.

Gasco Tantachuco, M. E. Efecto Diferencial de *Lepidium meyenii* (Maca Roja) y Finasteride Sobre Los Procesos Inflamatorios En La Hiperplasia Prostática Benigna Inducida Con Enantato de Testosterona En Ratas de La Cepa Holtzman, Universidad Peruana Cayetano Heredia, **2014**.

Gasco, M.; Aguilar, J.; Gonzales, G. F. Effect of chronic treatment with three varieties of *Lepidium meyenii* (Maca) on reproductive parameters and DNA quantification in adult male rats. *Andrologia* **2007**, *39* (4), 151–158.

Gasco, M.; Yucra, S.; Rubio, J.; Gonzales, G. F. *Lepidium meyenii* (Maca) varieties did not alter female reproductive parameters in adult intact rats. *Artic. J. Complement. Integr. Med.* **2008**, *5*(1).

Gaut, M. M.; Ortiz, J. Manejo Da Síndrome Do Compartimento Abdominal Pós-ressecção Transuretral de Próstata. *Brazilian J. Anesthesiol.* **2015**, *65* (6), 519–521.

Gonzales GF. Maca. Del Alimento Perdido de Los Incas Al Milagro de Los Andes; *Universidad Peruana Cayetano Heredia,* **2017**.

Gonzales G. F.; Alarcón-Yaquetto DE. Maca, A Nutraceutical from the Andean Highlands. In *Therapeutic Foods*; Elsevier: the Netherlands, **2017**; pp. 373–396.

Gonzales, G. F. Ethnobiology and ethnopharmacology of *Lepidium meyenii* (Maca), a plant from the Peruvian Highlands. *Evid.-Based Complement. Altern. Med.* **2012**, *2012*, 1–10.

Gonzales, C.; Cárdenas-Valencia, I.; Leiva-Revilla, J.; Anza-Ramirez, C.; Rubio, J.; Gonzales, G. F. Effects of different varieties of maca (*Lepidium meyenii*) on bone structure in ovariectomized rats. *Forschende Komplementärmedizin/Res. Complement. Med.* **2010**, *17* (3), 4.

Gonzales, G.; Cordova, A.; Gonzales, C.; Chung, A.; Vega, K.; Villena, A. *Lepidium meyenii* (Maca) improved semen parameters in adult men. *Asian J. Androl.* **2001**, *3* (4), 301–303.

Gonzales, G.; Córdova, A.; Vega, K.; Chung, A.; Villena, A.; Góñez, C.; Castillo, S. Effect of *Lepidium meyenii* (MACA) on sexual desire and its absent relationship with serum testosterone levels in adult healthy men. *Andrologia* **2002**, *34* (6), 367–372.

Gonzales, G. F.; Gasco, M.; Lozada-Requena, I. Role of Maca (*Lepidium meyenii*) Consumption on serum interleukin-6 levels and health status in populations living in the Peruvian Central Andes over 4000 m of altitude. *Plant Foods Hum. Nutr.* **2013**, *68* (4), 347–351.

Gonzales, G. F.; Gasco, M.; Malheiros-Pereira, A.; Gonzales-Castañeda, C. Antagonistic effect of *Lepidium meyenii* (red maca) on prostatic hyperplasia in adult mice. *Andrologia* **2008**, *40* (3), 179–185.

Gonzales, C.; Leiva-Revilla, J.; Rubio, J.; Gasco, M.; Gonzales, G. F. Effect of red maca (*Lepidium meyenii*) on prostate zinc levels in rats with testosterone-induced prostatic hyperplasia. *Andrologia* **2012**, *44*, 362–369.

Gonzales, G. F.; Rubio, J.; Chung, A.; Gasco, M.; Villegas, L. Effect of alcoholic extract of *Lepidium meyenii* (maca) on testicular function in male rats. *Asian J. Androl.* **2003**, *5* (4), 349–352.

Gonzales, C.; Rubio, J.; Gasco, M.; Nieto, J.; Yucra, S.; Gonzales, G. F. Effect of short-term and long-term treatments with three ecotypes of *Lepidium meyenii* (MACA) on spermatogenesis in rats. *J. Ethnopharmacol.* **2006**, *103* (3), 448–454.

Gonzales, G. F.; Ruiz, A.; Gonzales, C.; Villegas, L.; Cordova, A. Effect of *Lepidium meyenii* (maca) roots on spermatogenesis of male rats. *Asian J. Androl.* **2001**, *3* (3), 231–233.

Gonzales-Arimborgo, C.; Yupanqui, I.; Montero, E.; Alarcón-Yaquetto, D.; Zevallos-Concha, A.; Caballero, L.; Gasco, M.; Zhao, J.; Khan, I.; Gonzales, G. Acceptability, safety, and efficacy of oral administration of extracts of black or red maca (*Lepidium meyenii*) in adult human subjects: a randomized, double-blind, placebo-controlled study. *Pharmaceuticals* **2016**, *9* (3), 49.

Gormley, G. J.; Stoner, E.; Bruskewitz, R. C.; Imperato-McGinley, J.; Walsh, P. C.; McConnell, J. D.; Andriole, G. L.; Geller, J.; Bracken, B. R.; Tenover, J. S.; et al. The effect of finasteride in men with benign prostatic hyperplasia. *N. Engl. J. Med.* **1992**, *327* (17), 1185–1191.

Grino, P. B.; Griffin, J. E.; Wilson, J. D. Testosterone at high concentrations interacts with the human androgen receptor similarly to dihydrotestosterone. *Endocrinology* **1990**, *126* (2), 1165–1172.

Hajdu, Z.; Nicolussi, S.; Rau, M.; Lorántfy, L.; Forgo, P.; Hohmann, J.; Csupor, D.; Gertsch, J. Identification of endocannabinoid system-modulating N-alkylamides from heliopsis helianthoides Var. Scabra and *Lepidium meyenii*. *J. Nat. Prod.* **2014**, *77* (7), 1663–1669.

Huang, Y.-J.; Peng, X.-R.; Qiu, M.-H. Progress on the chemical constituents derived from glucosinolates in maca (*Lepidium meyenii*). *Nat. Products Bioprospect.* **2018**, *8*(6), 405–412.

INDECOPI. Resolución N°006065-2011/DSD INDECOPI http://www.indecopi.gob.pe/repositorioaps/0/11/jer/deno_deorigenotorgadas/R08_006065-**2011**.pdf.

Institute of Medicine (US) Panel on Micronutrients. Dietary Reference Intakes for Vitamin A, Vitamin K, Arsenic, Boron, Chromium, Copper, Iodine, Iron, Manganese, Molybdenum, Nickel, Silicon, Vanadium, and Zinc; *National Academies Press (US): Washington, D.C.*, **2001**.

Izumi, K.; Li, L.; Chang, C. Androgen receptor and immune inflammation in benign prostatic hyperplasia and prostate cancer. *Clin. Investig. (Lond).* **2014**, *4* (10), 935–950.

Khonche, A.; Fallah Huseini, H.; Abdi, H.; Mohtashami, R.; Nabati, F.; Kianbakht, S. Efficacy of Mentha Pulegium extract in the treatment of functional dyspepsia: a randomized double-blind placebo-controlled clinical trial. *J. Ethnopharmacol.* **2017**, *206*, 267–273.

Kim, W.; Kang, J.; Lee, S.; Youn, B. Effects of traditional oriental medicines as anti-cytotoxic agents in radiotherapy. *Oncol. Lett.* **2017**, *13* (6), 4593–4601.

Kronzucker, H.; Coskun, D.; Schulze, L.; Wong, J.; Britto, D. sodium as nutrient and toxicant. *Plant Soil* **2013**, *369* (1–2), 1–23.

Landén, N. X.; Li, D.; Ståhle, M. Transition from inflammation to proliferation: a critical step during wound healing. *Cell. Mol. Life Sci.* **2016**, *73* (20), 3861–3885.

Lebeda, A.; Dolezalova, I.; Valentova, K.; Dziechciarkova, M. Biological and chemical variability of maca and yakon. *Chem. List.* **2003**, *97*(7), 548–556.

León, J. The "maca" (*Lepidium meyenii*), a little known food plant of Peru. *Econ. Bot.* **1964**, *18* (2), 122–127.

León-Velarde, F.; Maggiorini, M.; Reeves, J. T.; Aldashev, A.; Asmus, I.; Bernardi, L.; Ge, R.-L.; Hackett, P.; Kobayashi, T.; Moore, L. G.; et al. Consensus statement on chronic and subacute high altitude diseases. *High Alt. Med. Biol.* **2005**, *6* (2), 147–157.

Lock, O.; Perez, E.; Villar, M.; Flores, D.; Rojas, R. Bioactive Compounds from Plants Used in Peruvian Traditional Medicine. *Nat. Prod. Commun.* **2016**, *11* (3), 315–337.

Marthe, F.; Schutze, W.; Kruger, H.; Scholze, P.; Kramer, R.; Rvschka, U. Maca (*Lepidium meyenii*) cultivation. Resistance and composition of secondary metabolites under European conditions. *Schriften zu Genet. Ressourcen* **2003**, 209–293.

McCollom, M. M.; Villinski, J. R.; McPhail, K. L.; Craker, L. E.; Gafner, S. Analysis of macamides in samples of maca (*Lepidium meyenii*) by HPLC-UV-MS/MS. *Phytochem. Anal.* **2005**, *16* (6), 463–469.

Meissner, H. O.; Mscisz, A.; Mrozikiewicz, M.; Baraniak, M.; Mielcarek, S.; Kedzia, B.; Piatkowska, E.; Jólkowska, J.; Pisulewski, P. Peruvian maca (Lepidium Peruvianum): (I) phytochemical and genetic differences in three maca phenotypes. *Int. J. Biomed. Sci.* **2015**, *11* (3), 131–145.

Melnikovova, I.; Havlik, J.; Fernandez Cusimamani, E.; Milella, L. Macamides and fatty acids content comparison in maca cultivated plant under field conditions and greenhouse. *Boletín Latinoam. Caribe Plantas Med. y Aromát.* **2012**, *11* (5), 420–427.

Méndez, L.; Rojas, J.; Izaguirre, C.; Contreras, B.; Gómez, R. Jatropha Curcas leaves analysis, reveals it as mineral source for low sodium Diets. *Food Chem.* **2014**, *165*, 575–577.

Nicholson, T. M.; Ricke, W. A. Androgens and estrogens in benign prostatic hyperplasia: past, present and future. *Differentiation.* **2011**, *82* (4–5), 184–199.

Nuñez, D.; Olavegoya, P.; Gonzales, G. F.; Gonzales-Castañeda, C. Red maca (*Lepidium meyenii*), a plant from the Peruvian Highlands, promotes skin wound healing at sea level and at high altitude in adult male mice. *High Alt. Med. Biol.* **2017**, *18* (4), 372–383.

Oshima, M.; Gu, Y.; Tsukada, S. Effects of *Lepidium meyenii* walp and jatropha macrantha on blood levels of estradiol-17 beta, progesterone, testosterone and the rate of embryo implantation in mice. *J. Vet. Med. Sci.* **2003**, *65* (10), 1145–1146.

Pan, Y.; Zhang, J.; Li, H.; Wang, Y.-Z.; Li, W.-Y. Simultaneous analysis of macamides in maca (*Lepidium meyenii*) with different drying process by liquid chromatography tandem mass spectrometry. *Food Anal. Methods* **2016**, *9* (6), 1686–1695.

Papaioannou, M.; Schleich, S.; Roell, D.; Schubert, U.; Tanner, T.; Claessens, F.; Matusch, R.; Baniahmad, A. NBBS isolated from Pygeum Africanum bark exhibits androgen antagonistic activity, inhibits AR nuclear translocation and prostate cancer cell growth. *Invest. New Drugs* **2010**, *28* (6), 729–743.

Parsons, J. K. Benign prostatic hyperplasia and male lower urinary tract symptoms: epidemiology and risk factors. *Curr. Bladder Dysfunct. Rep.* **2010**, *5* (4), 212–218.

Pino-Figueroa, A.; Vu, H.; Kelley, C. J.; Maher, T. J. Mechanism of action of *Lepidium meyenii* (Maca): an explanation for its neuroprotective activity. *Am. J. Neuroprot. Neuroregen.* **2011**, *3* (1), 87–92.

Pollycove, R.; Naftolin, F.; Simon, J. A. The evolutionary origin and significance of menopause. *Menopause* **2011**, *18* (3), 336–342.

Reinke, J. M.; Sorg, H. Wound repair and regeneration. *Eur. Surg. Res.* **2012**, *49* (1), 35–43.

Ruiz, H. Relación Histórica Del Viaje a Los Reinos Del Perú y Chile, 1777-1788.; *Acad. de Ciencias Exactas, Fis, y Nat: Madrid,* **1952**.

Ruiz-Luna, A. C.; Salazar, S.; Aspajo, N. J.; Rubio, J.; Gasco, M.; Gonzales, G. F. *Lepidium meyenii* (Maca) increases litter size in normal adult female mice. *Reprod. Biol. Endocrinol.* **2005**, *3* (1), 16.

Song, K.; Wang, H.; Krebs, T. L.; Kim, S.-J.; Danielpour, D. Androgenic control of transforming growth factor-beta signaling in prostate epithelial cells through transcriptional suppression of transforming growth factor-beta receptor II. *Cancer Res.* **2008**, *68* (19), 8173–8182.

Stojanovska, L.; Law, C.; Lai, B.; Chung, T.; Nelson, K.; Day, S.; Apostolopoulos, V.; Haines, C. Maca reduces blood pressure and depression, in a pilot study in postmenopausal women. *Climacteric* **2015**, *18* (1), 69–78.

Stone, M.; Ibarra, A.; Roller, M.; Zangara, A.; Stevenson, E. A pilot investigation into the effect of maca supplementation on physical activity and sexual desire in sportsmen. *J. Ethnopharmacol.* **2009**, *126* (3), 574–576.

Thigpen, J. E.; Haseman, J. K.; Saunders, H. E.; Setchell, K. D. R.; Grant, M. G.; Forsythe, D. B. Dietary phytoestrogens accelerate the time of vaginal opening in immature CD-1 mice. *Comp. Med.* **2003**, *53* (6), 607–615.

Valerio, L. G.; Gonzales, G. F. Toxicological aspects of the South American herbs cat's claw (Uncaria Tomentosa) and maca (*Lepidium meyenii*) : A critical synopsis. *Toxicol. Rev.* **2005**, *24* (1), 11–35.

Van Coppenolle, F.; Le Bourhis, X.; Carpentier, F.; Delaby, G.; Cousse, H.; Raynaud, J. P.; Dupouy, J. P.; Prevarskaya, N. Pharmacological effects of the lipidosterolic extract of Serenoa repens (Permixon) on rat prostate hyperplasia induced by hyperprolactinemia: comparison with finasteride. *Prostate* **2000**, *43* (1), 49–58.

Vasquez de Espinoza, A. Compendio y Descripción de Las Indias Occidentales. In *Biblioteca de Autores Españoles; Madrid,* **1969**; p. 330:355.

Wang, W.; Zhang, Z.; Yang, G.; Wang, Q. Health risk assessment of Chinese consumers to nickel via dietary intake of foodstuffs. *Food Addit. Contam. Part A* **2014**, *31* (11), 1861–1871.

Wu, X.; Gu, Y.; Li, L. The anti-hyperplasia, anti-oxidative and anti-inflammatory properties of Qing Ye Dan and swertiamarin in testosterone-induced benign prostatic hyperplasia in rats. *Toxicol. Lett.* **2017**, *265*, 9–16.

Yeganeh, M.; Afyuni, M.; Khoshgoftarmanesh, A.-H.; Khodakarami, L.; Amini, M.; Soffyanian, A.-R.; Schulin, R. Mapping of human health risks arising from soil nickel and mercury contamination. *J. Hazard. Mater.* **2013**, *244–245*, 225–239.

Zenico, T.; Cicero, A. F. G.; Valmorri, L.; Mercuriali, M.; Bercovich, E. Subjective Effects of *Lepidium meyenii* (Maca) Extract on well-being and sexual performances in patients with mild erectile dysfunction: a randomised, double-blind clinical trial. *Andrologia* **2009**, *41* (2), 95–99.

Zhang, J.; Wang, H.-M.; Zhao, Y.-L.; Zuo, Z.-T.; Wang, Y.-Z.; Jin, H. Comparison of mineral element content in a functional food maca (*Lepidium meyenii* Walp.) from Asia and South America. *J. Anal. Methods Chem.* **2015**, *2015*, 530541.

Zhang, Y.; Yu, L.; Jin, W.; Ao, M. Effect of *Lepidium meyenii* (maca) extracts on sexual performance in mice, and the seminal vesicle weight and serum testosterone levels in castrated mice. *Eur. J. Biomed. Pharm. Sci.* **2016**, *3* (5), 624–627.

Zhang, X.; Zhong, T.; Liu, L.; Ouyang, X. Impact of soil heavy metal pollution on food safety in China. *PLoS One* **2015**, *10* (8), e0135182.

Zhao, J.; Muhammad, I.; Dunbar, D. C.; Mustafa, J.; Khan, I. A. New alkamides from maca (*Lepidium meyenii*). *J. Agric. Food Chem.* **2005**, *53* (3), 690–693.

Zheng, B. L.; He, K.; Kim, C. H.; Rogers, L.; Shao, Y.; Huang, Z. Y.; Lu, Y.; Yan, S. J.; Qien, L. C.; Zheng, Q. Y. Effect of a lipidic extract from *Lepidium meyenii* on sexual behavior in mice and rats. *Urology* **2000**, *55* (4), 598–602.

CHAPTER 11

Nutraceutical Intervention for Treatment of Alcoholism and Drinking Problems

CHARU GUPTA* and DHAN PRAKASH

Amity Institute for Herbal Research and Studies, Amity University, Noida 201303, Uttar Pradesh, India

**Corresponding author. E-mail: charumicro@gmail.com*

ABSTRACT

Alcoholism is an obsession characterized by a lasting craving for alcohol. Unlike alcohol addicts, who drink too much but are not dependent on alcohol, alcoholics feel an obligation to drink overpoweringly. Despite its mental and societal sequelae, alcohol addiction is basically a mental ailment. Various medicinal plants and herbs such as *Silybum marianum*, *Pueraria lobata*, *Taraxacum officinale*, *Ganoderma lucidum*, *Passiflora edulis*, *Artemisia absinthium*, *Hypericum perforatum*, and *Tabernanthe iboga* including some other lesser known plants Ginseng and Cayenne pepper can be active either as a prime therapy or in combination with conservative remedies. Although various chemotherapies are available, the practice of herbs and plants for treating alcohol addiction has numerous benefits such as no side effects, no toxicity, natural, effective, and safe. The present review would focus on the health abuse of alcoholism and the role of herbs and medicinal plants (importance, medicinal value, benefits, and the mode of action) in handling the intoxication and drinking problem.

11.1 INTRODUCTION

Drunkenness remains a stern cause of illness and death, despite progress through neurobiological studies on finding new pharmacological approaches for its management and cure. Despite pronounced advancement made in the former two eras, the advancement in low-toxicity and high proficiency medications remains a perplexing job for liquor investigators. In Western nations, benzodiazepines, complex vitamins, cyanamide disulfiram, and aldehyde dehydrogenase inhibitors are directed to inebriating patients to subdue drawing and cut craving for drinking. However, it is known that these biochemical remedies lead to abuse, emotional dependency, and hostile behavioral effects. These drugs could also lead to physiological reliance when taken recurrently and result in dosage intensification (Dupuy et al., 1995). Aldehyde dehydrogenase inhibitors, especially disulfiram, have been stated to display stern side effects (Poulsen et al., 1992). Currently, a type of opioid receptors such as naltrexone is being used for treating drunkenness in individuals. Though the drug naltrexone effectively suppresses alcohol craving, it also bears some lateral effects such as an increase in withdrawal signs, biliousness, dysphoria, and tiredness (Froehlich et al., 1998; Jones and Dekker, 2000; O'Malley et al., 2000; Oncken et al., 2001).

Drunkenness is considered as a desire for liquor and incapability to discontinue consumption. It is conveyed by a bodily addiction and an augmented tolerance for liquor. Drunkenness involves a bodily dependency on liquor, but other prompting features include hereditary, mental, and social motives. The craving for alcohol increases day by day as the person consumes and their body starts developing acceptance and even demands for more to attain the desired effects. As the time progresses, the bodily addiction continues to increase. Even if a person wants to stop his drinking problem, the individual may experience painful indications of withdrawal that cause them to go back to alcohol for relief. As the habit grows, an individual suffering from intoxication will feel as if they need to drink to function and feel any sense of normality. Their health will begin to decline, they may have distress at family or work, there may be confusion of their assets, and they may even face legal difficulties as a result of their actions while drunk. Alcoholism also leads to family and work problems and can also cause grave health complications such as liver diseases (Jones and Dekker, 2000).

Alcohol addiction in the United States positions third on the list of avoidable causes of illness and death as per the records of the US Department of Health and Human Services (2000). In 2000, the United States had 20,687 alcohol-related deaths, excluding mishaps and murders, estimate the nation about $185 billion (US Department of Health and Human Services, 2000). It is estimated that about 17 million people in the United States misuse alcohol and more than 70 million Americans have faced drunkenness in their relations. Alcohol misuse is one of the four most public reasons for death in the United States, and it is also involved in almost half of all the traffic-related deaths in the country. Without a pharmacological aide to psychosocial treatment, the clinical consequence is poor, with up to 70% of patients restarting intake within 1 year (Swift, 1999; Finney et al., 1996). Therefore, psychosocial involvement alone is not an ideal treatment for liquor addiction.

The optimal treatment for subduing drinking problems in adults depends on the correct identification of chronic symptoms (early onset type) and a robust natural predisposition to the disease and for the late-onset type; it depends on the temperament signs typically brought on by psychosocial issues. Developing adults with binge-drinking patterns also might be helped by discerning managements. Although introductory effort on the pharmacogenetics of intoxication and its management has been hopeful, the task of handling still rests on medical valuation. Transitory communicative interventions that encourage the patient to set goals for a decrease in heavy consumption or self-discipline also are part of the ideal treatment (Johnson, 2010).

Western and Asian herbalists have used medicinal plants to treat liquor misuse, craving, and aftereffects for periods. Herbs are popular as a substitute to standard Western allopathic medicine for alcoholism. This is because mainstream pharmaceuticals have violent side effects and hard on the already overworked liver of those who have a drinking problem. Herbal remedies do not offer a magical panacea for the painful withdrawal from addiction but can help the healing process in two important ways: as a general strengthener of both the emotional and physical systems and in specific treatment of the side effects of withdrawal. Most of the herbs mentioned here help the recovery process by supporting and healing the liver and cleansing the body (Swift, 1999).

11.2 PHYTOTHERAPY: NATURAL TREATMENT OF ALCOHOLISM

The important part of phytotherapy in dealing the drunkenness is to use those plants that serve two main functions. First, the hepatoprotective activity that protects the liver against alcohol aggression, and second, the tranquilizer activity that reduces the anxiety of alcoholics, relaxes the mind, and aids in sleeping. The usage of plants/herbs is an age-old approach for supporting the body and handling illness. The herbal plants can sometimes elicit side effects and can interrelate with other herbs, supplements, or suppositories. Therefore, the herbal therapy should always be taken under the guidance of medical practitioners. Moreover, phytotherapy alone is not sufficient to treat drunkenness (Carai et al., 2000).

Presently, the research is more focused on some selected medicinal plants such as *Passiflora incarnata*, *Pueraria lobata*, *Salvia miltiorrhiza*, and *Salvia przewalskii*. These plants are a rich source of bioactive constituents, showing their likely help against alcohol addiction. The pharmacological studies of these plants have shown promising results in animal mockups of drunkenness, and insufficient research has been done on nicotine dependence using herbal plants. It has been shown that *P. incarnata* extract (and its benzoflavone derivative, BZF) possess good potential in the treatment of alcoholism problem and could also help in reducing the intensity of nicotine or alcohol withdrawal symptoms. Very limited work has been done in this area. Similarly, the root extract from the plant *P. lobata* (kudzu) has also been shown to exhibit the antialcohol effect. The clinical trials have shown the promising results of kudzu root extract in reducing alcohol intake in both animals and humans. The bioactive constituents present in the kudzu extracts are majorly isoflavones (daidzin, daidzein, and puerarin). These phytochemicals are chiefly responsible for the beneficial effects in reducing the craving of liquor drinking. However, the exact mechanism of action of kudzu's role in suppressing ethanol intake remains to be explained. It is well known that daidzin under in vitro studies is a strong, selective, and reversible inhibitor of aldehyde dehydrogenase. Moreover, studies on the central nervous system receptor gene expression have shown that the kudzu extract probably acts through the opioid system and shows adverse action by manipulating the opioid receptors (mi and delta) and the expression of endogenous opioid precursors (proopiomelanocortin), similarly to naltrexone. Besides kudzu, the preclinical studies also suggest the promising role of medicinal plants *S. miltiorrhiza* and *S. przewalskii* extracts in reducing voluntary alcohol intake in animal models of extreme liquor consumption. Their main bioactive constituents are tanshinones and miltiron (Ozarowski et al., 2013).

The other plants are *Tabernanthe iboga*, *Panax ginseng*, and *Hypericum perforatum* that are also found to be effective in decreasing liquor intake. The most common mode of action of all these medicinal plants is through their action on absorption from the gastrointestinal system (Carai et al., 2000).

The following medicinal plants are used to overcome alcoholism and drinking problem.

11.2.1 MILK THISTLE (*SILYBUM MARIANUM*)

Milk thistle is one of the best known herbal remedy for liver protection and is frequently used to treat liver complications. Milk thistle is inherent to the Mediterranean region and is now found all over the biosphere. This doughty thistle typically develops in arid, sunny parts. The spiky stalks branch at the top and reach an elevation of 4–10 ft. The leaves are wide, with white spots or veins. Milk thistle is so-called as it produces the milky white fluid that exudes out from their leaves when they are crushed. The florae are reddish purple. The small, hard-skinned drupe is brown, dotted, and glossy. Milk thistle spreads rapidly, that is why it is considered as a weed in some parts of the world, and it ripens quickly, in less than a year.

The active constituent is silymarin that is present in its seeds that helps the liver to eliminate the toxins, including alcohol. As alcohol is also a toxin, silymarin helps the liver to excrete alcohol. It is also recommended during undergoing alcoholism detox. The affected person is occasionally given milk thistle to get rid the liver of liquor. However, silymarin is not so effective in case of severe liver damage, such as alcoholic hepatitis or alcoholic cirrhosis.

Chemically, silymarin is a cluster of flavonoids (silibinin, silidianin, and silicristin), that aid in repairing damaged liver cells caused by alcohol and other noxious ingredients. Silymarin also helps in keeping new liver cells from being destroyed by the same substance(s), reduces inflammation (effective in liver inflammation or hepatitis), and is a powerful antioxidant. It also helps in the recovery process by supporting and healing the liver, detoxifying the body. It is typically used to treat liver cirrhosis, chronic hepatitis (liver inflammation), toxin-induced liver damage (including the prevention of severe liver damage) from *Amanita phalloides* (death cap) mushroom poisoning, and gallbladder disorders (Greenlee et al., 2007; Tamayo and Diamond 2007).

Numerous studies have been carried out in the recent past that has proved beneficial effect of milk thistle in the treatment of alcoholic liver disease and have found noteworthy developments in liver function. Individuals with the slightest form of alcohol-related liver injury have shown decent effects. Milk thistle is less effective for those suffering from severe liver disease, such as cirrhosis, which is characterized by scarring, everlasting, and irreparable injury to the liver. However, there are no studies on the role of milk thistle for liquor drawing. People who are sensitive to ragweed may have an allergic response to milk thistle. Milk thistle may interact with several medicines. However, some compounds in milk thistle resemble with estrogen in their composition, so people with hormone-sensitive conditions should consult with a physician before consuming.

Three cups of brew of a spoonful of crushed fruit per day for 1–2 months are generally recommended.

11.2.2 KUDZU (*PUERARIA MONTANA* VAR. *LOBATA*)

The anticraving and "antidrunkeness" effects of extracts of *P. lobata* Owhi (Fabaceae) have been well known to customary Chinese physicians for decades. Kudzu (*P. lobata*) is an ascending, semi-woody, perennial vine in the pea family Fabaceae, subfamily Faboideae. It is considered an invasive weed in the South and can help to dampen the desire to drink. It is a common herb in customary Chinese medication for giving liquor leftovers. This plant contains daidzin, a plant estrogen and an antioxidant, which impedes an enzyme known as mitochondrial aldehyde dehydrogenase that has been associated with alcoholism. According to research (American Journal of Drug and Alcohol Abuse, 2009), its roots have results comparable to the antialcoholism drug disulfiram. Kudzu is often used in dealing with drunkenness and stoppage of relapse in the start phases of retrieval.

There is stronger evidence to advocate that kudzu acts as a liver tonic, increasing its ability to fight off toxins. It is a likely way to give intoxicating cravings, through extracting the active constituent(s) and formulating it into an active drug (Associated Press, 2006). The mode of action for this is yet unknown but it is believed to be associated with alcohol metabolism and the reward circuits in the brain. Kudzu is a unique source of active phytochemicals such as isoflavones, including puerarin, daidzein (anti-inflammatory and antimicrobial), daidzin (anticancer), and genistein (antileukemic). The bioactive phytoconstituents are present in Kudzu roots, and they have the ability to affect neurotransmitters such as serotonin, gamma-aminobutyric acid (GABA), and glutamate. They are also effective in the treatment of migraine and cluster headache (Kudzu.Med-owl.com.2006). They are also suggested for allergies and diarrhea (Duke, 1997). The plant is also helpful in controlling some postmenopausal symptoms such as hypertension and diabetes type II. The plant is commonly known as gégen in the traditional Chinese medicine system. It is used to treat tinnitus, vertigo, and Wei syndrome, and the clinical studies data suggested that kudzu (*P. lobata*) is used to treat liquor misuse and also helps to decrease liquor cravings.

Numerous studies have been carried out in this way. An experimental study was carried out in alcohol-preferring Syrian Golden hamsters, and the results showed that the daily intraperitoneal administration of crude extract of *P. lobata* roots halved alcohol intake in alcohol-preferring Syrian Golden hamsters. They were given the choice between alcohol solution and water (Keung et al., 1993). The two bioactive constituents identified in *P. lobata* extracts were two major isoflavones (daidzin and daidzein); they reduced ethanol intake in Syrian Golden hamsters to a greater extent as observed with *P. lobata* extract (Keung et al., 1993).

In yet another study carried out on laboratory animals, *P. lobata* extract was used for treatment of the drinking problem. The herbal mixture of *P. lobate* is commonly used in China to prepare "tea of sobriety." The results showed that the administration of this mixture reduced alcohol consumption in alcohol-preferring rats (Carai et al., 2000).

However, clinical trials in humans have shown mixed results. One study on humans failed to show any advantage, while another found that heavy drinkers who had taken kudzu supplements for 7 days drank significantly less beer when given the opportunity than heavy drinkers who had taken a placebo. The other advantage of taking Kudzu is that it can interact with many medicines, including blood thinners, methotrexate, diabetes medicines, estrogens, and estrogen-modifying medicines, including birth control medicines and breast cancer medicines, such as Tamoxifen. There are also studies pointing out that Kudzu could deteriorate liver complaints and/or slow blood coagulation.

Through the above studies, it was however concluded that kudzu may be useful for treating alcoholism and averting relapse.

11.2.3 RED REISHI (GANODERMA LUCIDUM)

Red Reishi, normally known as Ling Zhi in China, is a herbal mushroom that possesses phenomenal health benefits. This has been used in traditional

Chinese medicine as a stimulant to check fatty liver and cirrhosis due to liquor abuse. Reishi is a basidiomycete, lamella-less fungus that belongs to the family of Polyporaceae. It commonly grows in nature in densely wooded mountains having high humidity and faint light. It is seldom found because it flourishes mainly on the desiccated trunks of dead plum, Pasonia (*Guercus serrata*) trees. It is believed that out of 10,000 aged trees, only 2 or 3 will have reishi growth; therefore, it is very rare indeed. The spores of reishi are enclosed within tough outer husks that make their germination tough, and this accounts for its infrequency.

Red Reishi is considered a much better medicinal herb than ginseng because of its nontoxic nature, no side effects, can be consumed habitually, can restore and rejuvenate the body to its natural healthy state, can enable all organs to function normally, and can act as an immunomodulator.

Red Reishi is chiefly composed of complex carbohydrates such as water-soluble polysaccharides, tri-terpenoids, proteins, and amino acids. Studies have shown that these water-soluble polysaccharides are actually the bioactive constituents that possess antitumour, immuno modulating, and blood pressure lowering effects. Another major active ingredient found triterpenes is called ganoderic acid. Primary studies showed that ganoderic acids are beneficial in improving common allergies through the inhibition of histamine release and also improve oxygen utilization and liver functions. Triterpenes are usually bitter in taste, and the amount of triterpene content present in a product can be determined by the degree of bitterness. Regular intake of red Reishi enhances body's immune system and improves blood circulation, thus cultivating better health conditions. Normally, Reishi is suggested as an adaptogen, immunomodulator, and a general stimulant. Red Reishi is also used to treat anxiety, hepatitis, bronchitis, insomnia, and asthma.

11.2.4 PASSION FLOWER (*P. incarnata*)

The passion flowers or passion vines (*Passiflora*) comprise of a genus of about 500 species of flowering plants and belong to the family of Passifloraceae. They are mostly vines, with some shrubs and a few herbs. The herbal plant is useful for treating nausea and vomiting resulting from withdrawal from Vicodin, cocaine, heroin, or opiate painkillers and may also help withdrawal from alcohol. The major bioactive constituents present in the passion flower are alkaloids and flavonoids, in combination with a variety of other constituents, such as amino acids, sugars, coumarins, and alcohols (sterols).

A decoction of the passion flower has been recommended for the treatment of bronchial asthma. The plant is also used as a topical treatment for burns in Europe and America as the compresses of herb have a marked effect on inflammations. The leaves of *Passiflora edulis* are used as a diuretic and for treatment of hemorrhoidal inflammations in South America. In Brazil, *Passiflora incarnata* is used as an antispasmodic and sedative. In North America, the passion flower is often used as an analgesic and anticonvulsant, with some achievement seen in case of tetanus also. In Italy, a combination of the passion flower, belladonna, and lobelia is used in the treatment of asthma. In Poland, one of the registered drugs used for treating excitability contains an extract of the passion flower. Several homeopathic drugs contain passion flower; thus, the main sedative activity of the plant is homeopathic in nature. The passion flower has been commonly used in the treatment of nervous, high-strung, easily excited children; cardiovascular neuroses; bronchial asthma; coronary illness; circulation weakness; insomnia; problems experienced during menopause; concentration problems in school children; and geriatrics. The passion flower is completely nontoxic and has been approved for food use by the FDA. The flower has been used in the treatment of opiate withdrawal. It helps to alleviate the nausea associated with withdrawal. Recent research has demonstrated that it can also help with alcoholism detox and related withdrawal indications.

11.2.5 WORMWOOD (ARTEMISIA ABSINTHIUM)

Wormwood is innate to temperate regions of Eurasia and Northern Africa. It is herbaceous silvery-green perennial plant, with a hard, woody rhizome, growing up to 1.5 m tall that contains the volatile oil thujone. It grows naturally on uncultivated, arid ground, on rocky slopes, and at the edge of footpaths and fields. Thujone, the intoxicating chemical in *Artemisia*, and tetrahydrocannabinol, the active ingredient of marijuana, have similar molecules, and both attach to the related receptor spots in the brain. It is added to distilled ethanol to create absinthe. Its effects alone are not well studied. However, the thujone content of alcoholic beverages containing *Artemisia* is less than one-twentieth of the amount needed for intoxication. It is a constituent in the liquor absinthe and also used for flavoring some other spirits and wines, such as bitters, vermouth, and pelinkovac. It is also used therapeutically as a liver tonic, in liver failure, stomachic, antiseptic, antispasmodic, carminative, cholagogue, febrifuge, and anthelmintic. It is used to reduce the inflammation of the gall bladder and to treat jaundice and hepatitis caused by the ingestion of chemicals.

In the Middle Ages, it was used to spice mead. Due to the plant's characteristic odor, its spray can be used as a biopesticide. The roots of the plant have an allelopathic effect and can be used in inhibiting the growth of surrounding plants, like weeds. It is also effective in repelling insect larvae, fleas, and moths indoors when planted on the edge of the area of cultivation.

11.2.6 ST. JOHN'S WORT (H. perforatum)

St. John's wort (*H. perforatum*) belongs to family Hypericaceae and is a perennial herb that is famous for its antidepressant effects. It is inherent to Europe but exists throughout Asia and North America. Clinical studies have shown that it is effective in the treatment of slight to moderately severe depressive ailments. It is beneficial in the treatment of liquor drawing symptoms as well as deterring its desire. People consuming St. John's wort experience reduced alcohol cravings, and its also aids them from relapsing during the beginning stages of recovery.

A pigment called hyperforin has been extracted from St. John's wort that imparts bright orange/yellow color to its flowers. Hyperforin is the main bioactive constituent that is responsible for its ability to decrease liquor consumption. It has also been shown to relieve depression. It has a comparable effect to that of Kudzu (URL Bradford School of Pharmacy).

In one of the studies, it was found that when St. John's wort extract was given by injection; it reduced voluntary ethanol intake in test animals that had a liking for liquor. The researchers also found that, gradually over time, St. John's wort reduced the desire for alcohol. These studies suggested that St. John's wort is useful in treating alcoholism and preventing a relapse in recovering alcoholics. However, results from human studies are sometimes different from the animal studies (Alcohol and Alcoholism, 2005).

St. John's wort is an effective herbal remedy that has been employed by the conventional health community.

11.2.7 DANSHEN (S. MILTIORRHIZA)

S. miltiorrhiza Bunge, also called as red sage or Danshen (Chinese Pinyin), is a perennial plant in the genus *Salvia* of the mint family, Lamiaceae

(Fang et al., 2008). Its roots are highly valued as a "supergrade herb" for the treatment of several pathological conditions such as insomnia.

In a study, acetone extract from the roots of *S. miltiorrhiza* was used to reduce alcohol fondness in Sardinian alcohol-preferring rats. In yet another study, it was found that the dose of *S. miltiorrhiza* extract, that is capable of reducing voluntary alcohol intake in rats, decreased blood alcohol levels by approximately 60% (Colombo et al., 1999). The modern research confirmed that a standardized extract of *S. miltiorrhiza* dose-dependently delayed the craving of alcohol-drinking behavior (Brunetti et al., 2003). Additional studies were also assumed to examine whether the Danshen extract was capable of averting the development of the alcohol deprivation effect (ADE) in rats (Serra et al., 2003). The acute, intragastric administration of 25, 50, and 100 mg/kg standardized extract resulted in the complete suppression of the extra amount of alcohol consumed during the first hour of re-access to alcohol after 7 days of deprivation. The results indicated that extract might possess antirelapse properties.

The chief bioactive constituents present in *S. miltiorrhiza* include phenanthrenequinones such as tanshinones I, II, cryptotanshinone, and miltirone that causes the inhibition of alcohol intake (Brunetti et al., 2003).

11.2.8 DANDELION (TARAXACUM officinale)

Dandelion (*Taraxacum officinale*) belongs to the family Asteraceae and is a herbaceous perennial plant that possesses various medicinal and culinary uses. Dandelion has been used as a therapy for anemia, blood purification, and immunomodulation. The folk medicines of China, India, and Russia have recognized dandelion's effect as a liver tonic. It is used conventionally for liver-related complications that justify its use in liquor drawing symptoms.

In a study, the defensive effects of *Taraxacum officinale* (dandelion) root extract against alcoholic liver damage was investigated in both HepG2/2E1 cells and ICR mice. On increasing the production of reactive oxygen species through administration of 300 mM ethanol in vitro, cell viability was significantly reduced by 39%. However, in the presence of hot water extract (TOH) from *T. officinale* root, no hepatocytic damage was observed in the cells treated with ethanol, while ethanol extract (TOE) did not show potent hepatoprotective activity. Mice, which received TOH (1 g/kg bw/day) with ethanol, revealed a complete prevention of alcohol-induced hepatotoxicity as shown by the noteworthy decline of serum aspartate aminotransferase, alanine aminotransferase, alkaline phosphatase, and lactate dehydrogenase enzyme activities as compared to ethanol-alone administered mice. On comparison with the ethanol-alone received group, the mice getting ethanol plus TOH showed noteworthy growths in hepatic antioxidant activities, including catalase, glutathione-S-transferase, glutathione peroxidase, glutathione reductase, and glutathione. Furthermore, the enhancement of malondialdehyde levels indicated TOH's protective effects against liver damage facilitated by liquor in vivo. These results suggest that the aqueous extract of *T. officinale* roots has defensive action against alcohol-induced toxicity in the liver by increasing antioxidative potentials and reducing lipid peroxidation (You et al., 2010).

It is often pooled with milk thistle or other herbs to treat hepatitis (Modaresi and Resalatpour, 2012). It is one of the best diuretics that increases urine output and cleanses the body to get rid of its toxins. Dandelion can react with many medicines, including lithium. Some people, who are allergic to ragweed, might have an adversative reaction to dandelion.

11.2.9 IBOGA (*T. iboga*)

Both *T. iboga* and *Voacanga africana* are perennial shrubs innate to Central Africa and belongs to family Apocynaceae. They are largely used in customary African medicine. The root bark of *T. iboga* comprises ibogaine as its major alkaloid, although Ibogaine is one of some naturally occurring alkaloids found in *V. africana*. The psychoactive indole alkaloid known as ibogaine is used to treat cravings of alcohol, cocaine, heroin, and methamphetamine. Ibogaine is present in the root bark of *Tabernanthe iboga* that acts on the serotonin, dopamine, and opioid receptors to decrease substance cravings. The stimulating, fatigue-, thirst-, and hunger-ameliorating effects of ibogaine are well known since times immemorial. The preclinical evidence indicates that ibogaine considerably affect morphine and cocaine after self-administration in rodents. Recently, ibogaine has been testified to markedly decrease voluntary liquor consumption in alcohol-preferring rats. The dropping effect of ibogaine on alcohol intake was detected only when ibogaine was injected intraperitoneally or intragastrically and not subcutaneously, suggesting that the bioactive principle of ibogaine could be a metabolite produced by the liver. It has been found that ibogaine can be toxic in high doses, so an ibogaine analog, known as 18-methoxycoronaridine, has been developed to produce the same antiaddiction effects as ibogaine but without the toxic side effects. Iboga acts on the neurotransmitters that control drinking behavior and helps to reduce cravings and subdue extreme drinking (Carai et al., 2000).

The potential of Ibogaine as "antiaddictive" effect has created a great deal of interest in recent years. It has been stated to antagonize the locomotor effects induced by cocaine, nicotine, and amphetamine in mice. Furthermore, ibogaine effectively blocks methamphetamine-induced hyperthermia in mice (Lu et al., 2009).

American and European addict self-help groups have claimed that ibogaine encourages long-term drug abstinence from heroin, psychostimulants, and cocaine. Anecdotal reports suggest that a single dose of ibogaine eradicates drawing indications and decreases drug desires for long periods. In a preliminary investigation of this effect, none of the seven opiate-dependent participants receiving ibogaine suggested important withdrawal symptoms, and three of the patients remained temperate for up to 14 weeks after the administration of ibogaine. Unluckily, a clinical trial that specifically examines ibogaine's beneficial benefits in drug-dependent patients has yet to be conducted (Lu et al., 2009).

Hence, a successive study proved that both the intraperitoneal and subcutaneous administration of noribogaine, a primary metabolite of ibogaine, considerably inhibited liquor intake in liquor preferring rats (Carai et al., 2000). It has been theorized that ibogaine may exert its reducing effects on voluntary ethanol consumption by interacting with the brain systems involved in the mediation of the reinforcing effects of liquor (Carai et al., 2000).

A study published in the June 2003 issue of "Pharmacology, Biochemistry, and Behavior" also established that ibogaine and its equivalent lessen cravings and subdue extreme drinking in test animals. The researchers attribute this effect to ibogaine's actions on the neurotransmitters that control drinking behavior.

11.2.10 GINSENG (*P. ginseng*)

P. ginseng (family Araliaceae) is well known for its revitalizing and energizing qualities and has been used in customary Korean medicine for numerous beneficial applications, including alcoholism. *P. ginseng* shows an intense influence on the central nervous and cardiovascular systems, and it stimulates endocrine secretions affecting immune function and metabolism (Attele et al., 1999). The chief active constituents in *P. ginseng* are called ginsenosides, and about more than 20 ginsenosides have been isolated, including Rb, Rc, Re, and Rg (Gillis, 1997). Another active constituent in *Panax quinquefolium* is pseudoginsenoside-F11 (PF11), an ocotillol-type saponin; which is absent in *P. ginseng* (Cheng et al., 1980).

P. ginseng has been confirmed to weaken the behavioral effects of drug abuse from morphine, methamphetamine, cocaine, and alcohol in both preclinical and clinical studies (Takahashi and Tokuyama, 1998). Ginsenosides significantly reduced the withdrawal syndromes precipitated by naloxone in morphine-dependent mice (Huang, 1991) and repressed the conditioned place preference and hyperactivity induced by morphine (Kim et al., 1998). However, *P. ginseng*'s ability to inhibit voluntary drug intake has not been examined in the animal model of drug self-administration, which makes it impossible to draw a definitive conclusion on its ability on alcohol withdrawal.

Initially, the mode of action of ginseng was proposed to be through accelerated alcohol metabolism and lowered blood alcohol levels by increasing antidiuretic hormone (ADH) activity and plasma clearance. However, a recent study established that when a red ginseng extract was given to the rats, it altered alcohol absorption from the gastrointestinal tract. Its main action on the body is to speed up metabolism, and by doing so, it can also help the body break down alcohol more quickly thus aiding detox from alcohol (Carai et al., 2000).

Ginseng's inhibitory effect on numerous drugs of abuse appears to be related to the modulation of dopaminergic transmission. Administration of the *P. ginseng* extract, ginseng total saponin, can significantly attenuate ambulatory hyperactivity to apomorphine in mice that had frequent contact with nicotine (Kim et al., 1999). This finding suggests that ginseng's total saponins successfully inhibit postsynaptic dopamine receptor super-sensitivity in nicotine-treated mice. A new study has also reported that ginseng total saponin greatly inhibited nicotine-stimulated dopamine release in the striatum and Fos protein expression in the nucleus accumbens of mice (Zhao et al., 2006). Taken collectively, these studies shed new light on *P. ginseng* as a possibly beneficial therapeutic agent for the management of drug dependence.

11.2.11 ASHWAGANDHA (WITHANIA somnifera)

W. somnifera, also known as Indian ginseng and Ashwagandha, is an important Ayurvedic medicinal plant because its biological effects are comparable to those of *P. ginseng* of Chinese origin. A few research studies have specially observed the effects of *W. somnifera* on addictive drugs. Kulkarni and Ninan (1997) reported that frequent administrations of *Withania somnifera* lessened the progress of tolerance to morphine's analgesic effects and inhibited morphine withdrawal obstacles in animals. The exact mechanism for *W. somnifera*'s effects on morphine involves multiple neurotransmitters (Andrade, 2006). However, in-depth studies are required to decide how *W. somnifera* can become used as a nonanalgesic herbal medicine to treat stress-induced relapse to drug misuse and addiction.

11.2.12 TOYEAM (HOVENIA dulcis)

Research has shown that an extract from the *H. dulcis* plant may treat alcoholism. The extracted compound is also present in the plant called Toyeam herb or Tengcha vine.

The active ingredient is dihydromyricetin, present in its herbal extract that is accountable for dealing with alcoholism.

The studies were also carried out on flavonoid extract derived from *Ampelopsis grossedentata*, *H. dulcis*, and a few other plants. The researchers gave this extract to rats that became addicted to alcohol intoxication. After treatment with dihydromyricetin, the rats reduced their consumption of alcohol with significantly fewer withdrawal symptoms. They exhibited less anxiety and fewer seizures. The treated group also exhibited significantly reduced levels of intoxication side effects when they did consume alcohol.

The researchers' in vitro (cellular) research found that dihydromyricetin antagonized GABA receptors while affecting the receptors' tolerance levels. This mechanism was expected to give dihydromyricetin its ability to alter the physical and mental issues of substance abuse.

The researchers concluded that pharmacological properties of dihydromyricetin are consistent with those expected to underlie successful medical treatment of alcohol use disorders.

Though dihydromyricetin is considered to be a flavonoid constituent of a number of medicinal herbs such as *Erica arborea, Glochidion sumatranum* (umbrella cheese tree) and others, it is most known as having been extracted from the *A. grossedentata* plant, a vine shrub known to grow in Asia. Ironically, the herb has a common name of Snake wine vine and a Chinese medicine name of Teng Cha. The preparation of *Ampelopsis* as a tea is sometimes referred to as vine tea (Overstreet et al., 2003).

11.2.13 KAVA (PIPER methysticum)

An extract of the *P. methysticum* plant is commonly known as Kava and is usually used by the people of Pacific Islanders and Indigenous Australians. Herbal preparations containing kava have been used for the psychiatric dealing of nervousness and sleeplessness (Cairney et al., 2002). Kava has also been used as a traditional medicinal support for smoking and liquor termination. To cite a working example, a village in Fiji developed and applied an effective community-based smoking termination program that comprised the usage of kava (Groth-Marnat et al., 1996). Clinical research has also provided subsidiary proof that the administration of kava reduces patients' desire for their drug of choice and substantially promotes self-discipline in drug-dependent patients (Steiner, 2001). The mode of action of kava is primarily due to the activation of a lipid-soluble group of compounds, called kava lactones (Bressler, 2005). Kava lactones bind to multiple sites in the brain and interact with numerous neurotransmitters. They significantly inhibit the update of noradrenaline but not serotonin (Seitz et al., 1997). Kava is natural and safe to consume with no side effects. Though, hepatotoxicity has been related to kava use due to mitochondrial toxicity in the liver (Lude et al., 2008).

11.2.14 THUNBERGIA laurifolia LINN

T. laurifolia Linn is commonly used as an antipyretic, antidote, and anti-alcoholism treatment in Thai customary medication (Ussanawarong and Thesiri, 2001). To date, no available clinical trial has examined *T. laurifolia* Linn in treating drug craving and only a few studies have discovered the plant's effects in alcohol pharmacotherapy for humans or in animal models. In a study by Chanawirat et al. (2000), the extracts of *T. laurifolia* Linn leaves protected mice from ethanol-induced hepatic injury. Similarly, Pramyothina et al. (2005) showed that the aqueous extract from *T. laurifolia*

Linn provided protection against alcohol-induced liver injury in both live rats and primary cultures of rat hepatocytes. A recent study showed that *T. laurifolia* Linn and amphetamine share some resemblance in potentiating potassium-stimulated dopamine release from rat striatal slices, suggesting that the efficacy of the plant's against addictive drugs is dopamine dependent (Thongsaard and Marsden, 2002; Lu et al., 2009).

11.3 OTHER PLANTS FOR DRINKING PROBLEM

11.3.1 CAYENNE PEPPER

Cayenne pepper is a good boost for the stomach. The alcoholic drinkers usually suffer from stomach problems in particular gastritis. Cayenne pepper provides relief in the pain and decreases the inflammation of the stomach lining that is usually damaged by too much liquor intake.

11.3.2 EVENING PRIMOSE (CENOTHERA BIENNIS)

It is particularly beneficial in dealing with binge drinking especially in regards to the negative symptoms that produce a hangover the next day. It also exerts a positive role in alleviating the negative effects of alcohol withdrawal syndrome. On the other hand, it also protects the liver from the negative effects of alcohol.

11.3.3 LICORICE (GLYCYRRHIZA GLABRA)

The ingestion of glycyrrhizin products with the licorice exerts a protective role in the liver, releasing those toxic substances that are harmful, and also protects and revitalizes the cells.

11.3.4 GREEN TEA (CAMELLIA SINENSIS)

Green tea protects the liver and exerts an antiviral action on the liver. It exhibits antioxidant activity and prevents the degeneration of liver and liver cancer or cirrhosis.

11.3.5 VALERIAN (VALERIANA OFFICINALIS)

It helps to reduce anxiety and is a sleep disorder remedy.

11.3.6 GINGER (ZINGIBER OFFICINALE)

It possesses antioxidant property, stimulates bile, and protects the liver by preventing the development of cancerous tumors.

11.3.7 SALVIA przewalskii

S. przewalskii is a herbaceous perennial plant native to the Chinese provinces of Gansu, Hubei, Sichuan, Xizang, and Yunnan, typically growing along stream banks, forest edges, among shrubs, and on granitic hillsides. It is also recommended by several herbalists as an effective treatment to prevent alcoholism or drinking problem. The mode of action is also unknown. Moreover, no scientific study or evidence is available to date this review was written.

Some other plants have also been listed as herbal medicine for alcoholism that includes angelica (*Angelica archangelica*), astragalus (*Astragalus membranaceous*), bupleurum (*Radix bupleurum*), cayenne pepper (*Capsicum annuum*), chamomile (*Matricaria recutita*), dandelion (*Taraxacum officinale*), ginkgo (*Ginkgo biloba*), goldenseal (*Hydrastis canadensis*), Siberian ginseng (*Eleutherococcus senticosus*), skullcap (*Scutellaria lateriflora*), and turmeric (*Curcuma longa*).

The brief descriptions of medicinal plants that show hepatoprotective activity in drinking problem are given in Table 11.1.

11.4 NUTRITION AND OTHER DIETARY SUPPLEMENTS

Because the continuing use of liquor drops hunger and averts the body from absorbing vital nutrients, there is deficiency of a number of vitamins and minerals. It is therefore essential to supplement the diet with vitamin B complex, vitamin C, selenium, magnesium, and zinc. A combination of amino acids, such as carnitine, glutamine, and glutathione, may help decrease longings, blood sugar variations, and anxiety that are related to liquor usage.

TABLE 11.1 Medicinal Plants With Hepatoprotective Activity (Adapted From Rajaratnam et al., 2014)

Medicinal Plant(s)	Plant Part Used	Type of Liver Protection in Drinking Problem
Camellia sinensis (green tea)	Leaves and seeds	NAFLD (nonalcoholic fatty liver disease), acute liver injury
Silybum marianum (milk thistle)	Seeds	Alcohol-induced cirrhosis
Combination of *Capparis spinosa, C. intybus, S. nigrum, Terminalia arjuna, Cassia occidentalis, Achillea millefolium, Tamarix gallica*	Mixture of leaves, roots, flowers, fruits and seeds of capers, blue daisy, black nightshade, arjuna, negro coffee, yarrow, and tamarisk	Alcoholic liver disease, liver cirrhosis
Combination of *Picorrhiza kurroa, Andrographis paniculata, Eclipta alba, C. intybus, S. nigrum, Phyllanthus niruri*	Mixture of rhizomes, roots, leaves of kutki, creat, false daisy, blue daisy, black nightshade, and stonebreaker.	Alcoholic liver disease, liver cirrhosis
Combination of *C. intybus, S. nigrum, P. kurroa, A. paniculata, Tephrosia purpurea)*	Mixture of flowers, rhizomes, leaves, and roots of kutki, creat, blue daisy, black nightshade and wild indigo	Alcoholic liver disease
Combination of *Thespesia populnea, Elettari cardamomum, Zingiber officinale, Glycyrrhiza glabra*	Mixture of leaves, roots, barks, fruits, seed pods and rhizome of portia tree, green cardamom, ginger and licorice	Alcoholic liver disease, liver cirrhosis
Combination of *E. alba, P. niruri, C. intybus, Rheum emodi, Tephrosia purpurea*	Mixture of leaves, roots, barks, and flowers of false daisy, stonebreaker, blue daisy, Himalayan rhubarb, and wild indigo	Toxic liver injury
Combination of *A. paniculata, P. niruri, E. alba,P. kurroa, Boerhaavia diffusa, Azadirachtica indica, Berberis aristata, Ipomoea turpethum, T. purpurea*	Mixture of leaves, roots, barks, and flowers of false daisy, creat, stonebreaker, kutki, red spiderling, Indian lilac, tree turmeric, St. Thomas lidpod, and wild indigo	Viral hepatitis, hepatomegaly
Combination of *P. kurroa, B. diffusa, Tinospora cordifolia, A. paniculata, P. emblica*	Mixture of leaves, roots, barks, and flowers of creat, red spiderling, giloy, kutki, and emblic	Hepatic dysfunction

TABLE 11.1 *(Continued)*

Medicinal Plant(s)	Plant Part Used	Type of Liver Protection in Drinking Problem
Combination of *A. paniculata, E. alba, Ocimum sanctum, P. niruri, P. kurroa, Piper longum, S. nigrum, Tephrosia purpurea, Terminalia chebula*	Mixture of leaves, roots, barks, fruits, seed pods, and rhizomes of creat, kutki, stonebreaker, false daisy, holy basil, Indian long pepper, black nightshade, wild indigo, and yellow myrobalan	Alcoholic liver damage, cirrhosis, Fatty liver, and drug-induced hepatotoxicity
Lycii fructus (wolfberry)	Dried berries and root bark	Hepatoma
Bacopa monnieri (brahmi)	Leaves	Ethanol-induced liver injury
Boerhaavia diffusa	Leaves	Oxidative stress of the liver
Vernonia amygdalina (bitter leaf)	Leaves	Chemical liver injury
Curcuma longa (turmeric)	Rhizomes	Acute liver damage by chemicals, for example, ethanol
Schisandra chinensis (five flavor berry)	Berries	Chemical or viral hepatitis

The supplementation of the diet with vitamin and mineral preparations is also known as "orthomolecular medicine" and has strong advocates in the addiction field, although these nutrients should be used carefully and in a pure form. A high-potency multivitamin preparation will assist detoxification and stabilize mood. Vitamins B are particularly important in this regard that are generally depleted after a period of drug or alcohol abuse, and 50 mg daily of a B complex is effective. Chromium plays an important role in stabilizing blood sugar and thus alleviates cravings. The amino acid L-glutamine will improve energy levels and stabilize mood, both of which can fluctuate for long periods, both during the period of immediate withdrawal and in the post-acute withdrawal period that can last for up to a year. The serotonin precursor 5-hydroxytryptophan will help with depression and anxiety (Hoffer and Saul, 2009).

Other important nutritional recommendations include phenylalanine for depression and fatigue; calcium and magnesium for tension, tremor, and muscle cramps and twitches; and niacin (vitamin B_3) is strongly recommended for alcoholism, along with inositol hexaniacinate.

N-acetyl-cysteine, a derivative of the amino acid cysteine, increases levels of glutathione, which is the most important antioxidant and has been a beneficial adjunct for dealing with drug dependence. Vitamin C also increases glutathione levels and augments antioxidant effects as well as detoxification. Many classic complications of alcoholism, particularly memory disorders, are due to depletion of thiamine (vitamin B_1), and supplementation with intramuscular or intravenous injection may be appropriate (Hoffer and Saul, 2009).

11.4.1 THIAMINE (VITAMIN B_1)

Intake of thiamine supplement is essential during withdrawal. Heavy use of alcohol causes thiamine deficiency, which can lead to a serious brain disorder called Wernicke–Korsakoff syndrome.

11.4.2 VITAMIN A

It has been commonly observed that generally people who are often deficient in vitamin A misuse liquor. The vitamin supplements should be taken only under a doctor's direction as high dosages of vitamin A can harm the liver, and may cause alcoholic liver disease to advance towards more quickly in people who drink heavily.

Some members of the alcohol recovery community endorse a heftier, nutritionally oriented attitude. This includes arterial dietetic remedies, along with targeted amino acid supplementation, to control mind function.

11.4.3 VITAMIN B_{12}

An important B-complex vitamin B_{12} helps regulate the function of the liver. Restoring the normal function of the liver can help ease alcohol cravings.

11.4.4 VITAMIN C

The entire body, but particularly the brain, requires high levels of vitamin C. This nutrient helps mediate the activity of the opiate receptors in the brain, which helps reduce an individual's interest in drugs and alcohol.

11.4.5 AMINO ACIDS

Amino acids, specifically L-glutamine, are responsible for regulating a number of processes in the body. L-glutamine, in particular, helps stabilize the mood, which can help interrupt the craving phase.

11.5 HOMEOPATHY IN ALCOHOLISM

There are a few studies conducted to assess the effectiveness of precise homeopathic therapies. Usually, professional homeopathic practitioners recommend a treatment for alcoholism based on their knowledge and clinical understanding. A homeopathic doctor generally considers the patient's inherent constitution before recommending a medication. In homeopathic terms, a patient's inherent constitution is his or her physical, psychological, and cerebral temperament. A skilled homeopathic doctor evaluates all of these aspects when defining the most suitable medicine for an actual individual. Homeopathy should not be used alone to treat alcoholism but can be a supportive treatment along with psychotherapy. There are a few instances of medications that an experienced homeopathic doctor will ponder for indicators related to liquor abuse or removal. *Arsenicum album* is used for worry and compulsiveness, with nausea, vomiting, and diarrhea; *Nux vomica* is used for irritability and compulsiveness with nausea, vomiting, and constipation; *Lachesis* is used for liquor desires, headaches, and strain in swallowing; and *Staphysagria* is used for irritated individuals who tend to overpower their sentiments and may have been ill-treated bodily, sexually, or emotionally in the past.

11.6 OTHER APPROACHES

11.6.1 ACUPUNCTURE

A form of ancient Chinese medicine acupuncture accesses specific pressure points to release endorphins, relieve pain and anxiety in the body, and stimulate wellness. As such, this form of treatment is also effective for curing alcohol addiction (Renaud and Renaud, 1998).

11.6.2 EXERCISE

Exercise is another important daily practice for managing alcoholism. Not only does exercise provide a release for stress and an opportunity to engage in an additional activity, but it also helps boost the endorphins in the body. Endorphins help improve the mood and decrease an individual's reliance on drugs to achieve the same result (Lu et al., 2009).

11.7 CONCLUSIONS

Alcoholism is a grave health, communal and fiscal problem facing nearly all human societies worldwide. Alcoholism has been established in an ~57% of the adult population in the Western countries, while a majority of the people have been categorized as "problem drinkers." From the above discussion, it can be safely concluded that traditional medicinal plants may establish unique and operative pharmacotherapies for drunkenness and liquor misuse. A balanced attitude to liquor retrieval is the best option. Pairing these treatments with an established recovery program will help you or your loved ones regain control over life and health. Summarizing, there is a necessity for more studies on the mechanisms of plant extracts and their bioactive compounds mode of action that are valuable unconventional modes for the stoppage and management of various drug addictions.

ACKNOWLEDGMENTS

The authors are grateful to Dr Ashok K. Chauhan, Founder President, and Mr Atul Chauhan, Chancellor, Amity University, Noida, India, for the encouragement, research facilities, and financial support.

KEYWORDS

- **alcoholism**
- **herbal treatment**
- **drinking problem**
- **medicinal plants**
- ***Silybum marianum***
- ***Pueraria lobata***
- ***Taraxacum officinale***
- ***Ganoderma lucidum***
- ***Passiflora edulis.***

REFERENCES

Addolorato, G.; Leggio, L.; Hopf, F. W.; Diana, M.; Bonci, A. Novel Therapeutic Strategies for Alcohol and Drug Addiction: Focus on Gaba, Ion Channels and Transcranial Magnetic Stimulation. *Neuropsychopharmacol* **2011**.

Ambrose, M. L.; Bowden, S. C.; Whelan, G. Thiamin Treatment and Working Memory Function of Alcohol-Dependent People: Preliminary Findings. *Alcohol Clin. Exp. Res.* **2001**, *25*(1), 112–116.

American Psychiatric Association. *Diagnostic and Statistical Manual of Mental Disorders*. 4th ed. Text Revision. Washington, DC: American Psychiatric Association, 2000.

Andrade, C. Indian Herbal Medicine and the Treatment of Addictive Disorders, *Proceedings of the UK/European Symposium on Addictive Disorders (UKESAD)*, London, 2006.
Assanangkornchai, S.; Srisurapanont, M. The Treatment of Alcohol Dependence. *Curr. Opin. Psychiatry* **2007**, *20*(3), 222–227.
Attele, A. S.; Wu, J. A.; Yuan, C. S. Ginseng Pharmacology: Multiple Constituents and Multiple Actions. *Biochem. Pharmacol.* **1999**, *58*, 1685–1693.
Bressler, R. Herb-Drug Interactions: Interactions Between Kava and Prescription Medications. *Geriatrics* **2005**, *60*, 24–25.
Brunetti, G.; Serra, S.; Vacca, G, et al. IDN 5082, a Standardized Extract of *Salvia miltiorrhiza*, Delays Acquision of Alcohol Drinking Behavior in Rats. *J. Ethnopharcol.* **2003**, *85*, 93–97.
Cairney, S.; Maruff, P.; Clough, A.R. The Neuro-Behavioural Effects of Kava. *Aust. N. Z. J. Psychiatry* **2002**, *36*, 657–662.
Carai, M.A.; Agabio, R.; Bombardelli, E.; Bourov, I.; Gessa, G. L.; Lobina, C.; Morazzoni, P.; Pani, M.; Reali, R.; Vacca, G.; Colombo, G. Potential Use of Medicinal Plants in the Treatment of Alcoholism. *Fitoterap.* **2000,** *71*(Suppl. 1), S38–S42.
Carr, G. Alcoholism: A Modern Look at an Ancient Illness. *Primary Care: Clin. Office Pract.* **2011**, *38*(1).
Chanawirat, A.; Toskulkao, C.; Temcharoen, P.; Glinsukou, T. *Protective Effect of Thunbergia laurifolia Extract on Ethanol-Induced Hepatotoxicity in Mice*; Faculty of Graduate Studies, Mahidol University: Bangkok, Thailand, 2000.
Cheng, R. S.; Pomeranz, B.; Yu, G. Electro-Acupuncture Treatment of Morphine Dependent Mice Reduces Signs of Withdrawal, Without Showing Cross Tolerance. *Eur. J. Pharmacol.* **1980**, *68*, 477–481.
Colombo, G.; Agabio, R.; Lobina, C. *Salvia miltiorrhiza* Extract Inhibits Alcohol Absorption, Preference, and Discrimination in sP Rats. *Alcohol.* **1999**, 18, 65–70.
Cooney, J.L.; Cooney, N.L.; Pilkey, D.T.; Kranzler, H.R.; Oncken, C.A. Effects of Nicotine Deprivation on Urges to Drink and Smoke in Alcoholic Smokers. *Addiction.* **2003**, 98(7): 913–921.
Correale, M.; Laonigro, I.; Altomare, E.; Di Biase, M. Alcohol-induced Cardiac Disease. *G Ital. Cardiol. (Rome)*. **2009**, 10(1): 18–27.
Dager, A.D.; Anderson, B.M.; Stevens, M.C. et al. Influence of Alcohol Use and Family History of Alcoholism on Neural Response to Alcohol Cues in College Drinkers. *Alcohol Clin. Exp. Res.* **2013**, 37 Suppl 1: E161–171.
Das, U.N. Essential Fatty Acids—A Review. *Curr. Pharm. Biotechnol.* **2006,** 7(6):467–482.
Duke, J. The Ultimate Compendium of Natural Remedies from the World's Foremost Authority on Healing and Herbs. *Green Pharm.* **1997**, 57: 281–282.
Dupuy, O.; Flocard, F.; Vial, C. et al. Disulfiram (Esperal) Toxicity. Apropos of 3 Original cases. *Rev. Med. Interne*. **1995**, 16: 67–72.
Fang, Z.Y.; Lin, R.; Yuan, B.X.; Yang, G.D.; Liu, Y.; Zhang, H. Tanshinone IIA Downregulates the CD40 Expression and Decreases MMP-2 Activity on Atherosclerosis Induced by High Fatty Diet in Rabbit. *J. Ethnopharmacol.* **2008**, 115: 217–222.
Ferri, F. *Ferri's Clinical Advisor 2016*. 1st ed. Elsevier Mosby: Philadelphia, PA. **2016**.
Ferri, M.; Amato, L.; Davoli, M. Alcoholics Anonymous and Other 12-step Programmes for Alcohol Dependence. *Cochrane Database Syst Rev*. **2006**, 3: CD005032.

Finney, J.W.; Hahn, A.C.; Moos, R.H. The Effectiveness of Inpatient and Outpatient Treatment for Alcohol Abuse: The Need to Focus on Mediators and Moderators of Setting Effects. *Addiction.* **1996**, 91: 1773–1796.

Froehlich, J.C., Badia-Elder, N.E.; Zink, R.W. et al. Contribution of the Opioid System to Alcohol Aversion and Alcohol Drinking Behaviour. *J. Pharmacol. Exp. Ther.* **1998,** 287: 284–292.

Garland, E.L.; Gaylord, S.A.; Boettiger, C.A.; Howard, M.O. Mindfulness Training Modifies Cognitive, Affective, and Physiological Mechanisms Implicated in Alcohol Dependence: Results of a Randomized Controlled Pilot Trial. *J Psychoactive Drugs.* **2010**, 42(2):177–192.

Gillis, C.N. *Panax ginseng* Pharmacology: A Nitric Oxide Link. *Biochem. Pharmacol.* **1997**, 54: 1–8.

Gonzalez-Reimers, E.; Santolaria-Fernandez, F.; Martin-Gonzalez, M.C.; Fernandez-Rodriguez, C.M.; Quintero-Platt, G. Alcoholism: A Systemic Pro-inflammatory Condition. *World J. Gastroenterol.* **2014**, 20(40): 14660–14671.

Greenlee, H.; Abascal, K.; Yarnel, E.; Ladas, E. Clinical Applications of *Silybum marianum* in Oncology. *Integr. Cancer Ther.* **2007**, 6: 158–165.

Groth-Marnat, G.; Leslie, S.; Renneker, M. Tobacco Control in a Traditional Fijian Village: Indigenous Methods of Smoking Cessation and Relapse Prevention. *Soc Sci Med.* **1996,** 43: 473–477.

Hall, M. Alcoholism and Depression. *Home Health Nurse.* **2012,** 30(9): 543–550.

Heilig, M.; Goldman, D.; Berrettini, W.; O'Brien, C.P. Pharmacogenetic Approaches to the Treatment of Alcohol Addiction. *Nat. Rev. Neurosci.* **2011**, 12(11): 670–684.

Hillemacher, T.; Kahl, K.G.; Heberlein, A.; Muschler, M.A.; Eberlein, C.; Frieling, H.; Bleich, S. Appetite- and Volume-Regulating Neuropeptides: Role in Treating Alcohol Dependence. *Curr. Opin. Investig. Drugs.* **2010**, 11(10): 1097–1106.

Hoffer, A.; Saul, A.W. The Vitamin Cure for Alcoholism: Orthomolecular Treatment of Addictions. *Health and Fitness*, Basic Health Publications, Inc., 2009, p. 134.

Huang, Z.M. Studies on Ginsenoside Inhibited Morphine-dependence. *World Phytomed.* **1991**, 6: 203–205.

Johnson, B.A. Medication Treatment of Different Types of Alcoholism. *Am. J. Psychiatry.* **2010**, 167(6): 630–639.

Jones, E.A.; Dekker, L.R. Florid Opioid Withdrawal-Like Reaction Precipitately By Naltrexone in a Patient With Chronic Cholestasis. *Gastroenterol.* **2000,** 118: 431–432.

Keung, W-M.; Vallee, B.L. Daidzin: A Potent, Selective Inhibitor of Human Mitochondrial Aldehyde Dehydrogenase. *Proc Natl Acacl Sci. USA.* 1993; 90:100.

Kim, H.S.; Hong, Y.T.; Jang, C.G. Effects of the Ginsenosides Rg1 and Rb1 on Morphine-induced Hyperactivity and Reinforcement in Mice. *J Pharm Pharmacol.* **1998**, 50: 555–560.

Kim, H.S.; Kim, K.S.; Oh, K.W. Ginseng Total Saponin Inhibits Nicotine Induced Hyper-activity And Conditioned Place Preference In Mice. *J Ethnopharmacol.* **1999**, 66: 83–90.

Kulkarni, S.K.; Ninan, I. Inhibition of Morphine Tolerance and Dependence by *Withania somnifera* in Mice. *J. Ethnopharmacol.* **1997**, 57: 213–217.

Kunz, S.; Schulz, M.; Lewitzky, M.; Driessen, M.; Rau, H. Ear Acupuncture for Alcohol Withdrawal in Comparison With Aromatherapy: A Randomized-Controlled Trial. *Alcohol Clin. Exp. Res.* **2007**, 31(3): 436–442.

Le Berre, A.P.; Pitel, A.L.; Chanraud, S.; et al. Chronic Alcohol Consumption And Its Effect On Nodes of Frontocerebellar And Limbic Circuitry: Comparison of Effects in France and the United States. *Hum. Brain Mapp.* **2014**, 35(9): 4635–4653.

Leone, M.A.; Vigna-Taglianti, F.; Avanzi, G.; Brambilla, R.; Faggiano, F. Gamma-hydroxybutyrate (GHB) for Treatment of Alcohol Withdrawal and Prevention of Relapses. *Cochrane Database Syst Rev.* **2010**, (2): CD006266.

Liu, Q.; Lawrence, A.J.; Liang, J.H. Traditional Chinese Medicine for Treatment of Alcoholism: From Ancient to Modern. *Am. J. Chin. Med.* **2011**, 39(1):1–13.

Lu, L.; Liu, Y.; Zhu, W.; Ling, W.; Kosten, T.R. Traditional Medicine in the Treatment of Drug Addiction. *Am. J. Drug Alc. Abus.* **2009**, 35:1–11.

Lude, S.; Torok, M.; Dieterle, S.; et al. Hepatocellular Toxicity of Kava Leaf and Root Extracts. *Phytomed.* **2008**, 15; 120–131.

Lukas, S. E.; Penetar, D.; Berko, J.; Vicens, L.; Palmer, C.; Mallya, G.; Macklin, E. A.; Lee, D. Y. An Extract of the Chinese Herbal Root Kudzu Reduces Alcohol Drinking by Heavy Drinkers in a Naturalistic Setting. *Alcohol Clin. Exp. Res.* **2005**, *29*(5), 756–762.

Lyndall, G. J.; Bass, N. J.; McQuillin, A. et al. Confirmation of Prior Evidence of Genetic Susceptibility to Alcoholism in a Genome-Wide Association Study of Comorbid Alcoholism and Bipolar Disorder. *Psychiatr. Genet.* **2011**, *21*(6), 294–306.

Lyon, J.E.; Khan, R.A.; Gessert, C.E.; Larson, P.M.; Renier, C.M. Treating Alcohol Withdrawal With Oral Baclofen: A Randomized, Double-Blind, Placebo-Controlled Trial. *J Hosp. Med.* **2011**, *6*(8), 474–479.

Malik, P.; Gasser, R. W.; Kemmler, G.; Moncayo, R.; Finkenstedt, G.; Kurz, M.; Fleischhacker, W. W. Low Bone Mineral Density and Impaired Bone Metabolism in Young Alcoholic Patients Without Liver Cirrhosis: A Cross-Sectional Study. *Alcohol Clin. Exp. Res.* **2009**, *33*(2), 375–381.

Mantere, O.; Suominen, K.; Valtonen, H.M. et al. Concomitants of Family Histories of Mood Disorders and Alcoholism in a Clinical Cohort of Patients With Bipolar I and II Disorder. *J. Nerv. Ment. Dis.* **2012**, *200*(5), 388–394.

Martinotti, G.; Andreoli, S.; Reina, D.; Di Nicola, M.; Ortolani, I.; Tedeschi, D.; Fanella, F.; Pozzi, G.; Iannoni, E.; D'Iddio, S.; Prof, L. J. Acetyl-l-Carnitine in the Treatment of Anhedonia, Melancholic and Negative Symptoms in Alcohol Dependent Subjects. *Prog. Neuro. Psycho. Pharmacol. Biol. Psychiatry* **2011**, *35*(4), 953–958.

Modaresi, M., Resalatpour, N. The Effect of *Taraxacum officinale* Hydroalcoholic Extract on Blood Cells in Mice. *Adv. Hematol.* **2012**, *2012*, 653412.

Narendran, R.; Mason, N. S.; Paris, J.; Himes, M. L.; Douaihy, A. B.; Frankie, W. G. Decreased Prefrontal Cortical Dopamine Transmission in Alcoholism. *Am. J. Psychiatry* **2014,** *171*(8), 881–888.

O'Malley, S. S.; Krishnan-Sarin, S.; Farren, C.; et al. Naltrexone Induced Nausea in Patients Treated for Alcohol Dependence. Clinical Predictors and Evidence for Opioid-Mediated Effects. *J. Clin. Psychopharmacol.* **2000**, *20*, 69–76.

Oh, S. H.; Soh, J. R.; Cha, Y. S. Germinated Brown Rice Extract Shows a Nutraceutical Effect in the Recovery of Chronic Alcohol-Related Symptoms. *J. Med. Food.* **2003**, *6*(2), 115–121.

Oncken, C.; Van Kirk, J.; Kranzler, H. R. Adverse Effects of Oral Naltrexone: Analysis of Data From Two Clinical Trials. *Psychopharmacology* **2001**, *154*, 397–402.

Otto, K. C. Acupuncture and Substance Abuse: A Synopsis, With Indications for Further Research. *Am. J. Addict.* **2003**, *12*(1), 43–51.

Overstreet, D. H.; Keung, W. M.; Rezvani, A. H.; Massi, M.; Lee, D. Y. Herbal Remedies for Alcoholism: Promises and Possible Pitfalls. *Alcohol Clin. Exp. Res.* **2003**, *27*(2), 177–185.

Ozarowski, M.; Mikołajczak, P. Ł.; Thiem, B. Medicinal Plants in the Phytotherapy of Alcohol or Nicotine Addiction: Implication for Plants in Vitro Cultures. *Przegl. Lek.* **2013**, *70*(10), 869–874.

Pilowsky, D. J.; Keyes, K. M.; Hasin, D. S. Adverse Childhood Events and Lifetime Alcohol Dependence. *Am. J. Public Health* **2009**, *99*(2), 258–263.

Poulsen, H. E.; Loft, S.; Andersen, J. R.; et al. Disulfiram Therapy-Adverse Drug Reactions and Interactions. *Acta Psychiatr. Scand.* **1992,** *86*, 59–66.

Pramyothin, P.; Chirdchupunsare, H.; Rungsipipat, A.; Chaichantipyuth, C. Hepato-Protective Activity of *Thunbergia laurifolia* Linn Extract in Rats Treated With Ethanol: In Vitro and in Vivo Studies. *J. Ethnopharmacol.* **2005**, *102*, 408–411.

Procopio, D. O.; Saba, L. M.; Walter, H.; et al. Genetic markers of comorbid depression and alcoholism in women. *Alcohol Clin. Exp. Res.* **2013**, *37*(6), 896–904.

Purohit, V.; Abdelmalek, M. F.; Barve, S.; Benevenga, N. J.; Halsted, C. H.; Kaplowitz, N.; et al. Role of S-adenosylmethionine, folate, and betaine in the treatment of alcoholic liver disease: summary of a symposium. *Am. J. Clin. Nutr.* **2007**, *86*(1), 14–24.

Rajaratnam, M.; Prystupa, A.; Lachowska-Kotowska, P.; Załuska, W.; Filip, R. Herbal medicine for treatment and prevention of liver diseases. *J. Pre-Clin Clin. Res.* **2014**, *8*(2), 55–60.

Rakel, D. *Integrative Medicine*. 3rd ed. Elsevier Saunders: Philadelphia, PA:, 2012.

Rakel, R. E.; Rakel, D. *Textbook of Family Medicine*. 8th ed. Elsevier Saunders: Philadelphia, PA, 2011.

Reddy, S.; Dick, A. M.; Gerber, M. R.; Mitchell, K. The Effect of a Yoga Intervention on Alcohol and Drug Abuse Risk in Veteran and Civilian Women With Post-Traumatic Stress Disorder. *J. Altern. Complement. Med.* **2014**, *20*(10), 750–756.

Renaud, J.; Renaud, M. *Big Publisher Explains Acu-Detox to Docs Guide Points: Acupuncture in Recovery.* **1998**, p 5.

Rezvani, A. H.; Overstreet, D. H.; Perfumi, M.; Massi, M. Plant Derivatives in the Treatment of Alcohol Dependency. *Pharmacol. Biochem. Behav.* **2003**, *75*(3), 593–606.

Roerecke, M.; Rehm, J. Ischemic Heart Disease Mortality and Morbidity Rates in Former Drinkers: A Meta-Analysis. *Am. J. Epidemiol.* **2011**, *173*(3), 245–258.

Ryan, M.; Merrick, E. L.; Hodgkin, D.; et al. Drinking Patterns of Older Adults With Chronic Medical Conditions. *J. Gen. Intern. Med.* **2013**, *8*(10), 1326–1332.

Seitz, U.; Schule, A.; Gleitz, J. [3H]-Monoamine Uptake Inhibition Properties of Kava Pyrones. *Planta Med.* **1997,** *63*, 548–549.

Serra, S.; Vacca, G.; Tumatis, S.; et al. Anti-Relapse Properties of IDN 5082, A Standardized Extract of *Salvia miltiorrhiza*, in Alcohol-Preferring Rats. *J. Ethnopharmacol.* **2003**, *88*, 249–252.

Shebek, J.; Rindone, J. P. A Pilot Study Exploring the Effect of Kudzu Root on the Drinking Habits of Patients With Chronic Alcoholism. *J. Altern. Complment. Med.* **2000**, *6*, 45–48.

Singh, A. K.; Jiang, Y.; Benlhabib, E.; Gupta, S. Herbal Mixtures Consisting of Puerarin and Either Polyenylphosphatidylcholine or Curcumin Provide Comprehensive Protection

Against Alcohol-Related Disorders in P Rats Receiving Free Choice Water and 15% Ethanol in Pure Water. *J. Med. Food.* **2007**, *10*(3), 526–542.

Sinha, R.; Fox, H. C.; Hong, K. I.; Hansen, J.; Tuit, K.; Kreek, M. J. Effects of Adrenal Sensitivity, Stress- and Cue-Induced Craving, and Anxiety on Subsequent Alcohol Relapse and Treatment Outcomes. *Arch. Gen. Psychiatry* **2011,** *68*(9), 942–952.

Steiner, G. G. Kava as an Anti-Craving Agent: Preliminary Data. *Pac. Health Dialog.* **2001**, *8*, 335–339.

Sukul, N. C.; Ghosh, S.; Sinhababu, S. P.; Sukul, A. Strychnos Nux-Vomica Extract and Its Ultra-High Dilution Reduce Voluntary Ethanol Intake in Rats. *J. Altern. Complement. Med.* **2003**, *7*(2), 187–193.

Swift, R. M. Drug Therapy for Alcohol Dependence. *N. Engl. J. Med.* **1999**, *340*, 1482–1490.

Takahashi, M.; Tokuyama, S. Pharmacological and Physiological Effects of Ginseng on Actions Induced by Opioids and Psychostimulants. *Methods Find Exp. Clin. Pharmacol.* **1998**, *20*, 77–84.

Tamayo, C.; Diamond, S. Review of clinical trials evaluating safety and efficacy of milk thistle (*Silybum marianum* [L.] Gaertn.). *Integrat. Canc. Ther.* **2007**, 6: 146–157.

Thongsaard, W.; Marsden, C. A. A Herbal Medicine Used in the Treatment of Addiction Mimics the Action of Amphetamine on in Vitro Rat Striatal Dopamine Release. *Neurosci. Lett.* **2002,** *329*, 129–132.

Trumpler, F.; Oez, S.; Stahli, P.; Brenner, H. D.; Juni, P. Acupuncture for Alcohol Withdrawal: A Randomized Controlled Trial. *Alcohol*, **2003**, *38*(4), 369–375.

US Department of Health and Human Services; National Institutes of Health; National Institute on Alcohol Abuse and Alcoholism. *10th Special Report to the U.S. Congress on Alcohol and Health*, 2000.

Ussanawarong, S.; Thesiri, T. Effect of *Thunbergia laurifolia* Linn. on Detoxification of Parathion in Rat. *Khon Kaen Univ. Res. J.* **2001,** *6*, 3–13.

Ventegodt, S.; Clausen, B.; Langhorn, M.; Kromann, M.; Andersen, N. J.; Merrick, J. Quality of Life as Medicine III. A Qualitative Analysis of the Effect of a Five-Day Intervention With Existential Holistic Group Therapy or a Quality of Life Course as a Modern Rite of Passage. *Sci. World J.* **2004**, *4*, 124–133.

Vieten, C.; Astin, J. A.; Buscemi, R.; Galloway, G. P. Development of an Acceptance-Based Coping Intervention for Alcohol Dependence Relapse Prevention. *Subst. Abus.* **2010**, *31*(2), 108–116.

Waldron, M.; Bucholz, K. K.; Lynskey, M. T.; Madden, P. A.; Heath, A. C. Alcoholism and Timing of Separation in Parents: Findings in a Mid-Western Birth cohort. *J. Stud. Alcohol Drugs* **2013**, *74*(2), 337–348.

Waldron, M.; Vaughan, E. L.; Bucholz, K. K.; et al. Risks for Early Substance Involvement Associated With Parental Alcoholism and Parental Separation in an Adolescent Female Cohort. *Drug Alcohol Depend.* **2014**, *138*, 130–136.

Wedekind, D.; Herchenhein, T.; Kirchhainer, J.; Bandelow, B.; Falkai, P.; Engel, K.; Malchow, B.; Havemann-Reinecke, U. Serotonergic Function, Substance Craving, and Psychopathology in Detoxified Alcohol-Addicted Males Undergoing Tryptophan Depletion. *J. Psychiatr. Res.* **2010**, *44*(16), 1163–1169.

Wilson, S. R.; Knowles, S. B.; Huang, Q.; Fink, A. The Prevalence of Harmful and Hazardous Alcohol Consumption in Older U.S. Adults: Data From the 2005–2008

National Health and Nutrition Examination Survey (NHANES). *J. Gen. Intern. Med.* **2014**, *29*(2), 312–319.

Xu, B. J.; Zheng, Y. N.; Sung, C. K. Natural Medicines for Alcoholism Treatment: A Review. *Drug Alcohol Rev.* **2005**, *24*(6), 525–536.

You, Y.; Yoo, S.; Yoon, H. G.; Park, J.; Lee, Y. H.; Kim, S.; Oh, K. T.; Lee, J.; Cho, H. Y.; Jun, W. *In vitro* and *in Vivo* Hepato-Protective Effects of the Aqueous Extract From *Taraxacum officinale* (dandelion) Root Against Alcohol-Induced Oxidative Stress. *Food Chem. Toxicol.* **2010**, *48*(6), 1632–1637.

Zhao, R. J.; Yoon, S. S.; Lee, B. H.; et al. Acupuncture Normalizes the Release of Accumbal Dopamine During the Withdrawal Period and After the Ethanol Challenge in Chronic Ethanol-Treated Rats. *Neurosci. Lett.* **2006**, *395*, 28–32.

CHAPTER 12

Beta Cells and Diabetes

SHIVANI SRIVASTAVA[1], DURGAVATI YADAV[1], KUMAR SANDEEP[2], HARSH PANDEY[1], SURYA KUMAR SINGH[3], and YAMINI BHUSAN TRIPATHI[1*]

[1]*Department of Medicinal Chemistry, Institute of Medical Sciences, Banaras Hindu University, Varanasi, Uttar Pradesh, India*

[2]*Dr Bhim Rao Ambedkar Institute Rotary Cancer Hospital, All India Institute of Medical Sciences, New Delhi, India*

[3]*Department of Endocrinology and Metabolism, Institute of Medical Sciences, Banaras Hindu University, Varanasi, Uttar Pradesh, India*

[*]*Corresponding author. E-mail: yaminibiochemist6@gmail.co, yaminiok@yahoo.com*

12.1 INTRODUCTION

Diabetes is generally considered a very complex disease. Because of having no clear idea of it, the permanent solution is absent. Even discrimination between Type 1 and Type 2 is also very difficult. Type 1 diabetes is the severe condition of diabetes mellitus (DM) but sometimes it mimics type 2 DM by performing insulin resistance. Mechanism of action involves hyperglycemia/glucotoxicity, lipotoxicity, autoimmunity, inflammation, adipokines, islet amyloid, incretins, and insulin resistance in DM (Cernea and Dobreanu, 2013). Insulin gets released from pancreatic β cells when circulating blood glucose reaches a threshold. Insulin performs two major functions: first, it stores glucose in skeletal muscles and liver in the form of glycogen and simultaneously inhibits glucose secretion from the liver and second, insulin preserves excess fatty acid in adipose tissue as triglyceride to be used as a reserve for energy release, in case of fasting. Excess intake of free fatty acids (FFAs), lack of exercise, glucocorticoid administration, and obesity are the main inducers of insulin resistance. In pathological conditions like obesity, these fatty acids get accumulated in the muscle, liver, and β cells, thus creating lipotoxicity. This ectopic accumulation results in insulin resistance and a loss of mass and function of β cells. Altogether, these conditions lead to hyperglycemia and type 2 diabetes (T2D) (Hajduch and Bourron, 2013) (Figure 12.1).

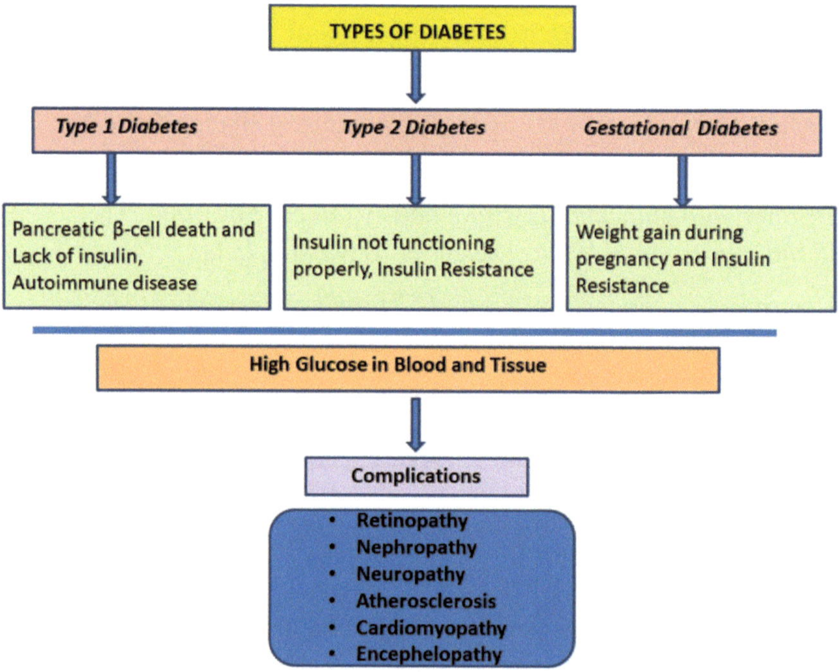

FIGURE 12.1 Types of diabetes and its complications (adapted from Pradhan et al., 2016).

12.2 AMYLOID DEPOSITION IN B CELLS

Many human disorders like Alzheimer's, Parkinson's, and T2D have a relation with amyloid deposition. It is the partially ordered fibrillar protein aggregated rich in beta sheets structure. In T2D, islets amyloid is usually formed by a neuropancreatic hormone islet amyloid polypeptide (IAPP), also known as amylin. This formation leads to β cells dysfunction and death. Interestingly, amyloid formation does not occur in all species. Cats, humans, and nonhuman primates develop islet amyloid formation but mice and rats do not. IAPP is secreted with insulin secretion, which means it is absent in TIDM. Reduced expression and reduction of IAPP in TIDM and later stage of T2DM usually leads to deleterious effect. But the case of T2DM is different, here the toxicity induced by amyloidosis is one of the major concerns. Heparansulfate proteoglycan perlecan, apolipoprotein E, and serum amyloid P component are the constituents of islet amyloid (Abedini et al., 2015).

12.3 INCRETINS

Diabetes is on the regular rise worldwide. It is due to progressive β cell damage present in the pancreatic islet of Langarhans. Intestinal function and incretin secretion are reported to directly regulate the health of the pancreas. One of them is glucagon-like peptide 1 (GLP-1), which regulates glucose-induced insulin secretion from pancreatic β cells. Its secretion and activity are governed by the activity of another intestinal enzyme dipeptidyl peptidase 4 (DPP-IV) and through regulation of GLP-1 receptors on target organs. In a report, it has been claimed that in the condition of compromised pancreas secretion, the enteroendocrine cells of the intestine get converted to insulin-positive cells (Suzuki et al., 2003; Talchai et al., 2012; Srivastava *et al*., 2019a) (Figure 12.2).

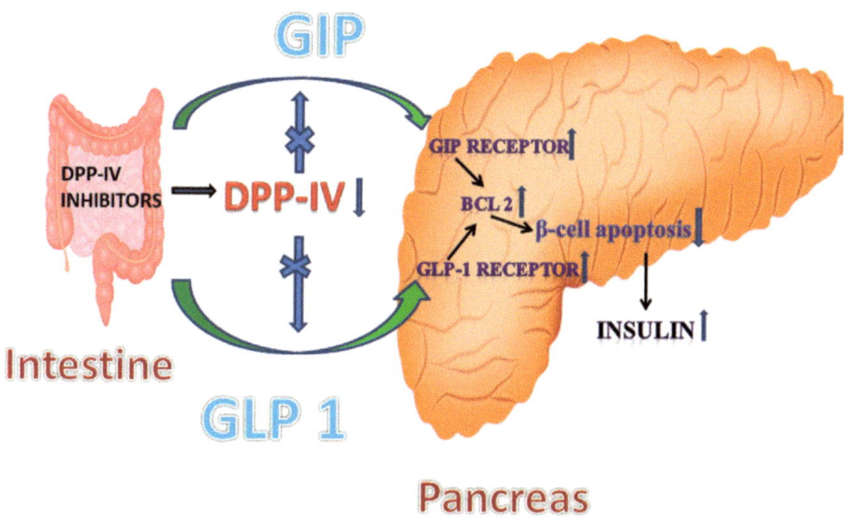

FIGURE 12.2 Mechanism of incretin pathways (adapted from Srivastava et al., 2018b).

12.3.1 GLUCOSE-DEPENDENT INSULINOTROPIC POLYPEPTIDE (GIP)

The location of 42-amino acid hormone GIP on the human chromosome is 17q21.3-q22 with six exons. This hormone is secreted from K cells of the upper small intestine (duodenum and jejunum) through proteolytic processing of pre-pro GIP. After secretion active GIP directly acts on the pancreatic islet to secrete insulin from β cells (Seino et al., 2010).

12.3.2 GLUCAGON-LIKE POLYPEPTIDE-1 (GLP-1)

The 31 amino acid GLP-1 is known to be secreted from the L cells of duodenum, ileum, and colon after processing from proglucagon present at the location of human chromosome 2q36–q37 with six exons. Like GIP it also acts on β cells directly to release insulin (Seino et al., 2010).

GLP-1 is one of the incretin hormones released in response to meal ingestion from L-cells of the intestine. Potentiation of glucose-stimulated insulin secretion (GSIS), enhancement of β-cell growth and survival, inhibition of glucagon release, gastric emptying, and food intake are some of the main physiological actions of GLP-1 (Lim et al., 2006). GLP-1 has a pleiotropic role in various body organs like heart, pancreas, brain, stomach, muscles, liver, kidney, and bones.

12.4 DIPEPTIDYL PEPTIDASE 4 (DPP-IV-EC 3.4.14.5)

DPP-IV, also known as CD 26 T-cell activating antigen, is a widely distributed proteolytic enzyme. It is highly expressed on the exocrine pancreas, kidney, gastrointestinal tract, biliary tract, thymus, lymph nodes, uterus, placenta, prostate, adrenal, sweat, salivary, and mammary glands. DPP-IV has a pleiotropic role in the body but it is highly considered in this era for its role in intestinal incretin hormones GLP-1 and GIP. It reduces the half-life of these hormones by cleaving dipeptides from their N-terminal side and makes them noninsulinotropic physiologically (Green et al., 2006).

DPP-IV (also known as CD26) is a ubiquitously expressed type II transmembrane glycoprotein (110 kDa) with dipeptidyl peptidase activity. As a member of the serine protease family, it removes dipeptides having proline or alanine from the NH_2-terminal penultimate position of various substrate hormones, chemokines, neuropeptides, and growth factors. GLP-1 and GIP secreted by enteroendocrine cells in response to postprandial hyperglycemia are the two incretin hormones, known as a substrate of DPP-IV enzyme (Aroor et al., 2014; Kim et al., 2014)). Thus, DPP-IV inhibitors also called "incretins" are a new class of oral antihyperglycemic drugs against T2DM. In addition to its contribution to the homeostasis of glucose and lipid metabolism in diabetic patients, it also improves endothelial function, reduces the pro-oxidative and the proinflammatory state, and exerts renal effects (Avogaro et al., 2014).

GLP-$1_{7\text{-}36}$ is released into circulation from the gut and is rapidly degraded to inactive GLP-$1_{9\text{-}36}$ by DPP-4. In the case of the classical mechanism, DPP-4 inhibitors inhibit DPP-4 in peripheral plasma. This prevents the inactivation of circulating GLP-1 by increasing the concentration of intact GLP-$1_{7\text{-}36.}$ An increase in GLP-$1_{7\text{-}36}$ stimulates the secretion of insulin, inhibiting glucagon secretion, which reduces hepatic glucose production and also increases fat and muscle glucose utilization. This, in turn, lowers the fasting and postprandial glucose in the body. While nonclassical mechanisms of DPP-4 inhibition involve inhibition of gut DPP-4 activity raising the intact GLP-1 at the tissue level in gut immediately after its release. This, in turn, activates gut autonomic nerves along with an increase in the portal concentration of intact GLP-1, thereby activating portal GLP-1 receptors. In another nonclassical mechanism of DPP-4 inhibitors is inhibition of DPP-4 in islets that prevents inactivation of islets GLP-1, thereby improving the function of islets in the pancreas. DPP-4 inhibitors also prevent the inactivation of other biologically active peptides such as SDF-1α, and PACAP that may result in improved function of islets (Omar and Ahrén, 2014).

12.5 GLUCAGON-LIKE PEPTIDE-1 RECEPTOR (GLP-1R)

GLP-1 R (64 kDa) was first cloned from rat pancreatic islets following human pancreatic insulinoma and a gut tumor cell line. It belongs to the class B family of seven-transmembrane receptor guanine nucleotide-binding protein (G protein) coupled receptor. In humans, its gene is located on the long arm of chromosome 6p21, having 90% identicality and 95% amino acid homology with rats (Doyle and Egan, 2008).

12.6 MICRO-RNAS AND THEIR ROLE IN DIABETES

miRNA are short endogenous noncoding RNA with approximately size of 22 nucleotides (~22 nt). They significantly regulate gene expressions through a post-translational mechanism. The disruption in miRNA levels leads to many diseases including DM. Till now > 2000 miRNAs have been discovered in the human genome. miRNAs cause pathogenesis of DM by affecting the functions of β cells, insulin resistance, or both. miRNAs often act as a negative regulator at the level of post-transcriptional state, leading to the degradation of mRNA due to the complimentary binding of the seed sequence region present at the 5′ end of the miRNA with 3′-untranslated region of mRNA (Feng et al., 2016).

12.7 DIABETOGENIC MODELS

Many chemicals are in use for the development of diabetogenic models such as alloxan, streptozotocin, dithizone and 8-hydroxyquinolone (Rees and Alcolado, 2005).

12.7.1 STREPTOZOTOCIN

STZ (IUPAC name: 2-deoxy-D-glucose derivative of *N*-methyl-*N*-nitrosylurea) was first discovered in 1959, isolated from soil micro-organism *Streptomyces acromogenes,* is a glucosamine nitrosurea compound resembles 2-deoxy-D-glucose (Goyal et al., 2016). Through glucose transporter (GLUT-2), STZ enters in β cells of the islet in the pancreas and induces DNA damage via its alkylation. This leads to ADP-ribosylation activation, the important step of diabetogenicity. This ADP-ribosylation further causes the depletion of NAD+ and ATP. This enhanced process of ATP dephosphorylation provides the substrate for xanthine oxidase resulting in superoxide radicals formation. As a result, hydrogen peroxide and hydroxyl radicals also get generated. In addition to this, STZ liberates nitric oxide in toxic amounts that participate in DNA damage through inhibition of aconitase activity. These pathways result in β cell necrosis (Szkudelski, 2001).

12.8 β CELL ADAPTATION TO STRESS

Studies have revealed the self-adaptive properties of β cells against stress conditions (Lacraz et al., 2009). β cells adapt themselves during diabetes progression through modulation of various markers of autophagy, apoptosis, and proliferation (Srivastava et al., 2018a).

12.9 FACTORS RESPONSIBLE FOR B CELLS AGING

Islet β cells from aged humans have reduced turn over as compared to young donors. 2–3 months old rats are known as an appropriate model to study hyperglycemia-induced β cells apoptosis (Maedler et al., 2006). Obesity and sarcopenia, mitochondrial dysfunction, hormonal changes, oxidative stress, inflammation, reduced physical activity, reduced GLUT-2 levels, accumulation of AGEs, telomeres shortening, impaired Ca^{2+} handling, and reduced response to GLP 1, are some of the age-related pathophysiological factors responsible for downregulation of glucose homeostasis (De Tata, 2014). Advanced age severely restricts cell cycle entry of β cells (Kushner, 2013) (Figure 12.3).

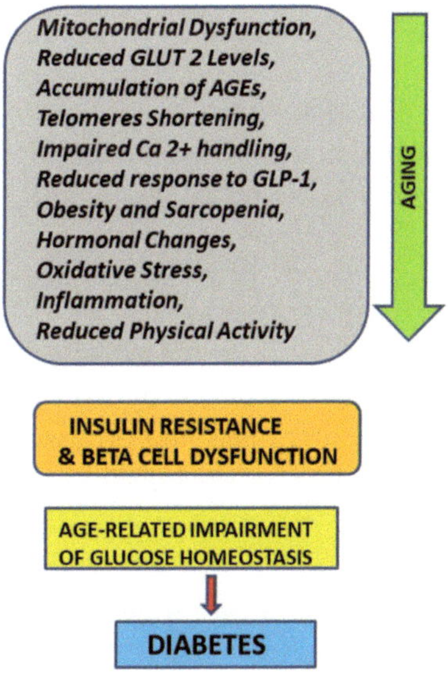

FIGURE 12.3 Pathophysiological factors responsible for the age-related failure of glucose homeostasis (adapted from De Tata (2014) published under the CC-BY licence).

12.9.1 REDUCED GLUT-2 LEVELS

GLUT-2 is essential for GSIS. GLUT-2-dependent glucose-sensing controls pancreatic islet cell mass and function through the nervous system. GSIS and glucose uptake get suppressed by genetic inactivation of GLUT in beta cells of rodents (Thorens, 2015).

12.9.2 ACCUMULATION OF AGES

Nuclear accumulation of Foxo1 is induced by AGEs, which leads to the downregulation of Pdx-1 expression at post-translational level via reducing its protein stability, ultimately leads to decreased synthesis of insulin in pancreatic β-cell (Shu et al., 2011).

12.9.3 TELOMERES SHORTENING

Telomere length acts as an important determinant of diabetes pathogenesis and β cell function even with an intact β-cell number (Guo et al., 2011).

12.9.4 IMPAIRED Ca^{2+} HANDLING

The immunosuppressive drug tacrolimus (FK506), also increased the risk of DM onset, because of its potential to alter mitochondrial Ca^{2+} fluxes, thus compromising the beta-cells metabolic secretion coupling (Lombardi et al., 2017).

12.9.5 REDUCED PHYSICAL ACTIVITY

β cells death were found to get reduced by exercise in T1D, with lower cleaved caspase 3, iNOS content and NO production. Such intracellular changes included IL-6 signaling, at least partly (Paula et al., 2015).

12.9.6 MITOCHONDRIAL DYSFUNCTION

The process like translation and transcription in mitochondria usually plays the determining role in β cell function (Mulder, 2017). Estrogen receptor α is known as a potential target for the T2D in women, having an important role in pancreatic β cell function and survival. Overexpressing ER α represses the Oma 1 and Chop promoters via binding to their consensus estrogen-response element sites, thus maintains mitochondrial–endoplasmic reticulum function and mitochondrial fission/fusion-mitophagy dynamics (Zhou et al., 2018).

12.10 CELLULAR SENESCENCE

Besides aging, cellular senescence also occurred in the case of obesity and diabetes. In T2D pathogenesis, senescence has a direct effect on pancreatic β cells, impact on adipose tissue dysfunction and tissue damage via senescence-associated secretory phenotype (SASP). Also, the diabetic factors like hyperglycemia, growth hormone axis perturbations, and altered lipid metabolism result in the formation of the senescent cell. The phenomenon like age-related chronic diseases, developmental factors, radiation exposure, as a defense mechanism against tumor development, DNA damage, oncogenic activation, developmental factors, the hormonal milieu, chronic infection (e.g., HIV), oxidative stress, certain medications (e.g., chemotherapy or certain HIV protease inhibitors), and telomere erosion causes irreversible growth arrest leading to cellular senescence. Senescent cells are metabolically active despite possessing the reduced capability to get divide. Due to high metabolic activity, these cells produce chemokines, growth factors, and proinflammatory cytokines collectively termed as SASP. Senescence signals possess heterogeneity and also have both positive and negative effects once formed. Aging and obesity pathogenesis accumulate senescent cells results in tissue dysfunction and diabetic complications as well as inflammation, metabolic dysfunction, insulin resistance, and progenitor cell dysfunction via the direct impact of senescent cells in tissues and SASP. This, in turn, leads to enhanced senescence cells through hyperglycemia, dyslipidemia, immune dysfunction, and GH/IGF 1 dysregulation. This pathogenic loop of diabetes, senescence, and tissue dysfunction can be broken with the help of glucose-lowering, senolytic or SASP-inhibiting therapies (Palmer et al., 2015).

12.11 UNUSED FREE FATTY ACIDS IN ISLETS AND B CELL APOPTOSIS

The esterification of unused FFAs in islets gets deposited over time and induces lipotoxicity, leading to increased iNOS expression, NO production as well elevated ceramide results in β cell apoptosis (Shimabukuro et al., 1998).

12.12 ROLE OF CERAMIDE AND AUTOPHAGY IN THE DEVELOPMENT OF TYPE 2 DIABETES

To regulate homeostasis under stress conditions, autophagy is essential. The autophagy acts through two main lysosomal degradation pathways, i.e., a general (nonspecific target) pathway and a selective (Specific target) pathway (Kang and Elledge, 2016). Thus, this process increases the survival rate of cells and makes them healthier by degrading the unwanted and toxic composition, results in recycling of energy and needed nucleotides, molecules, so forth, by cells. However, under certain stress conditions, persistent and/ or dysregulated autophagy enhances cellular death-causing molecular mechanisms (Jiang and Ogretmen, 2014). In diseases like diabetes, this homeostatic process gets disturbed due to some external factors like the Western diet or lack of exercise. These factors contribute to the accumulation of many harmful chemicals within the cells, making them susceptible to death.

Interestingly, ceramide is a key intermediate that links all the above conditions together, resulting in a state of insulin resistance (Galadari et al., 2013).

Ceramide, a central molecule of sphingolipid (bioactive lipid messengers) metabolism, has shown to trigger autophagy through signaling pathways involving mTOR and the dissociation of Beclin 1:Bcl 2 complex via JNK-1-mediated Bcl 2 phosphorylation in a dependent manner. More precisely, ceramide-synthase plays a vital role in inducing autophagy through the conversion of short-chain ceramide (C2 and C6) into long-chain ceramide (Pattingre et al., 2009).

Ceramide induces both cytoprotective as well as lethal autophagy (Jiang and Ogretmen, 2014). Therefore, the role of ceramide is paradoxical in nature, it is strongly known to induce autophagy, the clarification on how ceramide regulates autophagy-mediated cell protection versus death is not established yet in response to various alternative conditions, studied deeply in case of cancer. Also, the overall molecular detail of sphingolipid–protein binding, like ceramide-LC3B-II complex, might be essential to reveal the mechanisms by which lethal versus survival autophagy is being regulated by ceramide (Jiang and Ogretmen, 2014).

12.13 PUERARIA TUBEROSA AS GLP-1 R, GIP-R AGONIST AND DPP-IV INHIBITOR

According to our previous research, *Pueraria tuberosa* protects islets because of its anti-oxidant, anti-apoptotic, anti-hypoxic, anti-inflammatory and DPP-IV inhibitory potential. Along with these it also acts as GLP-1R and GIP R agonist, thus prevents β cells damage (Srivastava et al., 2015, 2017, 2018a, 2018b).

12.14 YOGA

Yoga has been reported as cost-effective as well as noninvasive adjuvant therapy against hyperglycemia in T2DM (Chimkode et al., 2015).

KEYWORDS

- **diabetes**
- **beta cells**
- **autophagy**
- **incretins**
- **aging**

REFERENCES

Abedini, A.; Schmidt, A. *et. al.* Mechanisms of islet amyloidosis toxicity in type 2 diabetes. *FEBS Lett.* **2015**, *587* (8), 1119–1127.

Aroor, A. R.; Sowers, J. R.et al. Pleiotropic effects of the dipeptidylpeptidase-4 inhibitors on the cardiovascular system. *Am. J. Physiol. Heart Circ. Physiol.* **2014**, *307* (4), H477-92.

Avogaro, A.; Kreutzenberg, S. De et al. Dipeptidyl-peptidase 4 inhibition : linking metabolic control to cardiovascular protection. *Curr. Pharm Des.* **2014**, *20* (14), 2387–2394.

Cernea, S.; Dobreanu, M. Diabetes and beta cell function: from mechanisms to evaluation and clinical implications. *Biochem. Medica* **2013**, *23* (3), 266–280.

Chimkode, S. M.; Kumaran, S. D.et al. Effect of yoga on blood glucose levels in patients with type 2 diabetes mellitus. *J. Clin. Diagn. Res.* **2015**, *9* (4), CC01-3.

De Tata, V. Age-related impairment of pancreatic beta-cell function: pathophysiological and cellular mechanisms. *Front. Endocrinol. (Lausanne).* **2014**, *5*, 138.

Doyle, M. E.; Egan, J. M. Mechanisms of action of GLP-1 in the pancreas.*Pharmacol. Ther.* **2008**, *113* (3), 546–593.

Feng, J.; Xing, W. et al. Regulatory roles of microRNAs in diabetes. *Int. J. Mol. Sci.* **2016**, *17* (10), 1–12.

Galadari, S.; Rahman, A. et al. Role of Ceramide in diabetes mellitus: evidence and mechanisms. *Lipids Health Dis.* **2013**, *12* (1), 98.

Goyal, S. N.; Reddy, N. M. et al. Challenges and issues with streptozotocin-induced diabetes—a clinically relevant animal model to understand the diabetes pathogenesis and evaluate therapeutics. *Chem Biol Interact.* **2016**, *244*:49–63.

Green, B. D.; Flatt, P. R.et al. Dipeptidyl Peptidase IV (DPP IV) Inhibitors: A newly emerging drug class for the treatment of type 2 diabetes. *Diab. Vasc. Dis. Res.* **2006**, *3* (3), 159–165.

Guo, N.; Parry, E. M.et al. Short telomeres compromise β-cell signaling and survival. *PLoS One* **2011**, *6* (3), e17858.

Hajduch, E.; Bourron, O. Type 2 diabetes: Ceramides as a therapeutic target? *Clin. Lipidol.* **2013**, *8* (6), 607–609.

Jiang, W.; Ogretmen, B. Autophagy paradox and ceramide. *Biochim. Biophys. Acta - Mol. Cell Biol. Lipids* **2014**, *1841* (5), 783–792.

Joseph, B.; Jini, D. Insight into the hypoglycaemic effect of traditional Indian herbs used in the treatment of diabetes. *Res. J. Med. Plant* **2011**, *5* (4), 352–376.

Kang, C.; Elledge, S. J. How autophagy both activates and inhibits cellular senescence. *Autophagy* **2016**, *12* (5), 898–899.

Kim, N. H.; Yu, T. et al. The nonglycemic actions of dipeptidyl peptidase-4 inhibitors. *Biomed Res. Int.* **2014**, *2014,* 1–10.

Kushner, J. A. The role of aging upon β cell turnover. *J. Clin. Invest.* **2013**, *123* (3), 990–995.

Lacraz, G.; Figeac, F. et al. Diabetic β-cells can achieve self-protection against oxidative stress through an adaptive up-regulation of their antioxidant defenses. *PLoS One* **2009**, *4* (8), e6500.

Lim, G. E.; Brubaker, P. L.; Glucagon-like peptide 1 secretion by the L-cell. The view from within. *Diabetes.* **2006**, *55* (2), S70–S77.

Lombardi, A.; Trimarco, B. et al. Impaired mitochondrial calcium uptake caused by tacrolimus underlies beta-cell failure. *Cell Commun. Signal.* **2017**, *15* (1), 47.

Maedler, K.; Schumann, D. M. et al. Aging correlates with decreased β-cell proliferative capacity and enhanced sensitivity to apoptosis. *Diabetes* **2006**, *55* (9), 2455–2462.

Mulder, H. Transcribing β-cell mitochondria in health and disease. *Mol. Metab.* **2017**, *6* (9), 1040–1051.

Omar, B.; Ahrén, B. Pleiotropic mechanisms for the glucose-lowering action of DPP-4 inhibitors. *Diabetes* **2014**, *63* (7), 2196–2202.

Palmer, A. K.; Tchkonia, T. et al. Cellular senescence in type 2 diabetes: A therapeutic opportunity. *Diabetes* **2015**, *64* (7), 2289–2298.

Pattingre, S.; Bauvy, C. et al. Ceramide-induced autophagy: To junk or to protect cells? *Autophagy* **2009**, *5* (4), 558–560.

Paula, F. M. M.; Leite, N. C. et al. Exercise increases pancreatic β-cell viability in a model of type 1 diabetes through IL-6 signaling. *FASEB J.* **2015**, *29* (5), 1805–1816.

Pradhan, P.; Upadhyay, N. et al. Genetic and epigenetic modifications in the pathogenesis of diabetic retinopathy: A molecular link to regulate gene expression. *New Front. Ophthalmol.* **2016**, *2* (6), 192–204.

Rees, D. A.; Alcolado, J. C. Animal models of diabetes mellitus. *Diabet. Med.* **2005**, *22* (4), 359–370.

Seino, Y.; Fukushima, M. et al. GIP and GLP-1, the two incretin hormones: similarities and differences. *J. Diabet. Invest.* **2010**, *00022* (10), 2040–1124.

Shimabukuro, M.; Zhou, Y. T. et al H. Fatty acid-induced beta cell apoptosis: A link between obesity and diabetes. *Proc. Natl. Acad. Sci. USA* **1998**, *95* (5), 2498–2502.

Shu, T.; Zhu, Y. et al. AGEs decrease insulin synthesis in pancreatic β-cell by repressing Pdx-1 protein expression at the post-translational level. *PLoS One* **2011**, *6* (4), e18782.

Srivastava, S.; Koley, T. K. et al. The tuber extract of pueraria tuberosa linn. competitively inhibits DPP-IV activity in normoglycemic rats. *Int. J. Pharm. Pharm. Sci.* **2015**, *7* (9), 227–231.

Srivastava, S.; Shree, P. et al. Active phytochemicals of pueraria tuberosa for DPP-IV inhibition: In silico and experimental approach. *J. Diabetes Metab. Disord.* **2017**, *16* (1), 46.

Srivastava, S.; Pandey, H. et al. Expression kinetics reveal the self-adaptive role of β cells during the progression of diabetes. *Biomed. Pharmacother.* **2018,** 106, 472–482.

Srivastava, S.; Pandey, H. et al. GLP 1 regulated intestinal cell's insulin expression and self-adaptation before the onset of type 2 diabetes. *Adv. Pharm. Bull.* **2019**, *9* (2), 325–330.

Srivastava, S.; Shree, P. et al. Incretin hormones receptor signaling plays the key role in antidiabetic potential of PTY-2 against STZ-induced pancreatitis. *Biomed. Pharmacother.* **2018**, *97*, 330–338.

Srivastava, S.; Yadav, D. et al. DPP-IV inhibitory potential of methanolic extract of pueraria tuberosa in liver of alloxan induced diabetic model. *Biosci. Biotechnol. Res. Asia* **2018**, *15* (1), 01–04.

Suzuki, A.; Nakauchi, H. et al. Glucagon-like peptide 1 (1–37) converts intestinal epithelial cells into insulin-producing cells. *PNAS* **2003**, *100* (9), 5034–5039.

Szkudelski, T. The mechanism of alloxan and streptozotocin action in B cells of the rat pancreas. *Physiol. Res* **2001**, *50*, 536–546.

Talchai, C.; Xuan, S. et al. Generation of functional insulin-producing cells in the gut by foxo1 ablation. *Nat. Genet.* **2012**, *44* (4), 406–412.

Thorens, B. GLUT2, glucose sensing and glucose homeostasis. *Diabetologia.* **2015**, 221–232.

Zhou, Z.; Ribas, V. et al. Estrogen receptor α protects pancreatic β-cells from apoptosis by preserving mitochondrial function and suppressing endoplasmic reticulum stress. *J. Biol. Chem.* **2018**, *293* (13), 4735–4751.

Index

B

C

D

O

S

T

U

V

W

X

Y

Z